“Climate change is one of the most significant challenges facing our planet in the 21st century. Addressing climate change is crucial for ensuring a sustainable future, yet it remains a contentious subject. *Climate Change Communication in Africa* offers valuable insights into the challenges and complexities of communicating climate change in Africa. Diverse in scope and rich with examples, the volume elucidates the trends, cultures and practices of communication climate change.”

Martin N. Ndlela, *University of Inland, Norway*

“Climate change fundamentally challenges humankind – and communication about it is a core influence on how societies around the world deal with it. *Communicating Climate Change in Africa* is a much-needed contribution to research on the topic: It compiles studies about public communication about the climate crisis in a range of African countries, some of which among the most affected by the climate crisis globally. Its broad vista – from stakeholder communication over climate journalism to social media such as TikTok – will help filling the gap of scholarship on countries from the Global South.”

Mike S. Schäfer, *Professor, University of Zurich, Switzerland*

“Without doubt, this book makes an important contribution to the transnational scholarship of climate change communication! It is a much-needed publication providing insight into a great variety of fascinating case studies – ranging from climate communication in local community radio, to national legacy media to social media and the role of influencer across the African continent. A must read for everyone interested in more balanced transnational debates of climate change across the Global North and Global South.”

Ingrid Volkmer, *Professor, University of Melbourne, Australia*

Climate Change Communication in Africa

This groundbreaking collection explores the complex intersections of communication and climate change in Africa.

From legacy media's framing of climate narratives to the mobilising potential of digital media, the volume examines the multi-faceted ways in which climate change is mediated through different communication channels and how different media can be harnessed to amplify African voices on climate change. Drawing on a broad range of theoretical and methodological approaches, case studies cover all regions of the continent, providing a nuanced examination of how the media shape our understanding of and responses to climate change in Africa. Contributions scrutinise different facets of climate change communication, providing illuminating accounts on pitfalls in climate change communication, climate justice communication strategies, social media activism and climate change, and datafication of climate change. This is the first collection to comprehensively address climate change communication across the continent, featuring perspectives from all regions in Africa.

An essential contribution to the global conversation on climate change communication, the collection will appeal to researchers, academics, policymakers, communication activists, and students at all levels in the fields of environmental communication, media studies, and journalism, as well as general readers keen to understand the complex dynamics of climate change in Africa.

Tendai Chari is Associate Professor of Media Studies and C1 National Research Foundation (NRF) Rated Researcher at the University of Venda, South Africa. He holds a PhD in media studies from the University of Witwatersrand, South Africa. Chari is widely published in the field of media and communication studies, and his research focuses on political communication with a broadened horizon on the interface between digital media and politics, media and conflict, and media ethics and popular culture. His other publications have appeared in the *Journal of African Media Studies*, *Global Media* (China Edition), *African Journalism Studies, International Communication Gazette*, *African Identities*, and *Journal of African Elections*. He is the co-editor of *Global Pandemics and Media Ethics: Issues and Perspectives* (Routledge, 2022, with Professor Martin N. Ndlela). Chari is a recipient of several grants and fellowships and is finalising his book *Diaspora Media and Homeland Conflict: Coloniality of Conflict Journalism in Zimbabwe* (Routledge).

Allen Munoriyarwa holds a PhD in Journalism from the University of Johannesburg. Currently, he is Associate Professor of Journalism at Walter Sisulu University (WSU) in South Africa, in the Department of Marketing, Public Relations and Communication. He is the former Coordinator of the Media Policy and Democracy Project (MPDP). This research project explored the growth of digital surveillance practices in Southern Africa. He is also a board member of Intelwatch, a civil society organisation. (It can be followed here: https://intelwatch.org.za/) Professor Munoriyarwa is also the Coordinator of the British Academy Research titled, "Watching the Watchers: Strengthening Public Oversight of Intelligence-Driven Surveillance". (It can be found here: www.gla.ac.uk/research/az/watchingthewatchers/about/ – PI Professor Jane Duncan). His research interests are in Artificial Intelligence (AI), digital surveillance, digital media ecosystems, and climate communication practices.

Routledge African Media, Culture and Communication Studies

This series features innovative and original research on African media, culture and communication from scholars both within and outside of Africa. With Africa still under-represented in media and cultural studies literature, this series provides a much-needed platform for comparative and interdisciplinary research that highlights the perspectives and needs of Africans as consumers and producers.

African Englishes
Contemporary Trends and Sociolinguistic Shifts
Edited by Collen Sabao and Esther Mavengano

Shifting Journalism Boundaries in Zimbabwe
Digital Disruption and Innovation at the Margins
Edited by Mphathisi Ndlovu and Lungile Tshuma

Nigeria's Cultural and Creative Industries
Perspectives, Problems, and Prospects
Edited by Ikechukwu Obiaya, Kizito Alakwe and Emmanuel Nzeaka

Generative AI in African News Media
Opportunities and Challenges
Edited by Trust Matsilele and Lungile A. Tshuma

Critiquing Generative AI in Africa's Media Ecosystems
Edited by Lungile A. Tshuma and Trust Matsilele

Climate Change Communication in Africa
Emerging Trends and Practices
Edited by Tendai Chari and Allen Munoriyarwa

For more information about this series, please visit: www.routledge.com/Routledge-African-Media-Culture-and-Communication-Studies/book-series/RAM

Climate Change Communication in Africa

Emerging Trends and Practices

Edited by Tendai Chari and Allen Munoriyarwa

Routledge
Taylor & Francis Group
LONDON AND NEW YORK

First published 2026
by Routledge
4 Park Square, Milton Park, Abingdon, Oxon OX14 4RN

and by Routledge
605 Third Avenue, New York, NY 10158

Routledge is an imprint of the Taylor & Francis Group, an informa business

British Library Cataloguing-in-Publication Data
A catalogue record for this book is available from the British Library

ISBN: 978-1-041-11482-6 (hbk)
ISBN: 978-1-041-11487-1 (pbk)
ISBN: 978-1-003-66021-7 (ebk)

DOI: 10.4324/9781003660217

Typeset in Times New Roman
by Newgen Publishing UK

We dedicate this book to our families, whose love, support and patience have been the wind beneath our wings throughout this project.

Contents

Figures

Tables

Contributors

Akwasi Bosompem Boateng (PhD, CPRP) is Senior Lecturer in the Department of Strategic Communication at the University of Education, Winneba, Ghana. Akwasi earned his Doctor of Philosophy (PhD) degree in Cultural and Media Studies from the University of KwaZulu-Natal, South Africa. He is a member of various academic and professional bodies, including the Public Relations Institute of Southern Africa (PRISA), French Institute for Research in Africa (IFRA), Nigeria, African Studies Centre in Leiden (ASCL), Collaborative Research Group African History, Leiden, African Network of Researchers on Inequality in Science (ANRIS), and Communication Educators Association of Ghana (CEAG). His research interests include strategic communication, public relations, advertising, media, digital (social) media, and political and environmental communication.

Nassir Bouali is Associate Professor in the Department of Mass Communication at the College of Communication at the University of Sharjah. He brings extensive professional experience, having worked for 8 years as a news reporter for *Al Wahda* Journal in Algeria. Dr. Bouali specialises in Media Communication and Journalism research, with over 40 published studies and seven authored books in the fields of Communication and Media Sciences, along with three book chapters. He is widely recognised for his academic contributions and has received numerous awards. Notably, he is one of the founding scholars behind the *Value Media Determinism Theory (VMDT)*, a significant contribution to the development of Arab media theories.

Albert Chibuwe is Research Fellow, Department of Communication Science, University of South Africa. His research interests are political communication, social media and politics, media framing, and journalism education and practice. He also occasionally researches on data journalism and intersections of citizen journalism and surveillance.

Gopolang Ditlhokwa (PhD) is Lecturer in the Department of Media Studies, Faculty of Humanities, University of Botswana. He teaches courses in communication, broadcasting, film, and digital media, and his research is on platform dependencies and creative industries.

Ogadimma C. Emenyeonu (PhD) is Faculty Member at the College of Communication, University of Sharjah. He has taught Mass Communication and Media Studies in Nigeria, Oman, and the United Arab Emirates. Dr. Emenyeonu has authored numerous articles in peer-reviewed journals and contributed several chapters to academic books. He serves on the editorial boards of various scholarly journals. A dynamic and forward-thinking academic, he is deeply committed to professional growth and actively engages in interdisciplinary research and collaborative projects. His research interests span new media, social media engagement, sociology of news production, health communication, environmental communication, peace journalism, and conflict resolution.

Uchenna Oluchi Eze is a graduate student at the School of Journalism and Mass Communication, University of Iowa, Iowa City, USA. She earned her Bachelor's degree in mass communication in 2017 from the University of Nigeria, Nsukka. Uchenna belongs to international associations, including National Communication Association (NCA), International Communication Association (ICA), International Association for Media and Communication Research (IAMCR), International Sociology of Sport Association (ISSA), North American Society for the Sociology of Sport (NASS), Association for Education in Journalism and Mass Communication (AEJMC), and The International Journal of Press/Politics (IJPP). She has previously worked with several media stations in Nigeria as a sports presenter and communication officer with the United Nations Office on Drugs and Crime (UNODC). She has some papers that are still under review for publication to her credit. Uchenna has reviewed for the National Communication Association, and her research interests lie in the intersection of media and sports, climate change communication, and advocacy journalism.

Victor C. Eze is Adjunct Lecturer at the Department of Mass Communication, Dominican University, Ibadan. He earned his Doctor of Philosophy and Master's Degree in Communication and Language Arts in 2021 from the University of Ibadan, Nigeria. Dr. Eze belongs to the following international academic associations: International Association for Media and Communication Research (IAMCR), International Communication Association (ICA), African Studies Association (ASA), and Chinese in Africa/Africans in China research network (CAAC). He has presented research papers at a number of international conferences hosted by IAMCR and ASA. In the past, he worked as a journalist with print media outfits and as a brand reputation expert in some integrated marketing communication (IMC) consultancies. He has published in a number of peer-reviewed journals and also reviews research papers for some journals. His research interests lie at the intersection of media framing, political communication, crisis/security communication, and public relations.

Jendele Hungbo (PhD) is Professor of Journalism and Media Studies at KolaDaisi University, Ibadan. He is also Research Associate in the Department of Media Studies at the University of the Witwatersrand, Johannesburg. His research interests cut across broadcast media, new (interactive) media, social media, postcolonial identities, and representations in the media in Africa, among others. He previously taught at North-West University, Mafikeng, South Africa. A former Cadbury Fellow at the Centre for West African Studies, University of Birmingham, Jendele was a recipient of the Volkswagen Stiftung Doctoral Fellowship and is widely published in reputable local and international outlets. His current research includes trauma and precarity among journalists in Nigeria and interactive media and the performance of identities in Nigeria and South Africa.

Leonard S. N. Kalindekafe (PhD) is Professor and former Executive Dean of Ndata School of Climate and Earth Sciences at the Malawi University of Science and Technology. He has over 30 years' experience in geology, geochemistry, mining policy, and climate change. He chaired the United Nations Intergovernmental Forum on Mining, Minerals, Metals and Sustainable Development (2009–2013), where he helped develop the Mining Policy Framework now being used globally. Currently, he is the Chief Executive Officer for the Malawi Mining Investment Company.

Shitemi Khamadi is a Kenyan journalist and a media development practitioner with over 15 years of experience. He has worked extensively in sub-Saharan Africa with different organisations, as a media trainer, project manager, and media specialist, including Aga Khan University Graduate School of Media and Communications (GSMC), Africa Uncensored, Internews, and the East African Editors Society (EAES). His research interests include environmental communication, science communication, information disorder, journalism studies, media effects, and migration. Shitemi holds two master's degrees: one in Digital Journalism from GSMC, Nairobi, and another in Peace and Conflict Management from Kenyatta University. Currently, he is a PhD in Media Arts and Sciences at The Media School, Indiana University-Bloomington.

Saizi Salim Katete Kimu holds a PhD in Communication Sciences from North-West University's Mahikeng Campus. His research interests are in development communication, environmental communication, indigenous communication systems, indigenous language media, discourse analysis, visual social semiotics, and grounded theory. Currently, he is also Lecturer at the Malawi University of Science and Technology's Language and Communication Studies Department.

Tumelo Koboyatau is Independent Researcher who holds a Master of Arts in Media and Communications from the University of Botswana. Her postgraduate research focused on the newsification of climate change in Botswana's local media, examining how environmental issues are framed and communicated to the public. Her current work centres on journalism covering climate, energy, mining, and construction – key sectors influencing Botswana's socio-economic development. Through a blend of academic inquiry and field reporting, she brings underreported narratives to light with depth, nuance, and critical insight.

Alex Lenferna is a Research Fellow at the Fort Hare Institute for Social and Economic Research. From 2020-2024, he also served as the co-founder and elected General Secretary of the Climate Justice Coalition - a South African coalition of civil society organisations working together to advance a transformative climate justice agenda. Alex has dedicated over sixteen years to research, writing, teaching, and activism on environmental and social justice with a primary focus on energy and climate justice. His work, writing, and research is freely available at AlexLenferna.com

Takudzwa Mandiringa holds a Master of Science in Journalism and Media Studies and a BSc (Honours) degree in Journalism and Media Studies both from the National University of Science and Technology. He is currently lecturing at Midlands State University, and his research interests include media, film, photography, culture, decolonisation, community development, and gender studies.

Takunda Maodza is Postdoctoral Research Fellow at Walter Sisulu University, Department of Marketing, Public Relations and Communications, East London, South Africa. He holds a PhD in Cultural and Media Studies with the Centre for Communication, Media and Society (CCMS) at the University of KwaZulu-Natal, Durban, South Africa. Maodza's research interests are in media and politics, investigative journalism, digital media, climate change journalism, and indigenous language media. Maodza has vast experience as a practising journalist spanning over 17 years having worked for both the private and state-controlled media in Zimbabwe, including the Zimbabwe Mirror Newspaper Group and Zimpapers.

Daniel Maposa holds a Master's in Arts and Culture Management from Wits University. He is the Founder and Director of Savanna Trust, an art-for-development organisation based in Harare, Zimbabwe. Currently, he is pursuing a PhD in Cultural Policy and Management at Wits University. His research interests are in the sustainability and resilience of arts and culture organisations.

Refilwe Whitney Mofokeng is nGAP Lecturer at Tshwane University of Technology in South Africa. She is also a PhD candidate at the University of Johannesburg, South Africa. Her study is centred around Artificial Intelligence (AI) practices in South African mainstream media. Her research interests are in AI practices in journalism and media ecosystems.

Nancy Mtambo holds a Bachelor's (Honours) degree in Film and Theatre Arts and a Master's degree in Media, Film, Television and Digital Technologies, both from the Midlands State University. She is currently Staff Development Fellow at the Midlands State University. Her research interests include community development through the arts, film, climate change, gender studies, and cultural sociology.

Lyton Ncube (PhD) is Senior Lecturer in the Department of Media Studies, Faculty of Humanities, University of Botswana. Ncube is also Research Associate in the Communication Science Department, University of South Africa. Ncube teaches media, journalism, and public relations courses. His research agenda is at the intersections of sociology of sport, digital cultural practices, and sports journalism in Africa.

Khulekani Ndlovu is a senior lecturer in the Journalism and Media Technology department at the Namibia University of Science and Technology. He is also a research fellow in the Department of Communication Science at the University of South Africa. He holds a PhD in Media studies from the University of Cape Town, South Africa. He has published journal articles and book chapters focusing on tabloid media; media and social justice; mediated discourse about the political agency of youth and the democratic implications of social media.

Kingsley L. Ngange (PhD) is Professor of Journalism and Communication Science and Deputy Vice-Chancellor for Research, Cooperation and Relations with the Business World, University of Buea (UB), Cameroon. He was Fulbright Visiting African Research Scholar at The Brechner Center for Freedom of Information Project, in the College of Journalism and Communications, University of Florida, for the 2023/2024 academic year and former Chair of the Department of Journalism and Mass Communication, University of Buea, for several years. He has been involved in journalism and mass communication training, practice, education, and management for the past 27 years. He has been Visiting Scholar in the USA (2010, 2015, and 2023/2024) and in different parts of Africa (Ethiopia, Uganda, Somaliland, South Africa, Rwanda, etc.) for over a decade now. Professor Ngange has supervised several master's and PhD theses and published extensively in reputable national and international journals and books. He has also attended and made top-notch academic presentations in national and international workshops and conferences in Cameroon, Rwanda, South Africa, Uganda, Ethiopia, Somalia, Somaliland, Sweden, and the USA. Professor Ngange's academic experiences and contributions spread across three continents: Africa, Europe, and North America, making him an international scholar of very high repute.

Jakub Siwak (PhD) is Senior Lecturer in the Department of Media and Communication at Nelson Mandela University, Gqeberha, South Africa, where he lectures in Cultural Studies and New Media/Media Studies. His research interests incorporate post-structural philosophy (most often the work of Michel Foucault and Gilles Deleuze), new media, and mediated representations of nature. More recently published articles in this specific regard include *Looking at animals through a digital lens – a general repetition of tendencies and a different take* (2023) in Stilet 35 (1) and *Affective Communication: Exchanges between People and Nature via Information Technology* (2019) in Communication 45 (4).

Thouraya Snoussi received her PhD from Stendhal University, France, in October 2000, and started her professional career as a radio and TV journalist in Tunisia and Canada, where she gained rich professional experience in journalism and communication. She moved to the teaching field in 2006 as Assistant Professor at Manouba University (Tunisia) and then at Emirates College of Technology in Abu Dhabi (UAE) from 2012 to 2015. Dr. Snoussi also served as Associate Professor, Head of Department, and then Acting Dean of the College of Mass Communication at Al Ghurair University in Dubai (UAE). She is now working as Associate Professor and Chair of Mass Communication Department in the College of Communication, at the University of Sharjah, in the UAE. Her area of research includes new media uses and impact. She published 6 books and more than 50 refereed research articles in indexed well-reputed international journals. Dr. Snoussi got international recognition as Leader in Humanities and Social Sciences in January 2018, in India; Woman in Education Leadership Award, in July 2018, in Singapore; Best Faculty Award during the International Academic and Research Excellence Awards (IARE), Noida, on 8 June 2019; and Woman Researcher Award from NESIN 2020 Awards, from Science Father in October 2020.

Melanie M. Tchewo (PhD) is part-time Lecturer in the University of Buea, where she teaches master's students in Corporate Communication. She also carries out part-time lectures at the Limbe Nautical Arts and Fisheries Institute in Rural Sociology. Melanie obtained a Bachelor of Science degree in Journalism and Mass Communication and a Master of Science in Journalism, all from the University of Buea, Cameroon. She holds a PhD in Sociology from the University of Buea. Her areas of research include the use of ICTs and community media as tools for sustainable development communication as well as health and environmental communication. She is currently the Director of Communication and Public Relations in the Cameroon Development Corporation (CDC), an agro-industry and second employer of labour after the State of Cameroon. She is Founder and CEO of Rural Communication for Community Engagement (RUCOCE).

Didunloluwa Tella is Lecturer in the Department of Communication and Language Arts, University of Ibadan, Ibadan, Nigeria. Her research interests cut across broadcast journalism, community radio, new media, development communication, and environmental and climate change communication. An early-career scholar in Media and Communication, Didunloluwa has presented her work at several academic conferences and is currently involved in the Journalism and Trauma Research Network.

Mmakgotso Theko is a Master's student in the Department of English, Media Studies and Linguistics in the Faculty of Humanities, Social Sciences and Education. Her research focuses on environmental communication with a particular emphasis on reporting and climate change.

Dércio Tsandzana holds a PhD in Political Science from Sciences Po (France) and Lecturer at Eduardo Mondlane University. Since 2013, he has been actively engaged in research and advocacy on youth, the Internet, social media, and political participation in Mozambique, including online activism for both national and international organisations. In addition to his PhD, Tsandzana has conducted research on digital rights and data privacy with organisations such as Freedom House, Meta/Facebook, Paradigm Initiative, CIPESA, Internews/USAID,

UNESCO, and the Centre for Human Rights at the University of Pretoria. He is a member of the Board of Directors of the African Digital Rights Network (ADRN) and an active member of the African Declaration on Internet Rights and Freedoms Initiative (AfDec). He is a member of the UNESCO AI Ethics Experts Without Borders (AIEB) network. He is also a member of UNESCO's AI Ethics Experts Without Borders (AIEB) Network.

Sehlisiwe Tshuma holds a Master's degree in Communication Science from the University of South Africa. She is currently lecturing at Midlands State University and pursuing a PhD in Communication Studies at the University of Johannesburg. Her research interests include media consumerism, cultural transformation, theatre for development, film, and gender studies.

Munetsi Zvavahera is a PhD holder in Zoology specialising in fisheries from the University of Fort Hare. He works as Research Fellow at Midlands State University (in the Tugwi-Mukosi Multidisciplinary Research Institute). His research interests include the impacts of climate change on biodiversity, alien invasive species, aquatic ecology, fisheries, and aquaculture.

Acknowledgements

Editing this volume was an extremely rewarding experience that we could never have imagined. There were many people without whose efforts this book could not have materialised. First and foremost, we would like to extend our deepest gratitude to all the contributors in this collection whose expertise, dedication, hard work, sacrifice, and passion for the subject matter have made this book a reality. Their commitment to sharing their knowledge and research has enriched the volume and will undoubtedly benefit readers. Our sincere gratitude also goes to all the anonymous reviewers of the book chapters in this volume, whose insightful comments and critiques significantly improved the quality and rigour of the contributions. The reviewers of the book proposal also deserve special thanks for insightful comments and constructive feedback, which helped in enhancing the quality of our manuscript. We are also thankful to the Senior Editor at Routledge, Helena Hurd, for believing in our project and for her unwavering support and guidance throughout the publication process. Her expertise and vision have been invaluable in shaping the book's direction. We would also like to appreciate Katerina Lade, Editorial Assistant at Routledge, whose organisational skills and attention to detail have been instrumental in keeping the project on track. The copyediting team also deserves special thanks for meticulously reviewing and refining the manuscript, ensuring consistency and clarity throughout the book. On a personal note, we would like to thank our family members, whose love, patience, and encouragement have sustained us throughout this journey. Their support has been a constant source of motivation and inspiration. Thank you all for your contributions to this book. We sincerely hope that it will have a meaningful impact in the field of media studies, environmental communication/journalism, and cognate fields.

Abbreviations

3D	Three-dimensional
4D	Four-dimensional
ACE	Action for Climate Empowerment
AI	Artificial Intelligence
AIC	African Instituted Churches
AKU-GSMC	Aga Khan University Graduate School of Media and Communication
AMARC	World Association of Community Radio Broadcasters
AP	Associated Press
APRI	Africa Policy Research Institute
ArtCOP21	Art Climate Change Conference (2015, Paris)
BBR	Bonakanda Rural Radio
BNL	Blantyre Newspapers Limited
CCBS	Community, Climate and Biodiversity Standard
CD	China Daily
CDC	Cameroon Development Corporation
CJC	Climate Justice Coalition
CJID	Centre for Journalism Innovation and Development
CO_2	Carbon dioxide
COP	Conference of the Parties
COP26	The 26th session of the Conference of the Parties (COP) to the United Nations Framework Convention on Climate Change (UNFCCC)
COP27	The 27th session of the Conference of the Parties (COP) to the United Nations Framework Convention on Climate Change (UNFCCC)
COVID-19	Coronavirus disease of 2019
CR	Community Radio
CSOs	Civil Society Organisations
DAPL	Dakota Access Pipeline
DDT	Dichlorodiphenyl-trichloroethane
DRC	Democratic Republic of Congo

EACOP	East African Crude Oil Pipeline Project
EMA	Environmental Management Agency
EPA	Environmental Protection Agency
EWEs	Extreme Weather Events
FAO Report, 2018	Food and Agriculture Organization
FGD	Focus Group Discussion
FSLC	First School Leaving Certificate
GDP	Gross Domestic Product
GES	Growth and Employment Strategy Programme
GHG	Greenhouse gas
ICPAC	IGAD Climate Predictions and Applications Center
IGAD	Intergovernmental Authority on Development
IMF	International Monetary Fund
IPCC	Intergovernmental Panel on Climate Change
IPS	Inter Press Services
MDG	Millennium Development Goals
MENA	Middle East and North Africa
MESTI	Ministry of Environment, Science, Technology and Innovation
NCCP	National Climate Change Policy
NCR	National Climate Reports
NDCC	National Observatory on Climate Change
NGOs	Non-governmental organisations
NOAA	National Oceanic and Atmospheric Administration
NPC	Nigerian Press Council
NPL	Nation Publication Limited
NYT	New York Times
PAR	Participatory action research
PC	Participatory Communication
REDD	Reducing emissions from deforestation and forest degradation in developing countries
TCCR	Tiko Community Radio
UN	United Nations
UNCCD	United Nations Convention to Combat Desertification
UNDP	United Nations Development Programme
UNFCCC	United Nations Framework Convention on Climate Change
US	United States
VCS	Verified Carbon Standard
WB	World Bank
ZANU–(PF)	Zimbabwe African National Union–Patriotic Front

Introduction

Climate Change and Communication in Africa – Trends and Practices

Tendai Chari and Allen Munoriyarwa

Introduction

> The information flow on climate change has potentially been from the rest of the world to Africa. It needs to be replaced by a sustained dialogue and two-way flow of information that empowers African citizens and ensures publics and politicians in countries mainly responsible for causing climate change are better informed of African realities and perspectives.
>
> (BBC World Trust, 2009: 2)

Why is a book on climate change communication in Africa needed? The answer to this question is partly encapsulated in the above excerpt, which foregrounds the motivation behind this edited volume. It is trite to note that climate change has become one of the most urgent and all-encompassing challenges confronting humanity and the environment in the 21st century. Marlies et al. (2022: 5) note that Africa is one of the most vulnerable regions in the world to climate change, with the continent experiencing increases in average temperatures, frequent and longer heat waves, severe droughts, a relative rise in sea levels, severe flooding and erosion. Socio-economic, political and environmental vulnerabilities have exacerbated Africa's exposure to climate change-related risks. The root causes of the continent's social vulnerabilities, such as poverty and structural inequalities, have been attributed to the continent's history of imperialism and colonialism "and are still influenced by colonial power relations" (Marlies et al., 2022: 5).

Despite Africa being the least responsible for emitting greenhouse gases attributable for global warming, it is the most vulnerable continent to climate change due to multi-layered historical, social, economic and geo-political factors (BBC World Service Trust, 2009). Consequently, climate change has become part of the quotidian discourses at the centre of any global debates, from hunger, famine, war, droughts, sustainable development, etc. Tadesse (2010) notes that African nations are among the least polluters, but they suffer from climate change the worst. Consequently, the continent hopes to lead in the carbon credit market by 2030. Home to some of the poorest populations of the world who already live on

DOI: 10.4324/9781003660217-1

the frontlines of disasters and degradation of natural resources, Africa is disproportionately affected by climate change, and the continent's ability to achieve the Millennium Development Goals (MDGs) is in jeopardy (Tadesse, 2010). Due to low resilience in impoverished communities on the continent, climate change has exacerbated existing socio-economic challenges such as poverty, hunger, inequalities, natural disasters, conflicts, political instability, human and security challenges linked to declining biodiversity, increased demographic pressures on agricultural and grazing land and other natural resources.

However, the threat posed by climate change to the African continent is not commensurate with the quality and quantity of information generated about climate change and the level of citizen awareness or understanding of this complex and multifaceted phenomenon (Tagbo, 2010). The aim of this book is to provide a comprehensive analysis of the key issues, trends and controversies around mediated climate change communication in Africa. Chapters in this book shed light on climate change communication practices in different regions of the African continent. Our central claims in this book are threefold: firstly, although Africa is the most vulnerable continent to climate change, its citizens are among the least informed about climate change, its causes, impacts, consequences and the ideological and geo-political contestations around it. This limits the capacity of African nations to respond and adapt to climate change and their ability to communicate their perspectives and experiences to those responsible for causing climate change (BBC World Service Trust, 2009; Tadesse, 2010). Secondly, climate change can be better tackled if communication and media are effectively and responsibly deployed to inform and educate African citizens about climate change. Thirdly, the quality of climate change information in the media is very much linked to the level of attention given to climate change issues in different communication and media platforms and the quality of information possessed by communicators. Contributions in this book represent diverse perspectives which deepen and broaden understanding of climate change to enable African citizens to communicate their perspectives and experiences about climate change to the global community. In this vein, we hope that the book will add to the existing voices interrogating the climate change discourse from a Global South perspective, thus enhancing the flow of information about climate change from the Global South to the rest of the world and empowering African citizens and policymakers to communicate climate change issues. Spotlighting climate change communication on the African continent is particularly important to make sense of the multiple narratives about climate change that African citizens are exposed to and the possible influence these narratives have on citizens' environmental behaviour and opportunities for climate adaptation (Okoliko & De Wit, 2020).

Mediated Communication and Climate Change

Despite the emergence of digital media, traditional media such as newspapers, radio and television still have a crucial role in Africa, where an estimated 70% of the population resides in rural areas. Mediated communication plays an important role in creating awareness about climate change, thereby influencing public policy and shaping citizen behaviours and action, particularly in a continent where a significant percentage of the population hardly hear about climate change (Okoliko & de Wit, 2020; John-Adeyemi, 2024). This edited volume mainly focuses on mediated climate change communication. The deliberate focus on mediated climate change communication stems from the fact that the mass mediation of climate change has a much longer history, dating as far back as the 1930s (although it has a much more recent history in Africa), and is more widespread than other modes of climate change communication. Scholars have noted the critical role played by mass media such as newspapers and broadcasting in disseminating information about climate change (Schafer, 2015; Schafer et al. 2016; Okoliko & de Wit, 2020; Okoliko & de Wit, 2021), particularly the way in which such media help societies to make sense of climate change science (Okoliko & de Wit, 2020). However, in contexts such as Africa, where climate change communication is met with numerous socio-cultural and political factors, effectively communicating climate change is a tough order (see Chapter 11 by Siwak, in this volume). Consequently, several roles have been attributed to the mass media to foster a better understanding of the phenomenon. These include being a source of information, awareness creation, shaping public and policy opinions through their framing practices, public sphere role and an interpretive role, to mention only a few. Broadcasting media such as radio and television play a crucial role in providing news and information about climate change, with community radio playing an instrumental role in raising awareness about climate change in rural communities in a context of low literacy (Kapinga et al., 2020; John-Adeyemi, 2024). This brings discussions on climate change closer to the people and mobilises local action – the point canvassed by Tchewo and Ngange (in this volume). While television may not be as omnipresent as radio in Africa (John-Adeyemi, 2024), it also has a critical role to play, particularly in urban areas.

Framing Climate Change

The perspective that the media contribute towards making citizens understand the science of climate change and the effects of climate change (Jungsberg et al., 2019) resonates with the notion of framing wherein the media select certain aspects of reality and amplify them while downplaying other aspects of reality (Entman, 1993). The concept of framing illuminates the way in which the media shape public opinions and perceptions by making some aspects of reality about climate change salient while "promoting a particular problem definition, causal interpretation, moral evaluation, and/or treatment recommendation for the item being described" (Entman, 1993:52). Understanding how the media frames climate change is therefore important because it influences the policy agenda, wherein policymakers may come up with interventions for climate change mitigation and adaptation, while citizens may adopt lifestyles that are compatible with the government's climate change mitigation agenda (see Chapter 7 by Eze and Eze, in this volume). As a major source of information on climate change, the media not only play a significant role in influencing citizens' knowledge, attitudes and behaviours about climate change but also directly influence public engagement and action about climate change (Okoliko, 2021). Sebastian (2021) notes that most people in South Africa heard about climate change in the mass media, but very little research has been directed at illuminating the efficacy of the mass media in fostering public engagement on climate change issues (Chari, 2016; Okoliko, 2021). While admitting that they argue that the mass media are critical channels for addressing climate change issues, particularly in raising "public awareness and support for addressing climate change challenges", they lament the fact that "not enough has been done to unravel the dynamics of communication as a conduct for getting people politically connected to the challenge of climate change" (Okoliko (2021: 3). While the media, including the Internet, have been considered significant sources of news and information about climate change, several shortcomings have been identified about the way in which the media frame climate change science (Boykoff, 2007; Bowen, 2011). These include journalists' lack of knowledge about climate change science, allegations of sensationalism, misinformation disinformation, malinformation, inadvertent misreporting, fearmongering or "alarmist reporting" what is referred to as "climate porn" by Erent and Segnit (cited in Boykoff & Boykoff, 2007: 3). In addition, underreporting of climate change occasioned by either lack of interest or lack of knowledge of complex scientific issues by journalists and episodic reporting of climate events are some of the shortfalls levelled against the media (Evans & Evans, 2016; Sithole 2023). Bowen (2011) identifies the inclusion of denier perspectives in the news and the tendency to focus on controversial aspects of climate change rather than consensus as some of the pitfalls of climate change reporting, which lead to public perceptions of unscientific certainty. This demonstrates that through their discursive practices and/or manipulation by powerful sources the media significantly affect how the public perceives climate change, and consequently, on what policy develops (Bowen 2011: 56).

Legacy Media and Climate Change: Trends and Practices

The bulk of research on mediated climate change communication has focused on traditional media, such as newspapers and television (Boykoff 2007: Boykoff & Roberts, 2007; Gonapathy, 2021), clearly positioning legacy media as the dominant sources of news and information about climate change. Such studies deploy both quantitative and qualitative research designs and a variety of theoretical frameworks to explore some aspects of climate change communication, such as the volume and frequency of coverage, selection of themes, journalistic norms, frames and agenda-setting effects (Boykoff & Boykoff, 2004; Boykoff, 2007; Boykoff & Roberts, 2007; Wang, 2020). The contested nature of climate change, particularly in relation to its existence, its extent, urgency and appropriate responses and measures (Schafer, 2016), stems from its unobtrusiveness and uncertainty, resulting in legacy media prioritising the polarisation between advocates of climate change and the so-called denialists (see Chapter 18 by Chari and Theko, in this volume).

Polarisation lends itself to normative traditional values and the commercial logic of the news production process wherein climate change is viewed through a conflictual lens. Earlier coverage of climate change in the developed world exhibits intense political polarisation along ideological lines, with one side questioning the validity of climate science and the urgency of the problem, while the other camp accepts climate change as a reality which needs urgent attention (Nisbet, 2009). These deep partisan divisions were fodder for the elite press, often resulting in the perpetuation of the discourse of "uncertainty" even in the face of what appeared to be a global consensus within the scientific community. Similarly, a study on media coverage of climate change in Australian newspapers (Wang & Richardson, 2020) suggests that despite assumed scientific consensus around climate science, politicisation narratives were dominant in news articles, constituting 57%, while climate change denialism articles constituted 8%. Wang and Richardson (2009: 4) contend that "economic interests" and a highly concentrated media ownership were the major driving force behind the intensity of political conflict over climate change in a country that is highly dependent on fossil fuels and has the highest levels of per capita greenhouse emissions.

Against the backdrop of the high-stakes debates, political polarisation has often triggered fear appeals in persuasive messages whereby climate change is depicted as an existential threat to livelihoods to raise awareness among citizens so that they can act against the threat (Bilfinger et al., 2024). Within the context of the developed world, Russil and Nyssa (2009) documented the use of "tipping points" in climate change discourse, enquiring whether tipping point warnings in prestige media induced unwarranted anxiety and fatalism or whether they helped to correct the "false sense of security" produced by smooth projections about climate change which tended to make society inactive. More recent studies in the Global South reveal that climate change communication tends to emphasise catastrophic and apocalyptic climate change scenarios that accentuate negative consequences of climate change (De Graaf et al., 2024). De Graaf et al. (2024) argue that such "doomsday" messages or "threat appeals" can yield either negative or positive effects on perception and action. The authors note that while threat can be a powerful motivator for adaptive action, it can have unintended effects that hinder behaviour change, adaptive action, as citizens remain uncertain about the effectiveness of behaviour in curbing the threat. A study analysing newspaper framing of climate change in Mexico (Pulver & Sainz-Sanatamaria, 2018) resonates with the dominant literature on framing and agenda-setting (see Chapter 1 by Kimu and Kalindekafe, in this volume, and Chapter 2 by Ogadimma and Bouali, also in this volume).

Despite the persistence of polarisation, there are significant shifts in the content and style of climate change reporting, with more voices and news sources joining the fray. These include scientists, industry, policymakers and non-governmental organisations (NGOs) who seek to assert their perspectives and agendas on climate change through the mass media and different communication strategies to attract the attention of the media (Schafer et al., 2016). With the increased mediatisation of climate change (Wang & Richardson, 2020), such stakeholders strategically position themselves by adopting the logic of the media to influence public perceptions about climate change and consequently policy. As amply demonstrated in this edited volume, new and diverse actors such as civil society, environmental activists, corporates, grassroots communities, social media influencers and celebrities have joined the climate change discourse and scientific sources and state actors no longer have monopoly of knowledge on climate science anymore. The book demonstrates the multifaceted and cross-cutting nature of climate change as encompassing such issues as communication of climate justice (see Chapter 5 by Lenferna, in this volume, and Ndlovu, also in this volume), Artificial Intelligence and climate change (see Chapter 12 by Mofokeng and Munoriyarwa, in this volume), social media influencers and climate change (see Chapter 18 by Chari and Theko, in this volume), datafication of news (see Chapter 13 by Chibuwe and Maodza in this volume), the role of visual arts in climate change (see Chapter 8 by Mtambo et al., in this volume) and various aspects of digital media and climate change. It is the contention of this book that the discourse of climate change communication is evolving. The emergence of digital media and its transformation of climate change communication is but one example of such evolution.

Digital Media and Climate Change Communication

Digital media now offers substantial climate change news coverage (Painter et al., 2016; Lee 2025). Research has shown that digital media offer climate change news in rich format and thematic diversity (Painter et al., 2016), which might not be available on legacy media spaces. Digital media also amplify climate change discourses on digital media (including social media) – and help build a sustainable climate change agenda – on digital media. Due to their technological affordances, digital media provide spaces for climate change communication debates, interaction and action, facilitating access for under-represented voices and alternative viewpoints which might be deliberately neglected in legacy media (Painter et al., 2016). Furthermore, digital media are known to provide space for informal climate communication discourses that have the potential to push, widen and sustain climate change communication narratives (see Painter et al., 2016). Lee (2025) notes that digital media offer a myriad of climatic and environmental representations and, in agreement with Painter et al. (2016), further notes that digital media challenge dominant media even though they still push the climate and environment narrative in three major ways. These are agenda setting, framing and tone and language. Lee (2025) notes that "The tone and language used in news coverage can affect the public's emotional response to environmental issues" (Lee, 2025, n.p.).

In addition, digital media challenge dominant narratives about climate change promote understanding and give access to multiple perspectives and experiences (see Chapter 18 by Chari and Theko, in this volume). There is also, an opportunity in digital media to provide an inclusive, equitable and diverse space for environmental climate change communication. Digital media are bound to assume an even more central role in climate change communication as the younger generation gradually detaches from legacy media like newspapers and radio to these platforms for information. In the United Kingdom and the USA (see Painter et al., 2016), digital media platforms are becoming the main source of news for people who seek to engage with climate change (O'Sullivan, 2024). In Africa, digital spaces like Ushahidi in Kenya and Econai+ have provided sustainable climate change news and awareness information (Elega et al., 2024). Econai+, the Nigeria-based digital platform, for instance,

> focuses exclusively on the environmental dimension of sustainability by sourcing, aggregating, and reporting climate-related issues, such as loss of biodiversity, erosion and depletion of raw materials, and acidification of waters in oil-producing states through interactive geodata on their website and mobile application.
>
> (Elega et al., 2024, p. 3)

Thus, an advantage of digital platforms in climate change is that they can be issue-based, focusing exclusively on climate change, and they can also be sustainable, considering that their sunk-in costs are less than legacy media.

Platforms and Climate Change communication

Platforms, like X (formerly Twitter), Meta (formerly Facebook), TikTok, LinkedIn and Instagram, have become important conduits in the climate change communication nexus. They have been able to set the climate change agenda. Schäfer (2025, p. 49) argues that, "Social media have transformed climate change communication and become central for sharing information, facilitating dialogue, and driving mobilisation". There is a famous global example of Greta Thunberg, the young Swedish climate activist. Greta has become a global climate change figure because platforms like X have made Thunberg's efforts to go viral. Greta Thunberg has mobilised millions to join the climate course. In Africa, climate activists like Susan Chomba, the Kenyan scientist and activist and the late Wangari Maathai from Kenya have made use of platforms to drive the message of climate change home. Other climate activists on the African continent, like Inna Modja from Mali and Vanessa Nakate from Uganda, have all used platforms to drive their activism messages and to raise awareness about climate change. For instance, Vanessa Nakate has created a popular LinkedIn page for her organisation the Tard Foundation. There are several climate change communication practices that have become all too banal on platforms and have largely been sustained. These include mobilising youth voices, centring them in policymaking, building climate change communication resilience and amplifying the climate change narratives of marginalised groups (O'Sullivan,

2024). It is instructive to note that one of the largest climate change movements on the African continent, the African Activists for Climate Justice (AACJ), has largely thrived on X and Meta and hosts a digital website as well. Hence, platforms are evidently widening the reach of climate change messaging on the African continent by bringing on-board, once-marginalised groups to assert their climate rights, for instance, indigenous groups (O'Sullivan, 2024). Elroy, Komendantova and Yosipof (2024) argue that platforms are at the forefront of creating cyber-echoes of the climate crisis.

As much as digital media and platforms are central to contemporary and possibly future climate change communication, we cannot ignore the potential harm they may wreak on climate change discourses. For a start, platforms have become notorious purveyors of disinformation, misinformation and malinformation (Forchtner, 2019; Mare & Munoriyarwa, 2026). Digital platforms have become, for instance, the "mainstream" spaces where climate denialist discourses find sanctuary (Mare & Munoriyarwa, 2026). In the process, these platforms can undermine our understanding of climate change, and they may also compromise existing (at least in some contexts) participatory climate communication cultures and how the public engages with climate change. Platforms' contribution to climate change discourse can be a double-edged sword, positively promoting and enhancing public engagement on climate change. Yet they may render their digital spaces open to misinformation and disinformation about climate change (Mare & Munoriyarwa, 2026). This would certainly undermine public trust in the platforms and their whole messaging of climate change communication. In more recent times, digital spaces have given room to far-right climate change deniers' articulations of denialism, providing these groups with a platform and setting an agenda for climate change denialism. Fringe platforms have been the most preferred spaces on which anti-climate discourses have circulated widely. On these fringe platforms, climate misinformation and disinformation are on steroids. Fringe platforms like Parler and Truth Social fuel disinformation and misinformation (Uyheng, Tyagi and Carley, 2021) and amplify climate distrust and uncertainty (Lee et al., 2024). It is on these platforms that climate change denialists keep asking, "Is climate change even real?" and foreground it as a massive global conspiracy theory, against all scientific evidence. It could be argued that the next frontier of the war against climate change will be in the production of AI-driven deepfakes that further sow doubt and distrust of pro-climate change voices. There is a risk that the anti-climate change crusaders who gain access to these platforms could polarise mainstream platforms with their climate denialism. Notwithstanding their side of platforms and digital spaces like podcasts and websites, they have been positively strategic by allowing more and diverse people to access information and to engage with climate change issues, further broadening not only the range of issues around climate change, but also giving a voice to marginalised groups, particularly the most affected by climate change, to put across their perspectives (see Chapter 15 by Snoussi, in this volume).

The Structure of the Book

This book comprises 18 chapters and an introduction. The chapters are categorised into four parts. Part I, which comprises six chapters, focuses on legacy media and change communication. Chapter 1 by Saizi Salim Katete Kimu and Leonard Kalindekafe focuses on coverage of climate change and extreme weather conditions in Malawian newspapers between 2014 and 2016. The chapter argues that during the period under study climate change stories were not only prominent, but also silent on causes, consequences and solutions to climate change as well as long-term adaptation. Like Saizi and Kalindekafe, in Chapter 2, Ogadimma C. Emenyeonu and Nassir Bouali also examine press coverage of climate change in Algeria and Nigera. Drawing from secondary literature and interviews, Ogadimma and Bouali's study reveals that effective communication of climate change remains a huge challenge for both countries because of multi-layered, socio-economic factors such as poverty, illiteracy, linguistic diversity, journalistic ineptitude, governmental interference and lack of inclusive and participatory communication strategies. In Chapter 3, Khulekani Ndlovu turns to ideological framing of Zimbabwe's Kariba REDD+ Carbon credit crisis. The chapter examines how carbon neutrality discourse is shaped by the political economy of the green economy industry, the extractivism and neo-colonial relations between the Global North and the Global South nations and the domestic politics of adopting countries. In Chapter 4, Tumelo Koboyatau, Lyton Ncube and Allen Munoriyarwa focus on newsroom factors that influence the inclusion of climate change into the news agenda in Botswana. Deploying the sociology of news theoretical framework, the authors demonstrate how lack of ready-made news discourse, particularly the fact that climate change does not naturally lend itself to the traditional news values of conflict and drama, journalists and editors' preferences and lack of knowledge and awareness of climate change issues hinder coverage of climate change. Like Saizi Salim Katete Kimu and Ndlovu, in Chapter 5, Alex Lenferna employs framing to explore how activists (and beyond) are building more support for climate action by challenging certain frames and narratives about climate change. Drawing from his experiences as a scholar-activist and employing intersectionality as a theoretical lens, Lenferna posits that framing often distorts understanding of climate change by highlighting certain benefits and costs while ignoring, exaggerating or downplaying many others. Chapter 6 by Shitemi Khamadi examines journalistic narratives on climate change in the Kenyan mainstream media. The study revealed that dependency on government sources and data from international organisations was significant in the media. Khamadi argues that grants played a key role in shaping climate change news agenda in newsrooms since media houses operated on shoestring budgets, which made it difficult to prioritise stories that required extensive travelling. Chapter 7 by Uchenna Oluchi Eze and Victor C. Eze carries forward the theme of media framing of climate change. The authors examine newspaper framing of climate change in selected Nigerian newspapers, specifically focusing on how the newspapers attributed responsibility to climate change. The study concluded that

Nigerian newspapers adopt more frames that tend to hold other actors like the international community and the United Nations more accountable than the Nigerian government officials.

Part II of the book, which focuses on visual arts, community media and climate change, opens with Chapter 8 by Nancy Mtambo, Sehlisiwe Tshuma, Takudzwa Mandiringa, Daniel Maposa and Munetsi Zvavahera, who explore the role of visual arts in climate change communication. The chapter argues that visual arts have the potential to bridge the communication gap between scientific knowledge and public understanding, by catalysing action towards climate change mitigation and adaptation. In Chapter 9, Melanie Tchewo and Kingsley Ngange investigated the role of community radios in the fight against climate change in Cameroon's Southwest Region. The chapter observes that there is a significant relationship between community radios and the promotion of easy-to-understand messages on climate change. However, challenges were observed with regard to climate change adaptation policies. Chapter 10 by Akwasi Bosompem Boateng employed a qualitative case study design anchored on interviews and document analysis to explore the use of Indigenous Media strategies to communicate climate change in Ghana. The study revealed that there was minimal use of Indigenous Media strategies in communicating the climate change crisis. Chapter 11 by Jacub Siwak carries forward the theme of obstacles on climate change communication explored by Ogadimma and Nassir Bouali in Chapter 2. Siwak examines how the "problematic Western genesis" of the concept of climate change hindered effective communication about climate change in the African context. He contends that the globalised, generalised communication on climate change stemming from Western actors is undermined by the noise of its association with a region which is often not trusted in Africa.

Part III, which focuses on digital media and climate change, opens with Chapter 12 by Refilwe Whitney Mofokeng and Allen Munoriyarwa, which explored the use of Artificial Intelligence in climate change communication in South African and Zimbabwean media. Using in-depth interviews with 11 journalists, the authors shed light on how the adoption of AI for climate change reporting altered news production and journalistic practices. The study found that AI affords journalists simple formats for presenting complex scientific data, while adoption of AI also enhances early warnings for natural disasters and breaks down scientific data into understandable texts. In Chapter 13, Albert Chibuwe and Takunda Maodza focus on the datafication of news in two Zimbabwean newspapers. The authors examined the deployment of big data in the selected newspapers, interrogating the extent to which big data is used in these newspapers in climate change reporting and whether use of big data had brought any meaningful changes in climate change journalism. The chapter concludes that Zimbabwean journalists in the selected newspapers do not seem to understand data journalism, yet it is a key ingredient in investigative journalism. Chapter 14 by Dércio Tsandzana focuses on Internet discourses on climate change in Mozambique. Drawing from qualitative content analysis of news articles, social media posts and blogs from Mozambique and international media outlets, Tsandzana examines how climate change is discursively constructed. Like Saizi Salim Katete Kimu's study, in Chapter 14, Tsandzana found that discourses on climate change in Mozambique are fixated on extreme weather conditions such as natural disasters like cyclones and floods, while the human impacts of climate change are downplayed. In Chapter 15, Thouraya Snoussi examined the role of environmental podcasts in the Middle East and North Africa (MENA). Through content analysis of three leading podcasts in the MENA region, the study assessed how these podcasts contribute to environmental discourse and public awareness through their framing practices. The study revealed that the intimate and engaging nature of podcasts and their strategic framing techniques help strengthen listeners' connection to nature, foster empathy and inspire action.

Part IV of the book focuses on platforms and climate change and comprises three chapters. In Chapter 16, Jendele Hungbo and Didunloluwa Tella explored how citizens in platformised mediated climate change communication imagine their participation through their contributions and messages from climate change advocacy. Drawing on quantitative content analysis of X (formerly Twitter) posts on the #GreenNigeria campaign, the chapter interrogated perceptions and reactions to the campaign and related issues. The study concludes that although platformisation provides for liberalisation and spontaneity in climate change communication, infomediation and greater consideration for the material and social conditions in Nigeria seem to drive discourses on climate change. In Chapter 17, Gopolang Ditlhokwa and Lyton Ncube examined climate change communication on TikTok. The study inquired how technological affordances of TikTok provide democratised climate change communication, foster awareness creation about climate change in a way that challenges dominant narratives about climate change. The authors, however, lament the potential of TikTok to promote misinformation and reinforcement of denialism because of its rapid and unfiltered content. Chapter 18 by Tendai Chari and Mmakgotso Theko explored the activism of two young African climate change activists on X. Drawing from netnography and interviews Chari and Theko examined how an Ugandan climate change champion, Leah Namugerwa and a Kenyan climate change skeptic use X to frame climate change, and how the two engage with their audiences on X. The chapter concludes that notwithstanding their contrasting views on climate change, they found some common ground in relation to issues of climate justice.

While chapters in this volume focus on different aspects of the climate change–communication nexus in different regions of the African continent, they thematically intersect in different ways. We reckon that no single edited volume can cover every aspect of a complex issue such as climate change communication. However our hope is that the book will be a launching pad for robust conversations about climate change communication from an African perspective.

References

BBC World Service Trust (2009). Least Possible, Most Affected, Least Informed: Public Understanding of Climate Change. Retrieved from https://assets.publishing.service.gov.uk/media/57a08b8b40f0b652dd000d30/PB_climatechange_web.pdf. Accessed 31 December 2023.

Bilfinger, L., Brummernhenrich, B., and Jucks, R. (2004). The Effects of Fear Appeals in Climate Change Communication. *Journal of Experimental Social Psychology,* 115:1–9.

Bowen, M.G. (2011). Insights on Media Practices and Representations (Global Warming) Science: Confusing the Public, Educating School Children? *Journal for Activism in Science & Technology Education*, 3(1): 53–79.

Boykoff, M.T. (2007). Flogging a Dead Norm? Newspaper Coverage of Anthropogenic Climate Change in the United States and United Kingdom from 2003 to 2006. *Area*, 39(4): 470–481.

Boykoff, M.T., and Boykoff, J.M. (2007). Climate Change and Journalistic Norms: A Case of US-Mass Media Coverage. *Geoforum*, 38: 1190–1204.

Chari, T. (2016). Rethinking Climate Change Communication in Africa: The Case for Indigenous Media. *Indilinga: African Journal of Indigenous Knowledge Systems*, 15(2): 217–232.

De Graaf, J., Bal, M., De Wit, J., and Stok, M. (2024). Climate Change Doom Communication from Fear Appeal Perspective. *European Journal of Health Communication*, 5(4): 26–50.

Elega, A.A., Munoriyarwa, A., Mesquita, L., Gonçalves, I., and de-Lima-Santos, M. F. (2024). Empowerment Through Sustainable Journalism Practices in Africa: A Walkthrough of EcoNai+ and Ushahidi Digital Platforms. *Journalism Practice*, 1–21. DOI: https://doi.org/10.1080/17512786.2024.2399567.

Elroy, O., Komendantova, N., and Yosipof, A. (2024). Cyber-Echoes of Climate Crisis: Unraveling Anthropogenic Climate Change Narratives on Social Media. *Current Research in Environmental Sustainability*, 7: 100256.

Entman, R.M. (1993). Framing: Toward Clarification of a Fractured Paradigm. *Journal of Communication.* 43(3): 51–58, Autumn, 1993.

Evans, H.C., and Musvipwa K. (2016). *News Coverage of Climate Change: Perspectives from Africa and Zimbabwe*. Africa Institute of South Africa: Pretoria.

Forchtner, B. (2019). Far-Right Articulations of the Natural Environment: An Introduction. In B. Forchtner (ed.), *The Far Right and the Environment* (pp. 1–17). London, UK: Routledge.

Gonapathy, D. (2021). *Media and Climate Change: Making Sense of Press Narratives*. London, UK: Routledge.

John-Adeyemi, E. (2024). Climate Action in Africa: The Role of Media Influence. Climate Action. https://climateaction.africa/climate-action-in-africa/. Accessed 20 May 2025.

Jungsberg, L., Turunen, E., Heleniak, T., and Wang, S. (2019). *Atlas Population on Society and Economy in the Arctic*. Nordregio Working Paper. Stockholm: Nordregio. Retrieved from www.diva-portal.org/smash/get/diva2:1352410/FULLTEXT03.pdf. Accessed 12 February 2025.

Kapinga, M.D., Siyao, P.O., Sife, A.S., and Silayo, D.S. (2020). Role of Community Broadcasting in the Dissemination of Climate Change Information among Small-Holder Framers in Isimani, Iringa Rural District. *Journal of Management and Development Dynamics*, 30(2): 1–18.

Lee, J., Booth, E., Farid, H., and Rizoiu, M.A. (2024). Misinformation is not about Bad Facts: An Analysis of the Production and Consumption of Fringe Content. *arXiv preprint arXiv:2403.08391*.

Lee, S. (2025). The Environment in Digital Media: Analyzing the Representation of Environmental Issues in Digital Media and Communication. www.numberanalytics.com/blog/environment-in-digital-media. Accessed on 12 May 2025.

Mare, A., and Munoriyarwa, A (2026). *Disinformation Campaigns in Africa: Actors, Consequences, and Responses*. London: Bloomsbury Publishing.

Marlies, C., Okem, A., Hunter, N., and North, M. (2022). *Climate Change in Africa: Findings of the Latest IPCC Reports*. Durban: Durban TSU Office. Retrieved from https://ipcc.ukzn.ac.za/wp-content/uploads/2022/10/Climate-change-in-Africa-Findings-of-the-latest-IPCC-AR6-reports.pdf. Accessed 13 December 2024

Nisbet, M.C. (2009). Communicating Climate Change: Why Frames Matter for Public Engagement: Environment: *Science and Policy for Sustainable Development*, 51(2): 12–23.

Okoliko, D.A., and De Wit, M.P. (2020). Media(ted) Climate Change in Africa and Public Engagement: A Systematic Review of Relevant Literature. *African Journalism Studies*, 41(1): 65–83.

Okoliko, D.A., and De Wit, M.P. (2021). From "Communicating" to "Engagement" Afro-Relationality as a Conceptual Framework for Climate Change Communication in Africa. *Journal of Media Ethics*, 36(1): 36–50.

Okoliko, D.A., and De Wit, M.P. (2020). Media(ted) Climate Change in Africa and Public Engagement: A Systematic Review of Relevant Literature: *African Journalism Studies*, 41(1): 65–83.

O'Sullivan, L. (2024). Just How Effective Is Social Media in Tackling the Climate Crisis. https://en.reset.org/just-how-effective-is-social-media-in-tackling-the-climate-crisis/.

Painter, J., Erviti, M.C., Fletcher, R., Howarth, C., Kristiansen, S., Bienvenido, L., and Schäfer, M.S. (2016). *Something Old, Something New: Digital Media and the Coverage of Climate Change*. Oxford: Reuters Institute for the Study of Journalism.

Pulver, S., and Sainz-Santamaria, J. (2018). Characterizing the Climate Issue Context in Mexico: Reporting on Climate Change in Mexican Newspapers, 1996-2006. *Climate and Development*, 10(6): 538–551.

Russill, C., and Nyssa, Z. (2009). The Tipping Point Trend in Climate Change Communication. *Global Environmental Change*, 19: 336–344.

Schafer, M.S. (2015). Climate Change and the Media. *International Encyclopaedia of the Social and Behavioural Sciences*, 3: 853–859.

Schäfer, M.S. (2025). Social Media in Climate Change Communication: State of the Field, New Developments and the Emergence of Generative AI. *Dialogues on Climate Change*, *2*(1): 49–55. https://doi.org/10.1177/29768659241300666

Schafer, M., Berglez, P., Wessler, H., Eide, E., Nerlich, B. and O'Neill, S. (2016). *Investigating Mediated Climate Change Communication: A Brest-Practice Guide*. Research Report No.6. Jonkoping University, School of Education and Communication. Retrieved www.diva-portal.org/smash/get/diva2:961854/FULLTEXT01.pdf. Accessed 17 November 2024.

Sebastian, L. (2021). Climate Change Communication in Africa. In *Oxford Research Encyclopaedia of Climate Science*. Oxford University Press. Retrieved from https://oxfordre.com/climatescience/display/10.1093/acrefore/9780190228620-e-851. Accessed 11 November 2025.

Sithole, E. (2023). Climate Change Journalism in South Africa: Noticeable Improvements Adequate. *FOJO Media Institute at Linnaeus University*. https://journalism.co.za/wp-content/uploads/2023/06/Climate-change-journalism-in-South-Africa-Enoch-Sithole.pdf. Accessed 12 May 2025.

Tadesse, D. (2010). *The Impact of Climate Change in Africa*. Pretoria: Institute for Security Studies.

Tagbo, E. (2010). Media Coverage of Climate Change in Africa: A Case Study of Nigeria and South Africa. Oxford: Reuters Institute Fellowship paper. https://reutersinstitute.politics.ox.ac.uk/our-research/media-coverage-climate-change-africa-case-study-nigeria-and-south-africa. Accessed 30 December 2023.

Uyheng, J., Tyagi, A., and Carley, K.M. (2021, June). Mainstream Consensus and the Expansive Fringe: Characterizing the Polarized Information Ecosystems of Online Climate Change Discourse. In *Proceedings of the 13th ACM Web Science Conference 2021* (pp. 196–204). New York, NY: Association for Computing Machinery.

Wang, G., and Richardson, L.M. (2020). *Media Coverage of Climate Change in Australian Newspapers: A Case Study of the State of the Climate Reports*. Melbourne: Monash Climate Change Communication Hub.

Part I

Legacy Media and Climate Change Communication

1 'Covering' the Changing Climate

Newspaper Reporting of Extreme Weather Events in Malawi

Saizi Salim Katete Kimu and Leonard S. N. Kalindekafe

Introduction

Since the turn of the 21st century, climate change has received a new sense of urgency at both national and international level due to the increase in daily experiences of extreme weather events (EWEs) (Carvalho and Burgess, 2005; Okoliko and de Wit, 2020; Legg, 2021). Despite their negligible contribution to climate change in comparison to the Global North, Africa is the most vulnerable due to "multiple interactions of biophysical, political and socio-economic stressors" (Okoliko and de Wit, 2021: 36). Malawi and Uganda are the two African countries that are highly vulnerable to climate change (Selormey et al., 2019). Scientific evidence in Malawi has shown that over the past two decades, there has been an increase in frequency, intensity and magnitude of EWEs such as drought, intense rainfall and flash floods (Government of Malawi, 2016, 2021, 2023, 2024; Mataya et al., 2020; Oxford Analytica, 2023). This has had a great impact on the country's rain-fed small-scale agro-based economy leading to government revising the country's Gross Domestic Product (GDP) in 2015 from the projected 5.1% to 3.1% due to 2014/2015 floods (Government of Malawi, 2016). This disaster was yet again followed by El Nino induced drought in 2015/2016 growing season leading to economic losses of US$199 million and left approximately 6.4 million people (one-third of the country's population then) without food (Botha et al., 2018). Such climate change-induced EWEs pose the greatest threat to the realisation of United Nations Sustainable Development Goals 1 and 2, Africa's Agenda 2063 and Malawi's Vision 2063. However, although there have been intense media polarisation about climate change in the West (Boykoff and Boykoff, 2007; Boykoff, 2007; Billett, 2010; Priest, 2016; Ashe and Poberezhskaya, 2022), that has not been the case in Malawi. Even political discourse regarding who should fix the problem between highest emitters in the Global North and the highly affected least emitters in the Global South (Billett, 2010; Priest, 2016) has not permeated into the media and everyday discourse.

Several geoscientific studies have unanimously agreed that climate change is the most serious environmental risk the world is facing in the 21st century (Boykoff, 2007; Ramanathan and Feng, 2009; Priest, 2016; Okoliko and de Wit, 2020; Legg, 2021), and about 80% of it is due to release of greenhouse gases (Ramanathan

DOI: 10.4324/9781003660217-3

and Feng, 2009). Climate change adaptation and mitigation requires a complete overhaul of people's life style, and mass media has a critical role (Bacot and Fitzgerald, 1999; Moser and Dilling, 2011; Selormey et al., 2019; Okoliko and de Wit, 2020) as it provides vital discursive platform where climate change issues can be deliberated (Okoliko and de Wit, 2020), which can inform and motivate public action. Global scholarship on media representation of climate change therefore reveals different discursive patterns in various societies, especially in Global North (Williams, 2000; Carvalho and Burgess, 2005; Boykoff and Boykoff, 2007; Billett, 2010; Holt and Barkemeyer, 2012; Danielsen, 2016; Engesser and Brüggemann, 2016; Elia, 2019; Ashe and Poberezhskaya, 2022).

In Africa, there is little climate change communication research (Tagbo, 2010; Okoliko and de Wit, 2020) and most of these studies utilise traditional theories that emphasise either impacts of media to the audience or how climate change media texts are produced and (or) consumed. However, no study from Africa to our knowledge has looked at how social construction of climate change issues by powerful people in society can lead to its prominence in the media or vice versa. The situation is even dire in Malawi despite the country's high level of vulnerability. Hence, this study used Bennett's (2001) news structural biases and Hilgartner and Bosk's (1988) public arenas model for a quantitative content analysis of Malawian newspaper's coverage of climate change from 1 October 2014 to 30 April 2016 in order to find out the messages they relayed to the public and their prominence, as well as how that relate to powerful people's claim-making activities. The study intended to answer the following questions:

1 How frequent did climate change-related stories appear in newspapers during the study period?
2 Did news articles mention causes of EWEs? If yes, what were the causes that news articles attributed to them?
3 Did news articles mention solutions to climate change? If yes, what type of solutions did they suggest?
4 How prominent was climate change as a social problem, depicted in the newspapers?

The next sections, therefore, discusses the analytical framework, studies on newspaper coverage of climate change, methodology, results, discussions and the study's conclusion and implications.

Analytical Framework

This study adopted Bennett's (2001) news structural biases and Hilgartner and Bosk's (1988) public arenas model. Bennett (2001) states that decisions regarding news content are dependent on three 'not-always-compatible' normative orders of journalistic norms (e.g. accuracy and objectivity), political norms (promoting political accountability by informing citizens about their elected leaders' actions) and economic norms (reporting stories efficiently and profitably). Balancing these three normative orders is delicate since the first two norms are usually in conflict with economic obligation of audience maximisation by reporting entertaining stories. This has led to news structural biases of personalisation, dramatisation, fragmentation and authority–(dis)order (Bennett, 2001).

Personalisation is the tendency to downplay the underlying social, economic, political or geo-physical reasons behind an event in favour of surface trials, tragedies and triumphs of actors (Bennett, 2001). Personalised stories are easily told within a short time without serious investigation(s) and offending some sectors of the audience with probing analysis. However, personalised news' focus on news actors encourages audience passivity (Bennett, 2001), thereby deactivating collective urgency needed to tackle climate change (Moser and Dilling, 2011; Priest, 2016). Meanwhile, dramatisation is an emphasis on "crisis over continuity, the present over the past or future, conflicts and relationship problems between the personalities at their center [sic] and the impact of scandals on personal political careers" (Bennett, 2001: 36). Melodramatic reporting of EWEs undermines analysis of the socio-political, economical and geo-physical contexts in which disasters occur. Due to dramatisation, news reports of EWEs will rarely discuss their origins, solutions or future projections.

Furthermore, news media's penchant emphasis on breaking news, dramatic structure and personalities leads to news fragmentation whereby events appear as self-contained incidents that are isolated from their context. This makes it difficult for the audience to assemble news fragments into a big picture. Traditional media's strict space limitations to avoid what they call, "boring readers and viewers with too much information", exacerbate the problem (Bennett, 2001: 38). The last structural bias is authority–(dis)order, which is journalists' inclination towards authoritative voices of officials to interpret the frightening experience at the expense of examining their underlying socio-cultural, geo-physical and economic context (Bennett, 2001). This undermines capacity for consensus building needed in climate change adaptation and mitigation.

In addition, this study also adopted Hilgartner and Bosk's (1988: 53) public arenas model which states that "social problems are not objective and identifiable societal conditions that have intrinsically harmful effects" but rather how they are conceived by groups with some power. Hence, social problems are assumed conditions that are labelled as such in the arena of public discourse such as courts, political campaign organisations, research community, professional societies, religious organisations and news media (Hilgartner and Bosk, 1988). A problem which has succeeded in one public arena tends to dominate agenda in other arenas. Hence, there is a huge population of potential problems competing with climate change for attention in the public arenas, and their success or failure in this competition is not an objective identifiable societal condition, but rather how it is conceived in society by powerful group(s) of people. However, this model has been criticised for ignoring the context of claim-making, and Ungar (1992) therefore extends that besides claim-making activities of some powerful groups, success in the public arena also depends on the audience's receptiveness, which is heightened by occurrence of some dreadful triggering events like floods and droughts. Hence, claim-makers' ability to piggyback on occurrence of EWEs due to the audience's perceived attentiveness may influence prominence of climate change news and its framing. Hence, while use of Bennett's (2001) news structural biases helps to explain newspaper coverage of climate change during the period, Hilgartner and Bosk's (1988) public arenas model elucidates that such coverage may result from success or failure of claim-making activity of influential people in the society.

Newspaper Coverage of Climate Change

Recent studies have consistently found that the public acquire knowledge of most social problems from the media (Moser and Dilling, 2011; Akpan et al., 2012; Priest, 2016; Okoliko and de Wit, 2020). A quick review of literature on newspaper coverage of climate change shows that various societies, especially in the Global North, are experiencing different discursive reconstitution of the issue (Billett, 2010; Okoliko and de Wit, 2020; Ashe and Poberezhskaya, 2022) revealing ideological contestations prevalent in these countries. Research in the UK reveal readings of uncertainty of climate change in various newspapers types (Carvalho, 2007). In some cases, uncertainty was a justification for inaction by de-legitimising scientific climate knowledge and de-authorising agents and institutions that mobilise the public to address the problem. In other cases, the same uncertainty was used as a precaution to emphasise the need for action against unpredictable climate change (Carvalho, 2007).

In the USA, studies show that climate change reporting has also been subjected to a great deal of contestation. Utilising framing theory and claims-making from social problems perspectives, McCright and Dunlap (2000) analysed the counter-claims that conservative movement in the USA used to de-problematise climate change. The results show that not only an emphasis on weak evidence and economic consequences of proposed solutions, but also climate change would also reduce heating bills, transport delays and accidents due to less snow. Hence, despite a growing scientific consensus on anthropogenic climate change, subsequent studies of the US press continue to show images of disagreement which assigned more attention to climate change contrarians and (or) sceptics than its advocates (Boykoff, 2007; Boykoff and Boykoff, 2007). This situation is similar in most industrialised countries. Recently, Ashe and Poberezhskaya's (2022) systematic review of media coverage of climate change in Russia as a major carbon energy exporter not only shows the media's weak public engagement on the issue but also expresses climate agnosticism which the authors attribute to an attempt to balance the country's environmental, economic and political needs.

However, a review of climate change reporting in Africa reveals that the field is still in its infancy with research noticed in only 9 out of 54 countries (Okoliko and de Wit, 2020). However, these studies depict a discursive orientation different from the one observed in the Global North. For instance, in Tanzania, Siyao and Sife's (2018) analysis of six private and public newspaper's coverage from 2006 to 2015 reveals a meagre 0.84% of all the articles for that period reported climate change suggesting that less attention was paid to the problem. Meanwhile, Ukonu et al.'s

(2012) examination of four Nigerian newspapers' reports on climate change reveal that although climate change stories were prominent as a human-induced problem, news articles linked EWEs to weather variability. In a follow-up to Siyao and Sife's (2018) findings, Elia's (2019) study confirmed Shanahan's (2009) findings that lack of training and editorial constraints are the main challenges that climate journalists face in Tanzania. Studies like these indicate the need for increased scholarship in climate change coverage in Africa to establish the continent's various discursive construction of climate change problem.

Research on climate change coverage in Malawian press is rare (Okoliko and de Wit, 2020). This is worrisome considering the country's vulnerability to climate change. However, one study available according to the researchers is a comparative multimodal analysis of discursive strategies in Malawi's print media and international online media in the representation of the catastrophic flooding disaster of January 2015 (Chikaipa, 2018). The results show negative representations of floods as destructive which embodied hegemonic, humanitarian and expertise discourses, reminiscent of Global North's view of Africa without mentioning its causes. Although Chikaipa's (2018) study emphasises on production and consumption of media texts, not only did it not investigate media texts' suggested solutions to such EWEs, but it also does not mention the role of claim-makers in such media representations. This study therefore intended to fill this gap by adopting Bennett's (2001) news structural biases and Hilgartner and Bosk's (1988) public arenas model for a quantitative content analysis of Malawian newspapers' coverage of climate change from 1 October 2014 to 30 April 2016 in order to find out the message(s) being relayed to the audience, their frequency and prominence while also highlighting how claim-making activity of powerful people in the country may have contributed to such coverage.

Methods

The dataset for the study comprised of news articles from Blantyre Newspapers Limited (BNL) and Nation Publications Limited (NPL). These two newspaper stables were chosen because they are the leading newspapers in Malawi (Kanyang'wa and Mkwaila, 2011). Although newspaper is an elitist medium in the country due to literacy and economic constraints (Chimombo, 2008; National Statistical Office, 2019), and with their circulation restricted to few literate urbanite dwellers, newspapers discourse was chosen due to their powerful influence over policy-makers (Kanyang'wa and Mkwaila, 2011). Furthermore, prominent stories from these newspapers are also broadcast in various radio and television stations in the country as press review programmes in both English and Chichewa (the country's lingua franca) which increase their audience reach and influence coverage in other media platforms. This study chose climate change coverage due to its far-reaching effects in Malawi's small-scale rain-fed agro-based economy, where EWEs seriously impact people's livelihood practices. Data was collected from 1 October 2014 to 30 April 2016. The target was two rainy (hot-wet) seasons: from November to April because as a rain-fed agro-based economy, the country experienced some of the worse impacts of climate change in the form of heavy rainfall leading to severe floods in 2014/2015 growing season and severe drought conditions in 2015/ 2016 season.

The study used quantitative content analysis to find out how climate change was broadly reported. Quantitative content analysis is a systematic and replicable examination of manifest and latent content of recorded communication in order "to describe the communication, draw inferences about its meaning, or infer from the communication to its context, both of production and consumption" (Riffe et al., 2024: 24). Since the audience is affected by certain statistical properties of mass-produced messages in a way that neither the producer nor the consumers may be conscious of (Bennett, 2001; Moser and Dilling, 2011; Priest, 2016; Riffe et al., 2024), this quantitative content analysis of climate change coverage should reveal the discursive construction of climate change in Malawian press during the study period. Prominence of news articles was measured based on placement on the cover page and presence or absence of a photograph as used by several scholars (Schooler et al., 1996; Ukonu et al., 2012; Siyao and Sife, 2018). Furthermore, availability of subheading and a frame encasing the report were also considered as these are the newspapers story enhancement mechanism (Serdali et al., 2016). Ideally, the most prominent news article would be placed on the cover page with a subtitle, photograph and encased in a frame. Such features create what Serdali et al. (2016: 4732) calls "an architectonic structure" that has expressive function of presenting important information while performing marketing function of optimising readers' perception of the content, thereby stimulating readers' emotional involvement not only in the article, but also towards the represented phenomenon.

News articles were found through manual identification of articles about climate change, EWEs, global warming, flooding or drought in their printed form since Malawi did not have an online newspaper archive during the period of data collection (i.e. November–December 2023). The last two terms were included in the search as they are the commonest climate change-related disasters in Malawi. A total of 415 news articles were found, from which, 203 came from NPL, while the remaining 212 were from BNL. These articles were analysed to find out the frequency and prominence of newspapers' coverage of climate change as well as the causes they attributed to EWEs and their suggested solutions. To minimise researcher bias, two independent and trained coders were used to code the data using the code form in Figure 1.1 on page 23.

1 was marked for a story that starts on the front page and 2 for a story that did not appear on the front page. This excluded the use of skybox to promote an item on a front page. If a story started on the front page and continued inside, it was still marked 1. For story origin, 1 was coded for stories from the newspaper's staff, 2 was used when it originated from wire services, and 3 when their origin was not stated. For geographical focus, 1 was used when the story focused on Malawi, 2 when it is not focusing on Malawi, and 3 was used when the location was not stated. A story that had a picture was coded 1 and the one without picture was coded

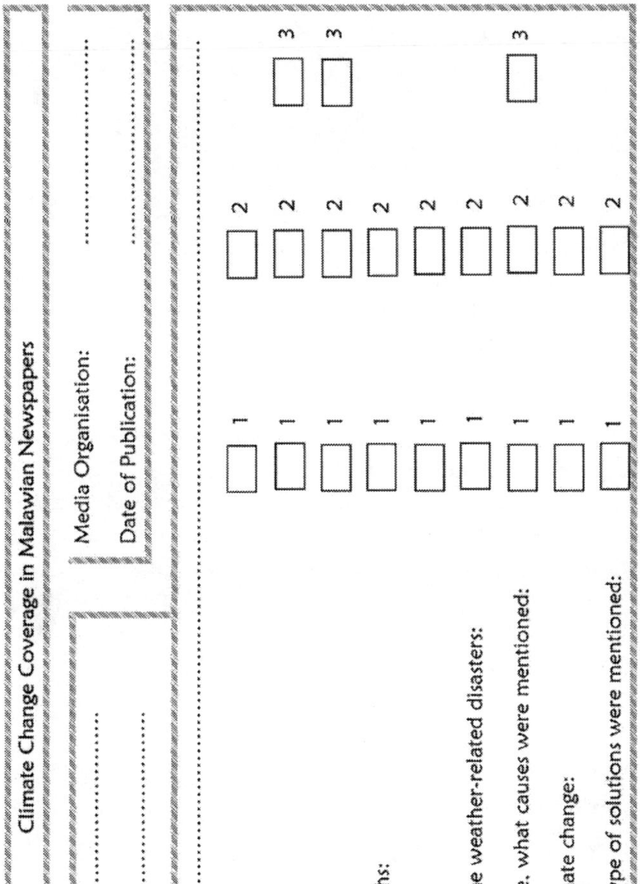
Climate Change Coverage in Malawian Newspapers

Coder's Initials: .... Media Organisation: ....

Date of Coding: .... Date of Publication: ....

Title of the article: ....

1. Placement of the story: ☐ 1 ☐ 2
2. Source of the story: ☐ 1 ☐ 2 ☐ 3
3. Geographical focus: ☐ 1 ☐ 2 ☐ 3
4. Availability of photographs: ☐ 1 ☐ 2
5. Availability of frame: ☐ 1 ☐ 2
6. Mention causes of extreme weather-related disasters: ☐ 1 ☐ 2
7. If yes to question 6 above, what causes were mentioned: ☐ 1 ☐ 2 ☐ 3
8. Mentions solution to climate change: ☐ 1 ☐ 2
9. If yes to 8 above, what type of solutions were mentioned: ☐ 1 ☐ 2

Figure 1.1 Coding form.

Source: Adapted from Lynch and Peer (2002).

2. A story with a frame was coded 1 and without a frame was coded 2. A story that linked climate change to EWEs was coded 1 and those that did not were coded 2. Furthermore, a story that provided solutions to climate change was coded 1 and those that did not were coded 2. A pilot content analysis of 30 news articles was done, and when the code book was reviewed by the two coders and the researcher, an average Holsti's intercoder reliability of 96% was found. Quantitative data was analysed based on descriptive statistics in form of percentages and frequencies.

Results

Frequency of Climate Change Reporting and Attribution of Causes to Climate Change

Regarding frequency of climate change coverage, results show that although there were floods and drought during the study period, there was high newspaper coverage during the period of floods (January–March 2015) than that of drought (January–March 2016) as depicted in the Figure 1.2 below.

As shown in Figure 1.2, there was a high coverage of climate change and EWEs in both newspaper stables between January and March 2015 as the country experienced severe flooding. This period was also followed by months of high coverage between May and December as the nation was grappling with the need to provide food and temporary shelter to the displaced flood victims and refurbishing destroyed public infrastructures. However, media coverage of the severe drought that followed from January to March 2016 slumped to almost non-existent as can be seen in February 2016, which is the middle of a typical growing season in Malawi. Coverage picked up a bit in April as people saw the destructive effects of the drought. However, even in this case, it was not as high as it was in January and March 2015.

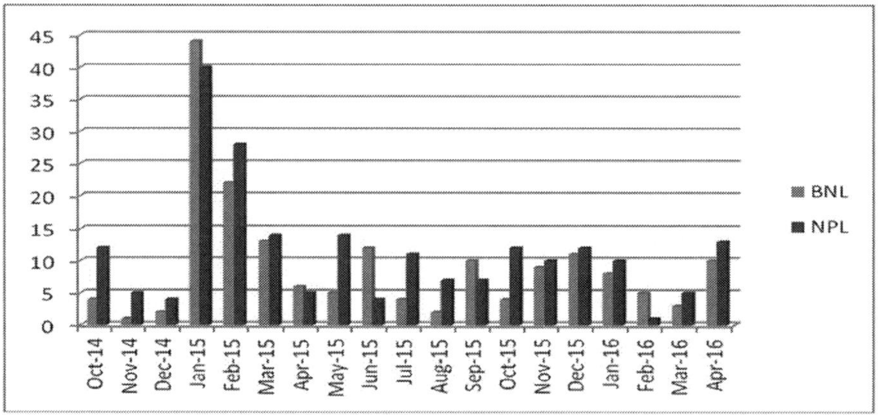

Figure 1.2 Frequency of climate change coverage in newspapers over the study period.

On whether news stories mentioned causes of EWEs, results show that 70.5% and 87.5% of BNL and NPL news articles, respectively, did not mention the causes. Instead, these news articles attributed the disasters to either heavy rains or El Nina, respectively. Furthermore, 63% and 58% of news articles from BNL and NPL, respectively, that showed the causes of EWEs were from wire services such as Inter Press Service (IPS), Reuters and Associated Press (AP) without Malawian context. Among the news articles that mentioned causes of EWEs (29.5% in BNL and 12.5% of NPL), 61.5% and 60% from BNL and NPL, respectively, indicated that EWEs were a result of climate change, which is caused by deforestation and overpopulation. The remaining news articles (38.5% for BNL and 40% from NPL) that mentioned greenhouse gas emissions as the main cause were from wire services.

Solutions to Climate Disasters and Prominence of Climate Reporting

Furthermore, results also showed that 73.4% and 66.8% of BNL and NPL news articles, respectively, did not provide solutions to climate change as they emphasised on how to help the affected people with short-term assistance like provision of food, plastic plates and buckets, plastic rolls for the constructions of temporary shelters as well as seeds and fertiliser for irrigation farming. Only 26.6% and 33.2% from BNL and NPL news articles reported on how the nation can contribute to climate change mitigation and adaptation. However, a majority of these news articles that provided solutions were from wire services like IPS, Reuters and AP with very little Malawian context. Furthermore, this study also tried to establish prominence of climate change and (or) EWEs news based on their placement, presence of photograph(s), subtitle and frame encasing the report, and Figure 1.3 on page 26.

As Figure 1.3 shows, less than 5% of all climate change news articles had all features of prominence and these were mostly stories featuring prominent political and religious figures. Hence, climate change news stories were less prominent in the newspapers analysed. Furthermore, Figure 1.3 also shows that use of photograph was the commonest prominence feature. However, there were more photographs depicting the disasters than those depicting short-term and long-term responses to the problem as Figure 1.4 illustrates.

Figure 1.4 on page 26 indicates that there were more photographs depicting the destruction that the floods and drought had caused such as flooded fields, wilting maize fields, flood-destroyed houses and roads. This was also followed by short-term responses where state and non-state actors were donating relief items such as food, plastic rolls for constructing temporary makeshift houses, plastic buckets and clothes. Meanwhile, long-term responses such as introducing irrigation farming and relocation of flood victims to upland areas and provision of social amenities in such newly relocated areas were scarcely reported.

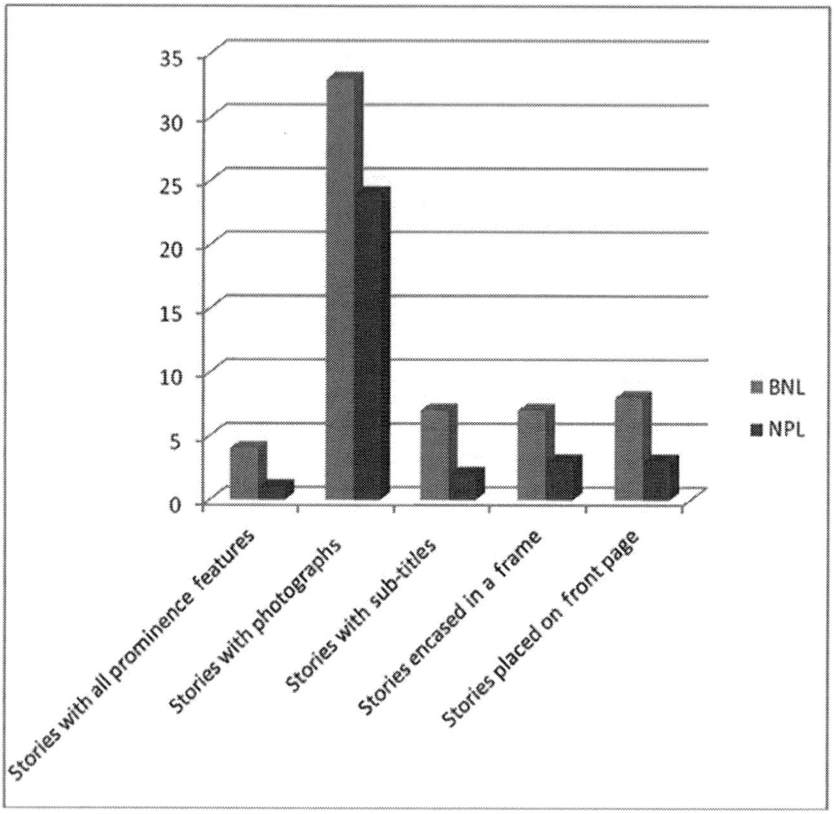

Figure 1.3 Prominence of climate change stories in the newspapers.

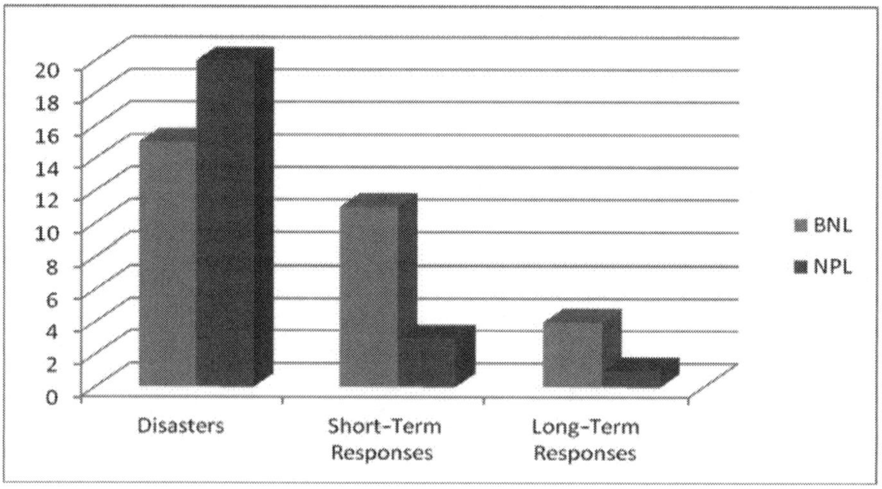

Figure 1.4 Messages in the photographs.

Discussions

This study has found that although there was minimal attention to climate change reporting during the period, the 2014/15 floods generated more attention than the 2015/16 droughts. However, in both cases, there was very little attempts to provide their causes and solutions. These findings may be viewed in the context of the four structural news biases and public arena model. First, although there were two different EWEs during the study period, floods were highly reported than droughts when coverage plummeted substantially downwards. However, despite such scanty newspaper coverage, scientific studies indicate that drought is the greatest risk that climate change poses to rain-fed agricultural economies in Africa like that of Malawi (Legg, 2021). Nonetheless, journalists in pursuit for melodramatic news found drought not news worthy as there was no sudden change in the status quo like that of floods. Although the drought satisfied Bednarek and Caple's (2017) timeliness news value as an on-going disaster, its lack of oddity as a gradual event could not fulfil unexpectedness news value for publication. Hence, events in the slow build up to worst drought of the current decade were rarely reported until when it reached frightening levels with large-scale impacts. In pursuit for 'alarmist' reporting, the shocking effects of the drought, with its vivid images of wilting crops, then started to appear in April of 2016 towards the end of Malawi's growing season. However, even then, such stories rarely mentioned origins, consequences and solutions. Although this could be due to the shocking loss of lives and property that the floods caused which created what Bennett (2001: 57) calls "hyper-dramatism", this coverage has the potential to lead to decline in the public's interest on drought as a serious environmental problem since media attention have been found to influence audience's concerns on social problems (Shanahan, 2009; Priest, 2016; Riffe et al., 2024).

Furthermore, contrary to Shanahan's (2013) recommendation to African journalists, the results show that most articles neither mentioned causes of climate change nor did they show any relationship between climate change and EWEs. Instead, climate change reports were personalised to emphasise on affected people and their properties. Although readers are free to interpret news in their own way, as Bennett (2001: 145) tells us, "they cannot interpret what they don't see [or read]" unless in rare cases where the reader has a direct experience with an issue or is so interested that she/he searches for additional information. Furthermore, in line with Bennett's (2001) news fragmentation, reports on impacts of climate change such as floods or drought in different parts of the country were depicted as episodic climatic hazards, usually caused by the affected peoples' actions like deforestation without any bearing to negative ecological activities occurring in other parts of Malawi or the world. Such sporadic reporting could not "look beyond human drama to explore origin of problems and the larger socio-cultural, [geo-physical,] economic and political context in which the immediate story has developed" (Bennett, 2001: 41). This de-contextualised coverage which fails to adequately report the cumulative, non-visible and long-term environmental problems makes the drought and (or) flood affected people more susceptible to blame for either their choice of settlement site, refusal to relocate to a safer location or their environmental behaviour.

Furthermore, the media's preoccupation with authority and (dis)order made them devote more space to authority figures like government officials, members of the academia and business gurus making pronouncements for 'normalising' the flood or drought situations through either short-term humanitarian efforts or doubting the government's ability to do so. While this heightened the story's dramatic tensions which could then attract readers especially where the authorities fail to resolve the problem, it has elsewhere led to the promotion of political voices with economic links to carbon-based industry (Carvalho and Burgess, 2005; Boykoff and Boykoff, 2007). Furthermore, by privileging expert voices, local knowledge about climate change and people's experiences and actions are backgrounded, which encourages a passive attitude among the general public as they might feel that their actions are not connected to the disaster being reported. Although authority and (dis)order bias is due to journalists' preoccupation to maintain social order after threatening crises (Boykoff and Boykoff, 2007), it may have a numbing effect on mobilising the much-needed collective efficacy which is essential to mitigate and adapt to climate change.

By 'covering' the link between EWEs and climate change, as well as the causes (especially fossil fuels), solutions and ways of mitigating and adapting to it, it makes the mass media silent accomplices to climate change problem. Although "exposing the impacts of big businesses, especially transnational corporations on climate change, is the necessary [though insufficient] condition" for ecological citizenship (Carvalho, 2007: 181), this study reveal that deforestation is a synecdoche sign for climate change in Malawi. Since accurate knowledge of climate change is the strongest predictor of the audience's decision to adopt positive ecological behaviour (Boykoff, 2007; Selormey et al., 2019), any efforts to devise a panacea to the problem is likely to remain ineffective if the causes and consequences of the problem and its solutions are not well understood by the public. On the other hand, while international media may be better resourced than their counterparts in Malawi, relaying on their climate change reports to explain causes of climate change in the country's press de-contextualises the problem, thereby making it difficult for the local readers to appreciate its relevance to them.

These findings are contrary to Government of Malawi's (2016) and Shanahan et al.'s (2013) recommendation for reporters to reduce disaster narratives in their climate change reports and emphasise on prevention, preparedness, mitigation and recovery. However, the results coheres with Williams' (2000: 68) findings of newspapers coverage of global warming between 1976 and 1998 in the USA where the "proposed solutions to global warming made up of only a small portion of the overall discourse." Hence, structural news bias of dramatisation took centre stage with shocking photographs as witnesses to EWEs. Nonetheless, studies have shown that the same images can also paralyse and demobilise the public from exploring possible actions to combat climate challenges as they feel overwhelmed and therefore helpless (O'Neill et al., 2013). This is even more pertinent in Malawian situation where newspapers' coverage of EWEs paid very little attention to their causes and solutions.

Lastly, results also reveal that climate change newspaper stories were not prominent during the study period. Out of the four features of prominence that this study used, i.e. availability of a photograph, subheading, encasement in a frame and placement on front page that together provide what Serdali et al. (2016: 4731) call "stimulus-signal" capable of arousing reader's interest, the use of photographs was the commonest. Hence, news stories reporting the extent of the destruction the floods or drought had caused were made prominent through use of vivid images. However, the stories rarely appeared on the cover page unless when they were connected to some prominent people, thereby making them susceptible to being easily glossed over by most unengaged readers. This is similar to Tagbo's (2010) findings that Nigerian and South African newspaper editors unanimously indicated that climate change is not a front page subject unless it has a strong political and economic aspect. Bennett (2001: 41), therefore, argues for the replacement of the current information system of news with perspectives that are "institutional, analytical, historical and reflectively critical" in their orientation. Considering the enormity of the challenge Bennett (2001) is proposing, the current authors are of the view that Hilgartner and Bosk's (1988) public arena model may still be useful in the current information system. Through the claim-making activity of various environmental stakeholders, it is possible to make climate change dominant in the news to warrant front page coverage by roping in powerful politicians, traditional authorities and religious leaders while lobbying for proper framing of climate change reporting in terms of its causes, solutions and linkage with EWEs.

From a public arenas' point of view, climate change and (or) EWEs were unsuccessful in the competition for attention in newspapers space during the study period in terms of prominence and framing. The minimal prominence was even more pronounced during the 2015/2016 drought. Unlike during the 2014/2015 floods where prominent politicians, religious leaders and business captains were involved due to dramatic nature of the destruction and the need to restore social order, there was very minimal prominent figure involvement in monitoring and therefore raising awareness of the drought situation in different parts of the country, and as such, these stories held less news value for journalists. This concurs with Williams (2000) and Boykoff and Boykoff (2007) conclusion that climate change may not be automatically at the top of media agenda basing on scientific reasons and (or) its destructive effects, but rather basing on how it is identified and defined by influential stakeholders in climate change mitigation and adaptation campaign. As Boykoff and Boykoff (2007) and Holt and Barkemeyer (2012) indicate, maintaining a higher level of interest and salience in environmental news is not a simple matter since journalists keep on looking for new dramatic events. However, reinvigorating the climate debate through subtle shift of the topic by influential stakeholders may still be possible (Holt and Barkemeyer, 2012) to motivate continued media coverage that may shape public opinion and motivate positive ecological behaviour.

Second, climate change and EWEs also lost the competition for attention in the newspapers through its framing which left out causes, consequences as well as mitigation and adaptation mechanism although floods and drought disasters had already increased people's attentiveness. It is therefore not surprising that recent Afro-barometer survey found only 43% of Malawians as climate literate; for example, they know the causes and its negative consequences (Selormey et al., 2019). However, by not mentioning the causes and solutions, such reporting might lead to Elisabeth Noelle-Neumann's spiral of silence among the general public, thereby leading to unabated continuation of the problem. Based on public arena model, climate change stakeholders would have mobilised for alternative news framing that emphasises its causes, consequences and solution amidst media's space limitations. This may help overcome what Ungar (1992) calls 'psychic numbing' and stimulate government and public support for sustainable disaster recovery and management since once a problem and its framing gains 'celebrity status' in one public arena like media, it has the possibility to dominate discussions among family members, acquaintances, religious gatherings and parliament deliberations (Hilgartner and Bosk, 1988; Boykoff and Boykoff, 2007). To achieve proper climate change framing in the press, stakeholder in climate change campaign in the country needed to build climate change editorial support, train journalists on climate change reporting (Shanahan, 2009) and create an award for media houses that provide quality coverage of climate change, among others.

Conclusion and Implications

The findings of this study reveal a bleak situation in the coverage of climate change and (or) EWEs in Malawian newspapers between October 2014 and April 2016 as it hides causes of climate change, its role on the emergence of EWEs and also its adaptation mechanisms, as well as long-term solutions. Worse still, reports of the 2015/2016 drought also slumped despite drought being the most serious effect of climate change in Africa (Lisk, 2009). This study attributes such reporting to news media's emphasis on economic norm of reporting stories efficiently and profitably which has led to structural news biases of personalisation, dramatisation, fragmentation and authority–(dis)order (Bennett, 2001). Since the general public get most of their knowledge about climate change from the mass media (Priest, 2016; Selormey et al., 2019; Riffe et al., 2024), such newspaper coverage of climate change and (or) EWEs has the possibility of blurring the public's understanding of the problem in the country leading to the continuation of negative ecological behaviours. This study therefore argues that informing the general public about climate change is not the responsibility of the press alone, but rather a coordinated effort of all climate change stakeholders. Public attention is a scarce resource allocated through competition of powerful claim-makers (Hilgartner and Bosk, 1988), and as Birks (2010) observes, in newspapers, such competition is so intense on the front page due to its function as the newspaper's billboard for advertising articles that the producers consider important to the audience. Hence, all stakeholders in climate change mitigation and adaptation campaign should have collaborated to raise its prominence in terms of its causes, effects and solutions as an urgent social problem that warrant not just newspaper coverage but also its appearance on front page.

The findings of this study may be limited in its emphasis on newspaper coverage of climate change in Malawi in a two-year period. Future studies may examine radio and television reporting of climate change in the country to find out if they adopt the same episodic reporting or not. Furthermore, while this study reports newspaper coverage within the 19-month period between October 2014 and April 2016, other studies may consider adopting a longitudinal approach to find out if the event-centred sporadic climate change reporting that prioritises risk and crisis over their causes, consequences and solutions persists. Other studies may also consider finding out from reporters why they did not prioritise causes, consequences, solutions and adaptation mechanism in their coverage of climate change. However, despite all these, results of this study still imply that research on media coverage of climate change in democratic societies like that of Malawi may be better understood if viewed in relation to climate change stakeholders' claim-making activities during the study period.

References

Akpan C, Anorue L and Ukonu M (2012) An analysis of the influence of the Nigerian mass media on public understanding of climate change. *Journal of Alternative Perspectives in the Social Sciences* 4(4): 688–710.

Ashe T and Poberezhskaya M (2022) Russian climate scepticism: An understudied case. *Climatic Change* 41: 41–20.

Bacot H and Fitzgerald MR (1999) Issue salience, news coverage, and attention cycles of environmental problems. *Southeastern Political Review* 27(3): 525–543.

Bednarek M and Caple H (2017) *The discourse of news values: How news organisations create newsworthiness.* New York: Oxford University Press.

Bennett WL (2001) *News: The politics of illusion.* Chicago: Longman.

Billett S (2010) Dividing climate change: Global warming in the Indian mass media. *Climatic Change* 99(1–2): 1–16.

Birks J (2010) Press protest and publics: The agency of publics in newspaper campaigns. *Discourse and Communication* 4(1): 51–67.

Botha BN, Nkoka FS and Mwumvaneza V (2018) Hard Hit by El Nino. DOI: https://doi.org/10.1596/30037.

Boykoff MT (2007) Flogging a dead norm? Newspaper coverage of anthropogenic climate change in the United States and United Kingdom from 2003 to 2006. *Area* 39(4): 470–481.

Boykoff MT and Boykoff JM (2007) Climate change and journalistic norms: A case-study of US mass-media coverage. *Geoforum* 38(6): 1190–1204.

Carvalho A (2007) Communicating global responsibility? Discourses on climate change and citizenship. *MCP* 3(2): 180–183.

Carvalho A and Burgess J (2005) Cultural circuits of climate change in UK broadsheet newspapers, 1985–2003. *Risk Analysis: An International Journal* 25(6): 1457–1469.

Chikaipa V (2018) *A critical analysis of media discourse after a natural flooding disaster in Malawi, in 2015*. Stellenbosch University. PhD Thesis, Stellenbosch, South Africa.

Chimombo M (2008) Government journalism in Malawi: From totalitarianism to democracy. In: Phiri KM and Ross KR (eds.) *Democratisation in Malawi: A stocktaking.* Zomba: CLAIM, pp.217–236.

Danielsen A (2016) Climate activism and the mass media: Potentially a politically challenging debate. In: Almlund P, Jespersen PH and Riis S (eds.) *Rethinking climate change research*. Surrey, United Kingdom: Routledge, pp.279–294.

Elia EF (2019) Media coverage of climate change information in Tanzania. *Global Knowledge, Memory and Communication* 68(4/5): 258–274.

Engesser S and Brüggemann M (2016) Mapping the minds of the mediators: The cognitive frames of climate journalists from five countries. *Public Understanding of Science* 25(7): 825–841.

Government of Malawi (2016) National climate change management policy. Lilongwe: Ministry of Environment and Climate Change Management.

Government of Malawi (2021) Third national communication of Malawi. Lilongwe: Ministry of Natural Resources.

Government of Malawi (2023) Malawi 2023 tropical cyclone Freddy post-disaster needs assessment. Lilongwe: Department of Disaster Management Affairs.

Government of Malawi (2024) National El Niño induced prolonged dry spell and floods response appeal. Lilongwe: Department of Disaster Management Affairs.

Hilgartner S and Bosk CL (1988) The rise and fall of social problems: A public arenas model. *American Journal of Sociology* 94(1): 53–78.

Holt D and Barkemeyer R (2012) Media coverage of sustainable development issues–attention cycles or punctuated equilibrium? *Sustainable Development* 20(1): 1–17.

Kanyang'wa M and Mkwaila S (2011) Media markets and commodification of news. In: Kishindo PJ, Kondowe EBZ and Mkandawire LF (eds.) *Journalism practice in Malawi: History, progress, and prospects*. Lilongwe: Malawi Council for UNESCO, pp.163–194.

Legg S (2021) IPCC, 2021: Climate change 2021-the physical science basis. *Interaction* 49(4): 44–45.

Lisk F (2009) Overview: The current climate change situation in Africa. In: Besada H and Sewankambo N (eds.) *Climate change in Africa: Adaptation, mitigation and governance challenges*. Ontario: Centre for International Governance Innovation, pp.8–15.

Lynch SM and Peer LJ (2002) *Analysing newspaper content: A how–to guide*. Chicago: Readership Institute.

Mataya DC, Vincent K and Dougill AJ (2020) How can we effectively build capacity to adapt to climate change? Insights from Malawi. *Climate and Development* 12(9): 781–790.

McCright AM and Dunlap RE (2000) Challenging global warming as a social problem: An analysis of the conservative movement's counter-claims. *Social Problems* 47(4): 499–522.

Moser SC and Dilling L (2011) Communicating climate change: Closing the action-science gap. In: Dryzek JS, Norgaard RB and Schlosberg D (eds.) *The Oxford handbook of climate change and society*. Oxford: OUP, pp.161–174.

National Statistical Office (2019) 2018 Malawi population and housing census: Main report. Zomba: National Statistical Office.

Okoliko DA and de Wit MP (2020) Media(ted) climate change in Africa and public engagement: A systematic review of relevant literature. *African Journalism Studies* 41(1): 65–83.

Okoliko DA and de Wit MP (2021) From "communicating" to "engagement": Afro-relationality as a conceptual framework for climate change communication in Africa. *Journal of Media Ethics* 36(1): 36–50.

O'neill SJ, Boykoff M, Niemeyer S, et al. (2013) On the use of imagery for climate change engagement. *Global Environmental Change* 23(2): 413–421.

Oxford Analytica (2023) Aftermath may add to Malawi-Mozambique cyclone damage. Emerald Expert Briefings (oxan-es). https://doi.org/10.1108/OXAN-ES276725

Priest S (2016) *Communicating climate change: The path forward.* Washington: Springer.

Ramanathan V and Feng Y (2009) Air pollution, greenhouse gases and climate change: Global and regional perspectives. *Atmospheric Environment* 43(1): 37–50.

Riffe D, Lacy S, Watson BR, and Lovejoy, J. (2024) *Analyzing media messages: Using quantitative content analysis in research.* New York and London: Routledge.

Schooler C, Shyam Sundar S and Flora J (1996) Effects of the Stanford five-city project media advocacy program. *Health Education Quarterly* 23(3): 346–364.

Selormey E, Dome MZ, Osse L, and Logan, C. (August 2019) Change ahead: Experience and awareness of climate change in Africa. Afrobarometer Policy Paper No. 60. Accra, Ghana: Afrobarometer.

Serdali BK, Ashirbekova GS, Isaeva Z, et al. (2016) Newspaper headings as a means of presenting priority and secondary information. *International Journal of Environmental and Science Education* 11(11): 4729–4738.

Shanahan M (2009) Time to adapt? Media coverage of climate change in nonindustrialised countries. In: Boyce T and Lewis J (eds.) *Climate change and the media.* London: Peter Lang, pp.145–157.

Shanahan M, Shubert W, Scherer C, and Corcoran T. (2013) *Climate change in Africa: A guidebook for journalists.* Paris: UNESCO.

Siyao PO and Sife AS (2018) Coverage of climate change information in Tanzanian newspapers. *Global Knowledge, Memory and Communication* 67(6/7): 425–437.

Tagbo E (2010) *Media coverage of climate change in Africa.* UK: Reuters Institute for the Study of Journalism.

Ukonu MO, Akpan CS and Anorue LI (2012) Nigerian newspaper coverage of climate change, 2009–2010. *New Media and Mass Communication* 5: 22–33.

Ungar S (1992) The rise and (relative) decline of global warming as a social problem. *Sociological Quarterly* 33(4): 483–501.

Williams J (2000) The phenomenology of global warming: The role of proposed solutions as competitive factors in the public arenas of discourse. *Human Ecology Review* 7(2): 63–72.

2 Pitfalls in Climate Change Communication

Insights from Nigeria and Algeria

Ogadimma C. Emenyeonu and Nassir Bouali

Introduction

Climate change is one of the most urgent global challenges, marked by long-term shifts in temperature, precipitation, and more frequent extreme weather events (Moser, 2010; Emenyeonu et al., 2023; Ripple et al., 2023). The Intergovernmental Panel on Climate Change (IPCC) confirms that human activities, especially fossil fuel use and deforestation, are primary drivers (IPCC Sixth Assessment Report, 2021). The socioeconomic impacts are unevenly distributed, with vulnerable populations in the Global South facing the greatest risks due to limited adaptation resources. Climate change intensifies food insecurity, water scarcity, health crises, and displacement (FAO Report, 2018). Addressing these challenges requires coordinated global and local action, including a shift to renewable energy, ecosystem protection, and support for affected communities.

The media plays a vital role in raising awareness and shaping policy responses. Against this backdrop, this chapter explores climate change communication in Nigeria and Algeria, two African countries with distinct cultural and geographic contexts but shared climate vulnerabilities. Nigeria faces communication barriers rooted in traditional beliefs, low literacy, poverty, and an economy reliant on agriculture and oil. Algeria also struggles with reconciling economic dependence on hydrocarbons with environmental sustainability, further complicated by political instability and limited freedom of expression.

Through interviews and data analysis, this chapter identifies key obstacles to effective climate communication in both countries and proposes practical strategies to overcome them. It highlights the need for localized, inclusive communication approaches that reflect societal realities and actively engage communities. The findings contribute to global climate discourse by offering insights for improving communication in similarly affected regions.

The Media's Role in Shaping Climate Change Communication in Africa: A Comparative Study of Nigeria and Algeria

Climate journalism plays a crucial role in addressing the global climate crisis by informing, educating, and mobilizing public action. As a specialized field, it

DOI: 10.4324/9781003660217-4

focuses on climate change – its causes, impacts, and potential solutions. However, as climate change reshapes societal priorities and the ethical scope of journalism (Zelizer, 2017), the media's ability to engage the public and promote meaningful action is increasingly scrutinized (Hackett, 2017; Schäfer et al., 2017). Media influences how people perceive climate change (Boykoff & Roberts, 2007). Guided by framing theory, this process helps the public interpret diverse perspectives, enhancing communication effectiveness (Nisbet, 2009; Scheufele & Iyengar, 2014). Media representations also shape public discourse, determining whether climate change is viewed as an urgent crisis or a distant concern (Olausson, 2009; Takahashi & Meisner, 2013).

As the primary source of climate science and policy information, the media is key to raising awareness and framing public discourse (Schäfer & Schlichting, 2014). Campaigns can galvanize support and push climate change to the forefront of policy agendas (Moser, 2010). Their effectiveness, however, depends on journalists' ability to report climate issues accurately and compellingly. Research shows that training and capacity-building can significantly improve climate reporting by making scientific information more accessible and engaging (Painter & Ashe, 2012). Yet, this effectiveness varies by region, with Africa facing unique climate communication challenges.

In Africa, the media contends with numerous barriers to comprehensive climate coverage. Socioeconomic, political, and scientific factors limit the depth and consistency of reporting (Sharif & Medvecky, 2018; Emenyeonu et al., 2023). Commercial pressures push sensationalism and breaking news over substantive climate stories (Jan et al., 2020). Additionally, climate change is often framed as a Western-centric issue, disconnecting African audiences from relevant scientific narratives (Batta, Ashong & Bashir, 2013). Limited data access and underrepresented African voices in global discourse further hinder effective communication (Nyasulu & Koteyko, 2020).

Localized content is essential for making climate communication relevant to African communities (Jegede, 2016). Yet, many media outlets fail to report in local languages, creating a gap between climate discourse and daily realities. Bridging this divide requires greater use of community radio, local media, and social platforms to educate rural populations (Manyozo, 2012). Social media, in particular, has become a powerful tool for reaching younger audiences, who increasingly engage in climate activism and online discussions (Bosch, 2017).

Nigeria highlights both the promise and the challenges of climate communication. The Nigerian media has helped raise awareness through public engagement, though coverage remains sporadic and often tied to specific events (Emenyeonu et al., 2023; Oyebode & Adeyemi, 2017). Efforts to train journalists have improved the environment for in-depth reporting (Nwankwo & Okafor, 2017). In some cases, media campaigns have influenced policy, especially in urban areas where climate issues receive more attention (Akpan & Etuk, 2015). However, rural communities remain largely excluded due to infrastructural and linguistic barriers (Onyisi, 2021).

In contrast, Algeria faces challenges rooted in media control. Government restrictions limit the scope of climate reporting, though independent and digital media are emerging as alternatives (Hamouchene, 2016). In the absence of robust journalism, environmental NGOs and international organizations have taken the lead in raising awareness, often using media channels to communicate their messages (Lounis, 2017). Media efforts in Algeria frequently target youth, promoting awareness and environmental activism (Bencherif, 2018). Digital platforms have enabled activists to bypass traditional restrictions and reach wider audiences (Earl, 2022).

In both Nigeria and Algeria, media plays a vital role in shaping public understanding of climate change. Despite differences in media systems and political environments, both countries reflect broader African challenges in climate communication. While media has the potential to inform, educate, and mobilize, its impact is constrained by structural and contextual factors. Strengthening journalist training, promoting localized content, and leveraging digital tools are essential for improving communication across the continent.

In conclusion, Nigeria and Algeria face significant hurdles in climate change communication that hinder public understanding and limit effective action. Despite global urgency, limited resources, political constraints, linguistic diversity, and lack of specialized training complicate climate reporting. In Nigeria, coverage tends to be fragmented and driven by disasters rather than long-term trends or mitigation. In Algeria, tight government control restricts media freedom and narrows climate narratives. To improve communication, both countries need specialized journalist training, access to localized information, and a shift toward inclusive, context-aware storytelling. Engaging underrepresented communities through local languages, trusted leaders, and indigenous knowledge is vital for fostering a more informed and active response to climate change.

Environmental Degradation and Climate Challenges in Nigeria and Algeria: Governance, Corporate Accountability, and Media

The Niger Delta in Nigeria presents a stark case of environmental degradation, primarily caused by the activities of multinational oil companies. Once rich in biodiversity, the region is now plagued by oil spills, gas flaring, and deforestation, which have led to severe socio-economic and health consequences. Research by Nriagu et al. (2016) and the UNDP (2006) highlights the devastating impact of oil exploitation on the health, environment, and livelihoods of local residents. Among the most pressing challenges are oil spills, which contaminate water sources, destroy aquatic ecosystems, and decimate fishing and agriculture which are the key income sources. Gas flaring contributes to severe air pollution, acid rain, respiratory issues, and degradation of ecosystems, particularly mangrove forests (Amnesty International, 2018). The loss of mangroves, vital to ecological balance, intensifies environmental damage and increases community vulnerability to climate change. The IPCC projects sea level could rise by up to one meter by century's end, severely impacting Nigeria's coastal communities and compounding existing problems.

Scholars and environmental groups consistently emphasize these issues, especially their effects on health and livelihoods. Okonta and Douglas (2003) describe the socio-environmental devastation from oil extraction and the lack of corporate accountability. Watts (2008) links petro-politics to regional instability. Oil companies' failure to address these challenges, as noted by Amnesty International (2018), reveals the need for improved governance, transparency, and corporate responsibility. Yet, media coverage often focuses on isolated incidents like oil spills, rather than the broader socio-economic and political dimensions, limiting public understanding of the systemic nature of the crisis.

Similarly, Algeria faces severe environmental challenges, driven by its arid climate. Desertification and water scarcity are major threats. Climate change, combined with human activities such as overgrazing and deforestation, accelerates desert expansion, reducing arable land and threatening agriculture. Research by Bencherif and Bensalem (2013) and the UNCCD (2014) shows rapid soil degradation and declining crop yields, contribute to food insecurity. Water scarcity, worsened by population growth and climate change, affects both rural and urban areas. Reports such as Water Crisis in Algeria: A Thirsty Future (The National) highlight the impact on agriculture and domestic supplies, calling for better water management and international cooperation. The growing frequency and intensity of extreme weather events, such as heatwaves, also pose rising health risks (World Meteorological Organization, 2021).

Environmental degradation in both Nigeria and Algeria illustrates the consequences of fossil fuel dependence. Both are major fossil fuel producers – Nigeria reliant on oil and Algeria on natural gas. This reliance drives environmental damage: oil spills, gas flaring, and deforestation in Nigeria and pollution and unsustainable land use in Algeria. Weak governance, corporate impunity, corruption, and bureaucracy worsen these issues in both nations.

In both countries, the media plays a vital role in raising awareness on climate change issues. However, journalists often report on discrete events including oil spills or water shortages without linking them to broader environmental and political systems. Governance failures and corporate negligence further complicate mitigation efforts, as noted by Okonta and Douglas (2003) for Nigeria and Schilling et al. (2012) for Algeria. The absence of deeper coverage limits public understanding and pressure on decision-makers. In Algeria, restricted press freedom makes it especially difficult for journalists to critique environmental policies or government responses to climate change, reducing the media's effectiveness in advocacy.

While both countries face climate vulnerabilities: Nigeria with flooding, desertification, and biodiversity loss, and Algeria with water scarcity and extreme heat yet important differences shape their environmental responses. Nigeria's diverse climate results in varied impacts, while Algeria's arid terrain prioritizes desertification and water conservation. Moreover, Nigeria's more pluralistic media allows some independent reporting, whereas Algeria's state-controlled press limits environmental discourse, often framing climate issues as matters of national security (IMF, 2024; World Food Program, 2024).

Despite extensive documentation of environmental crises, key barriers remain. Fragmented media coverage, weak regulation, poor journalistic training, and inadequate oversight hinder progress. This is seen in Nigeria's slow cleanup and enforcement efforts in the Niger Delta and Algeria's limited success in combating desertification and managing water shortages. Climate change-driven food insecurity and displacement, as highlighted in FAO's 2018 report, disproportionately affect vulnerable groups, deepening social inequalities.

Comparing Nigeria and Algeria reveals how governance, corporate responsibility, and media ecosystems shape climate action in resource-rich African nations. The environmental challenges faced by both underscore the need for urgent reforms in policy, regulation, and communication. Despite mounting evidence, media often fails to provide sustained, contextualized coverage that informs and mobilizes the public. Addressing climate change in these contexts requires not only global cooperation and a shift to renewable energy, but also locally grounded journalism, stronger institutions, and targeted support for affected communities. Comprehensive media engagement and proactive governance are crucial for advancing meaningful climate action in Nigeria, Algeria, and beyond.

Drawing from the preceding literature discussion, this study addresses these research questions:

1 What are the primary challenges impeding effective climate change communication in Nigeria and Algeria, and how do socio-economic, political, and media-related factors contribute to these challenges?
2 How do cultural and linguistic diversity, as well as indigenous knowledge systems, influence the dissemination and public understanding of climate change information in both countries?
3 What role does the media play in shaping public perceptions of climate change in Nigeria and Algeria, and how can media strategies be improved to enhance awareness and engagement, particularly in rural and underrepresented areas?
4 What are the most effective approaches for fostering climate action in Nigeria and Algeria, considering governance, corporate accountability, and community engagement?

Research Methodology

This study adopts a qualitative approach, integrating in-depth interviews and document analysis to explore how journalists in Nigeria and Algeria communicate climate change. This mixed-method design offers a cross-national perspective, revealing how political, socio-economic, and cultural factors shape media narratives on climate issues.

Data collection methods were tailored to the distinct media environments in both countries. In Nigeria, semi-structured interviews were conducted with three journalists, two from private newspapers and one independent freelancer. They were selected for their expertise in climate reporting. Although five journalists were initially invited, only three responded, possibly reflecting limited engagement with climate communication due to disinterest, time constraints, or institutional barriers. Interviews were conducted by telephone, recorded, and transcribed for analysis. To maintain confidentiality, participants were labeled J1, J2, and IJ (Independent

Journalist). Interview questions aligned with the study's objectives, exploring journalists' experiences, challenges, and strategies in reporting climate change. Given the limited responses, document analysis was employed to supplement the findings, enriching insight into Nigeria's climate communication landscape.

In Algeria, interviews proved unfeasible due to political sensitivities and media restrictions. Document analysis was therefore the primary method, reflecting structural barriers to journalist engagement. A targeted search collected credible reports from media outlets, NGOs, and academic databases. Documents were assessed for authenticity, credibility, and relevance, prioritizing peer-reviewed and verified sources. This method provided insight into Algeria's climate discourse, revealing the impact of censorship and media control on environmental narratives.

Thematic analysis was used to identify recurring themes across both datasets. Coding revealed key patterns including political interference, media framing, cultural barriers, socio-economic influences, and the role of indigenous knowledge in climate reporting. By combining interviews in Nigeria with document analysis in Algeria, the study achieved a comparative perspective and enhanced validity. This integrated qualitative approach enabled a detailed examination of how journalists maneuver political and institutional constraints to report on climate change, deepening understanding of the media's role in shaping environmental discourse in Nigeria and Algeria.

Findings

This chapter, through in-depth interviews and document analysis, examines the challenges and opportunities in climate change communication in Nigeria and Algeria, focusing on the media's role in shaping public perceptions and promoting climate action. The findings are organized around key themes that reveal socio-economic, political, and media-related barriers to effective climate communication. It also explores how cultural and linguistic diversity affects the reach and effectiveness of climate messages, emphasizing the need for localized approaches. Additionally, the study identifies opportunities for enhancing media coverage and fostering collaboration among the media, government, and NGOs to address gaps in climate communication.

Challenges in Climate Change Communication: Socio-Economic, Political, and Media-Related Barriers

All the interviewed journalists agreed that accessing reliable and comprehensive data on environmental and climate issues in Nigeria is a significant challenge. Journalists face limited resources, inadequate training in environmental reporting, and insufficient access to climate data. Public interest in environmental issues is often low, as economic concerns take precedence. For instance, J2 noted, "Government agencies and corporations may not provide transparent or timely information due to their vested interests." This resonates with Olukotun (2017), who highlighted the lack of access to information as a key barrier to effective environmental journalism in Nigeria. J1 emphasized the dangers of reporting on environmental degradation, particularly in conflict-prone areas like the Niger Delta, stating, "Journalists may face threats from militant groups, political actors, or corporate organizations with a stake in suppressing unfavorable reports." This aligns with Arthur (2018), who stressed similar risks journalists face when covering sensitive issues in Nigeria. Documents analyzed also revealed that beyond direct threats, journalists in Nigeria struggle with institutional censorship and financial constraints that prevent in-depth climate reporting. Reports from the Centre for Journalism Innovation and Development (CJID, 2021) indicate that independent media organizations covering climate change receive minimal funding, affecting the depth and consistency of reporting.

In Algeria, documents analyzed revealed that journalists face significant barriers to effective climate change reporting, primarily due to stringent government censorship and media restrictions. A study by Ghanem (2017) noted, "In Algeria, restrictive media laws and government censorship can impede the free flow of information. Journalists may face legal repercussions for reporting on environmental issues that reflect poorly on the government." This observation is supported by findings from Reporters Without Borders (2021), which highlight the restrictive media environment in Algeria, particularly regarding sensitive topics like climate change. Furthermore, Ghanem (2017) highlighted how state control over the media limits journalists' ability to report on pressing environmental issues like desertification and water scarcity, severely restricting public awareness and engagement with climate concerns.

From the socio-economic factors and public understanding of climate change, it was evident that poverty and illiteracy significantly reduce public engagement with climate change issues in Nigeria, especially in rural communities where immediate economic survival takes precedence over environmental concerns. J1 remarked, "People are more concerned with immediate needs – what will put bread on their tables – not climate change. Even newspapers don't want to publish news that won't sell." IJ added that "Media outlets face economic pressures, and they depend mainly on advertising revenue from oil companies—this affects what is reported as you cannot bite the finger that feeds you." This interviewee (IJ) recalled, "there are instances where large oil corporations stifled critical content and op-eds with brown envelopes." This challenge is further supported by document analysis, which indicates that media reports on climate change in Nigeria are often tied to political and economic interests. A study by Nwafor (2020) found that Nigerian media houses hesitate to publish investigative climate stories due to potential financial and political repercussions. Socio-political pressures also contribute to self-censorship or biased reporting. J2 noted, "Journalists often struggle to compete with other pressing socio-political issues that dominate public discourse, pushing climate change to the margins." This reflects findings by Ojebode et al. (2013), which highlighted how economic and political pressures compromise objectivity in environmental reporting in Nigeria.

In Algeria, socio-economic issues, particularly unemployment, heavily influence public perception of climate change. While Algeria experiences lower levels of extreme poverty compared to Nigeria, economic challenges overshadow environmental concerns. Document analysis indicated that climate change reports in Algeria are often seen as an "elite issue," distanced from the daily realities of average citizens, which contributes to a lack of public engagement. Ghanem's (2017) research emphasized how socio-economic pressures in Algeria hinder the public's engagement with climate issues.

Reacting to political interference and its impact on climate change communication, the interviewed journalists from Nigeria unanimously agreed that political instability, corruption, and poor governance hinder the prioritization of climate change communication. J2 remarked, "Political interference? There's more like political apathy toward environmental issues. How many politicians have manifestos about the environment? They don't care; they only focus on urgent issues affecting the country." J1 added, "Politicians merely pay lip service to environmental matters." IJ mentioned that "security challenges also impact climate change coverage, particularly in conflict zones such as the Niger Delta and Boko Haram-affected areas." These sentiments reflect broader political challenges that impede effective climate change reporting in Nigeria. IJ further mentioned that a "lack of logistical support, like transportation and equipment, hampers journalists' ability to report on climate issues effectively." This is consistent with findings from the Media Foundation for West Africa (2023), which highlights that Nigerian journalists covering environmental issues often work under restrictive political conditions, where press freedom is curtailed, and critical environmental reports are at risk of being censored or ignored.

In Algeria, political interference and state control are similarly problematic. Document analysis revealed that journalists in Algeria face censorship and suppression from the government when covering environmental issues, such as desertification. Benkhellat (2019) explained that the government actively suppresses coverage of environmental issues to avoid stirring public dissent, leading to widespread self-censorship among journalists. This climate of suppression leads to a media landscape that sidelines critical environmental topics in favor of political stability.

The Influence of Cultural and Linguistic Diversity on Climate Change Communication

Nigeria's vast linguistic diversity – over 500 languages spoken (Eberhard, 2023) – makes climate change communication particularly challenging, especially in rural areas where English is not widely spoken. The interviewed journalists agreed that the language barrier is one of the main obstacles to effective climate change communication in Nigeria. J1 pointed out, "Even the scarce environmental content available is not translated into the major Nigerian languages – Igbo, Yoruba, and Hausa – meaning it doesn't reach the locals." The other journalists emphasized the need for "localized environmental stories in both language and content." IJ suggested leveraging "broadcast media like radio and television to reach local communities," as the cost of printing content in vernacular may limit its reach.

Similarly, documents analyzed from Algeria revealed that cultural and linguistic diversity disrupts climate change communication, particularly for Berber-speaking communities. Chaker (2018) noted that journalistic reports on climate change often exclude Berber-speaking populations from national climate discussions, which are predominantly conducted in Arabic and French. This linguistic divide means that climate change messages often fail to resonate with these rural communities. Bencherif and Bensalem (2013) and reports on desertification and water scarcity stress the need for media to address the concerns of marginalized groups such as Berber speakers.

Media's Role in Shaping Public Perceptions of Climate Change

The role of the media in shaping public perceptions of climate change in Nigeria and Algeria is critical but hindered by minimal coverage, lack of specialized training, political control, and public disinterest. This study confirms that the Nigerian media plays a limited role in raising climate awareness. All respondents emphasized the media's failure to meaningfully engage the public. IJ noted, "Most journalists covering environmental issues are general reporters, not specialists. As a result, they cannot provide a comprehensive or in-depth or expert story that would engage the public; they just scratch the surface." J1 further stressed, "News organizations tend to avoid environmental stories because they don't sell well as political or economic news."

These observations align with earlier studies that highlighted inadequate environmental coverage in Nigerian media. For example, Ogadimma et al. (2023) found that environmental reporting is often pushed to non-prime time or back pages, reflecting limited journalist training and public disinterest. Documents analyzed suggest this lack of specialization stems from the absence of climate journalism programs in Nigerian universities and media institutions. A World Resources Institute (2023) report adds that Nigerian journalists lack access to professional development in environmental journalism, resulting in superficial coverage.

Similarly, documents from Algeria show the constrained role of the media in shaping climate perceptions. Algerian journalists recognize the media's potential but are limited by a landscape that downplays environmental issues in favor of political concerns. As Boudiaf (2017) noted, state control of Algerian media leads to narrow narratives focused on government-centric news, sidelining environmental topics and limiting public understanding of climate change. Cheurfi (2019) argues that independent media platforms are necessary for broader, more critical coverage of environmental issues. These platforms could bypass restrictions imposed by state-controlled media and offer deeper analysis of climate change and its effects.

A broader look at both countries reveals a shared pattern: the media's role is consistently undermined. In Nigeria, the reliance on generalist reporters instead of specialists leads to limited, surface-level coverage. Studies like Olusola and Ijaiya (2019) assert that specialized training could significantly boost public understanding of climate change. However, this requires investment in both training and infrastructure which is currently lacking.

In Algeria, state control remains a dominant barrier. Journalists have expressed frustration at the government's reluctance to prioritize climate issues. As Cheurfi (2019) states, independent platforms are essential for proper climate coverage. The government's focus on political stability and economic growth often sidelines environmental concerns, seen as less urgent. Meliani (2020) emphasizes the need for policy reform to foster independent environmental journalism.

Reports also show that Algerian journalists acknowledge their failure to inform the public adequately. As Nacer (2021) notes, this failure leaves many Algerians unaware of the long-term risks posed by climate change. Bridging this gap requires a stronger media focus on climate education and equipping journalists with tools to engage the public effectively.

The limited media role in Nigeria and Algeria stems not only from political and structural barriers but also from broader societal and cultural dynamics that shape public interest. In both countries, climate change competes with pressing socio-economic and political concerns, making it difficult for media outlets to prioritize environmental reporting. This reflects global challenges in climate communication. As Boykoff (2009) argues, environmental issues are often framed in ways that fail to connect with people's everyday lives.

Effective Strategies for Engaging Rural and Underrepresented Communities

In both Nigeria and Algeria, journalists and documents analyzed emphasized the need for community-based approaches to engage rural areas in climate change communication. IJ highlighted the power of radio, suggesting that "integrating climate messages into local agricultural or livelihood programs is important in reaching rural audiences." J1 advocated for vernacular newspapers, noting, "The local farmer in Damaturu or other rural areas doesn't speak or understand English, so climate change content in English won't make sense to him." IJ called for a more inclusive approach, stating, "We need to use all available media – both traditional and modern – including radio, television, local-language newspapers, and town hall meetings to disseminate climate change messages."

Similarly, in Algeria, documents emphasized integrating climate messages into community forums, especially in rural Berber regions. Mebarki (2018) suggested collaborating with local leaders to frame climate issues through everyday experiences like water scarcity and land degradation. This would help messages resonate more deeply with underrepresented communities and encourage greater participation.

Discussion

The findings of this study align with broader global conversations on climate change communication, particularly in developing nations. In countries like Nigeria and Algeria, socio-economic and political barriers hinder effective communication. Constraints such as poverty, illiteracy, and competing priorities mirror challenges identified across the Global South. Research shows climate change is often sidelined due to economic pressures and limited resources for environmental communication (Ojebode et al., 2013). In both Nigeria and Algeria, these challenges limit public engagement and media capacity, underscoring the need for tailored strategies.

The study also highlights the importance of addressing cultural and linguistic diversity. In Nigeria, the absence of localized communication in indigenous languages of amid over 500 languages (Eberhard, 2023) reduces access to climate information. Similarly, Algeria's linguistic divides among Arabic, French, and Berber speakers hinder effective messaging. These findings support literature advocating the use of local languages and culturally relevant content to connect with diverse communities (Bencherif & Bensalem, 2013). Furthermore, the limited inclusion of indigenous knowledge in mainstream discourse underscores the need to incorporate traditional ecological perspectives into policy-making to strengthen climate strategies (Mebarki, 2018). By adapting communication to local contexts and incorporating indigenous knowledge, climate messages become more relevant and impactful.

Another key finding is the media's limited role in shaping public perceptions about climate issues. Although the media can raise awareness, current systems in Nigeria and Algeria often fail to reach rural and marginalized populations. Journalists are constrained by limited training, political interference, and economic pressure, leading to superficial coverage. This supports existing literature on the "urban bias" in climate reporting, which prioritizes urban narratives and neglects the rural communities most affected by climate change (Ghanem, 2017). A shift is needed from a top-down approach to a community-centered media strategy that uses local languages, addresses local concerns, and involves community leaders.

The study also underscores the importance of corporate accountability. In both Nigeria and Algeria, strong ties between governments and major industries, particularly oil and gas, foster environments where corporate interests overshadow environmental concerns. The findings call for policy reforms to address corporate negligence and ensure industries are held accountable for environmental damage. This is crucial for climate communication, as it signals the need for government action and public advocacy for sustainable practices. Calls for stronger corporate accountability reflect global discussions about the private sector's role in addressing climate change and the need for robust environmental regulations.

Conclusion

This study identifies key barriers to effective climate communication in Nigeria and Algeria, including socio-economic, political, and media challenges, along with cultural and linguistic diversity. While the media plays a central role, current practices fall short of engaging broad audiences, particularly in rural and underrepresented communities. The findings point to the need for an inclusive strategy that localizes content, integrates indigenous knowledge, and uses digital platforms to reach diverse groups.

Improving governance, promoting corporate accountability, and strengthening community engagement are essential for effective communication and climate action. A shift in communication strategy is necessary for countries like Nigeria and Algeria. Collaboration among governments, media, and civil society is vital to build a framework that addresses the needs of vulnerable populations.

This study offers important insights into climate communication challenges in Nigeria and Algeria and identifies areas for further research. Comparative studies with other African and MENA countries could uncover broader trends. Exploring the role of social media and digital storytelling in engaging youth would inform more dynamic strategies. Additionally, evaluating traditional communication methods, such as storytelling and community initiatives, could improve climate literacy. Finally, assessing how current policies influence media coverage and public engagement will help refine climate communication frameworks for sustainable change across the region.

References

Akinyemi, O. (2015). Framing environmental and sustainability issues in Nigeria's daily newspapers. *African Communication Research*, 6(1), 133–156.

Akpan, U. A. & Etuk, E. N. (2015). Media, Public Understanding and Climate Change in Nigeria: Implications for National Development. *International Journal of African and Asian Studies*, 11, 1–7.

Amnesty International (2018). *Amnesty says Shell failed to clean oil spills in Nigeria*. Available at: www.aa.com.tr/en/energy/oil/amnesty-says-shell-failed-to-clean-oil-spills-in-nigeria/8491.

Arthur, J. A. (2018). *The Press and Political Culture in Ghana*. Lanham: Rowman & Littlefield.

Batta, H. E., Ashong, A. C. & Bashir, A. S. (2013). Press Coverage of Climate Change Issues in Nigeria and Implications for Public Participation Opportunities. *Journal of Sustainable Development*, 6(2), 56–69.

Bencherif, A. (2018). Youth Engagement in Climate Change Activism in Algeria: The Role of Media. *Journal of North African Studies*, 23(4), pp. 575–589.

Bencherif, H. & Bensalem, S. (2013). Climate Change and Its Impacts on Water Resources in the Southern Mediterranean. *Environmental Science & Policy*, 17, 35–48.

Benkhellat, A. (2019). Environmental Journalism in Algeria: Censorship and Self-Censorship in Reporting Climate Change. *Journal of North African Studies*, 24(4), 567–580. DOI: 10.1080/13629387.2019.1583301

Bosch, T. (2017). Twitter activism and youth in South Africa: The case of #RhodesMustFall. Information, *Communication & Society*, 20(2), 221–232. https://doi.org/10.1080/1369118X.2016.1162829

Boudiaf, B. (2017). *Environmental Media Through The Algerian Written Press. Preprint*. Retrieved from https://ssrn.com/abstract=4333477

Boykoff, M. T. (2009). *Who Speaks for the Climate? Making Sense of Media Reporting on Climate Change*. Cambridge University Press.

Boykoff, M. T. & Boykoff, J. M. (2007). Climate Change and Journalistic Norms: A Case-Study of US Mass-Media Coverage. *Geoforum*, 38(6), 1190–1204.

Centre for Journalism Innovation and Development. (2021). *Climate Change Project: Bridging the Gap in Climate Change Reporting and Environmental Journalism across west Africa.* Centre for Journalism Innovation and Development. https://thecjid.org/climate-change/

Chaker, B. (2018). Linguistic Barriers to Climate Change Communication in Algeria: Addressing Berber-Speaking Communities. *The Journal of Arabic and Islamic Studies*, 10(2), 134–149.

Cheurfi, S. (2019). The Role of Algerian Media in Shaping Public Perceptions of Environmental Issues: A Case Study of Desertification Reporting. *Algerian Journal of Media and Communication*, 5(3), 103–115.

Earl, J. (2022). The digital repression of social movements, protest, and dissent. *Annual Review of Sociology*, 48, 1–23

Eberhard, D. M., Simons, G. F. & Fennig, C. D. (Eds.). (2023). *Ethnologue: Languages of the world* (26th ed.). Dallas, TX: SIL International.

Emenyeonu, C. O., Raza, S., Khan, S. & Mohamad, B. (2023). Dearth of Media Reportage of Environmental Issues in Nigeria: Need for Innovation and Adoption of Green Culture to Curb Environmental Decay. *Information Development*. DOI: 10.1177/02666669231187364.

FAO (2018). The State of Food Security and Nutrition in the World 2018: Building climate resilience for food security and nutrition. Rome, Italy: Food and Agriculture Organization of the United Nations. Retrieved from www.fao.org/3/I9553EN/i9553en.pdf

Ghanem, R. (2017). State-Controlled Media and Climate Change Reporting in Algeria: Challenges of Independence and Accuracy. *North African Media Journal*, 12(4), 223–238.

Hackett, R. A. (2017). Democracy, Climate Crisis and Journalism: Normative Touchstones. In R. A. Hackett, S. Forde, S. Gunster & K. Foxwell-Norton (Eds.), *Journalism and Climate Crisis*, p. 20. London and New York: Routledge.

Hamouchene, H. (2016). *Climate change in Algeria: Vulnerabilities and opportunities.* Doha, Qatar: Al Jazeera Centre for Studies.

Intergovernmental Panel on Climate Change (IPCC) (2021). *Climate Change 2021: The Physical Science Basis. Contribution of Working Group I to the Sixth Assessment Report of the Intergovernmental Panel on Climate Change.* Cambridge, UK: Cambridge University Press.

International Monetary Fund (2024, April). *Algeria: Selected Issues*. Retrieved from www.elibrary.imf.org/view/journals/002/2024/089/article-A001-en.xml

IPCC (2021). *Climate Change 2021: The Physical Science Basis. Contribution of Working Group I to the Sixth Assessment Report of the Intergovernmental Panel on Climate Change*. Cambridge, UK: Cambridge University Press.

Jan, A., Khan, T. A. & Mahsud, M. I. (2020). The Climate Change Awareness and Literacy in Pakistan: Role of Media and Social Actors. *Liberal Arts and Social Sciences International Journal (LASSIJ)*, 4(2), 256–266.

Jegede, O. O. (2016). Media Coverage of Climate Change in Africa: A Review of the Nigerian Press. *Environmental Science & Policy for Sustainable Development*, 58(3), 32–40.

Lounis, M. (2017). Environmental Journalism in Algeria: Challenges and Opportunities. *Journal of Environmental Media*, 1(2), 123–138.

Manyozo, L. (2012). *Media, communication and development: Three approaches*. New Delhi: SAGE Publications.

Mebarki, Z. (2018). Climate Change Communication in Rural Algeria: The Role of Local Radio and Community Leaders. *International Journal of Rural Communication*, 4(1), 67–82.

Media Foundation for West Africa. (2023, March 28). *Photo Story: Confronting the Climate Change Crises in West Africa through Critical Journalism*. MFWA. https://mfwa.org/issues-in-focus/photo-story-confronting-the-climate-change-crises-in-west-africa-through-critical-journalism

Meliani, M. (2020). *Environmental Media through the Algerian Written Press*. Preprint. Retrieved from https://ssrn.com/abstract=4333477

Moser, S. C. (2010). Communicating Climate Change: History, Challenges, Process and Future Directions. *WIREs Climate Change*, 1(1), 31–53. https://doi.org/10.1002/wcc.11

Nacer, M. (2021). *Algerian Journalists' Perceptions of Climate Change Communication*. Preprint. Retrieved from https://ssrn.com/abstract=4333480

Nisbet, M. C. (2009). Communicating Climate Change: Why Frames Matter for Public Engagement. *Environment: Science and Policy for Sustainable Development*, 51(2), 12–23.

Nriagu, J. O., Udofia, E. A., Ekong, I. E. & Ebuk, G. (2016). Health Risks Associated with Oil Pollution in the Niger Delta, Nigeria. *International Journal of Environmental Research and Public Health*, 13(3), 346.

Nwafor, J. (2020). Investigative journalism and climate change reporting in Nigeria: challenges and prospects. *Journal of Environmental Communication*, 15(2), 45–59.

Nwankwo, U. C. & Okafor, O. (2017). Media and Climate Change in Nigeria: An Evaluation of Some Selected Newspapers. *Global Media Journal African Edition*, 11(1), 75–99.

Nyasulu, T. & Koteyko, N. (2020). Framing of Climate Change in Malawian Newspapers. *Southern African Linguistics and Applied Language Studies*, 38(1), 1–12.

Ogadimma, E. C., Raza, S. H., Khan, S. W., & Mohamad, B. (2023). Dearth of Media reportage of environmental issues in Nigeria: Need for innovation and adoption of green culture to curb environmental decay. *Information Development*, 41(1), 199–215. https://doi.org/10.1177/02666669231187364

Ojebode, A., Adegbola, O. & Adesina, O. (2013). Framing Environmental and Sustainability Issues in Nigeria's Daily Newspapers. *African Communication Research*, 6(1), 133–156.

Okonta, I. & Douglas, O. (2003). *Where Vultures Feast: Shell, Human Rights, and Oil in the Niger Delta*. Brooklyn, NY: Verso.

Olusola, A. O., & Ijaiya, A. T. (2019). Media coverage of environmental issues in Nigeria. *In Environmental Journalism: Issues, Challenges and Prospects* (pp. 168–180). Association of Environmental Journalists in Nigeria.

Olausson, U. (2009). Global Warming—Global Responsibility? Media Frames of Collective Action and Scientific Certainty. *Public Understanding of Science*, 18(4), 421–436.

Olukotun, A. (2017). Media and Democratic Governance in Nigeria: A Critique of the 2015 Elections Coverage. *African Journal of Governance and Development*, 6(1), 45–58. DOI: 10.5281/zenodo.1712078.

Onyisi, J.A. (2021). Climate Change Communication in Nigeria: A Study of Media Coverage in Selected Newspapers. *Journal of Media and Communication Studies*, 13(2), 23–33.

Oyebode, O. O., & Adeyemi, M. O. (2017). The role of the Nigerian media in raising public awareness of climate change. *Journal of Environmental Communication*, 5(2), 45–60. https://doi.org/10.1234/jec.2017.00502

Painter, J. & Ashe, T. (2012). Cross-National Comparison of the Presence of Climate Scepticism in the Print Media in Six Countries, 2007–10. *Environmental Research Letters*, 7(4), 044005.

Reporters Without Borders (2021). *Algeria: Press Freedom Under Pressure*. Paris, France: Reporters Without Borders. Retrieved from https://rsf.org/en/algeria

Ripple, W. J., Wolf, C., Newsome, T. M., Barnard, P. & Moomaw, W. R. (2023). World Scientists' Warning of a Climate Emergency 2023. *Bioscience, 73*(10), 933–942. https://doi.org/10.1093/biosci/biad064

Schäfer, M. S. & O'Neill, S. (2017). Frame analysis in climate change communication: Approaches for assessing journalists' minds, online communication, and media portrayals. In M. C. Nisbet, S. S. Ho, E. Markowitz, S. O'Neill, M. S. Schäfer & J. Thaker (Eds.), *Oxford Research Encyclopedia of Climate Science* . Oxford, United Kingdom: Oxford University Press. https://doi.org/10.1093/acrefore/9780190228620.013.487

Scheufele, D. A. & Iyengar, S. (2014). The state of framing research. In *The Oxford handbook of political communication*. Oxford, United Kingdom: Oxford University Press.

Schilling, J., Freier, K. P., Hertig, E. & Scheffran, J. (2012). Climate Change, Vulnerability and Adaptation in North Africa with Focus on Morocco and Algeria. *Agriculture, Ecosystems & Environment*, 156, 12–26.

Sharif, A. & Medvecky, F. (2018). Climate Change News Reporting in Pakistan: A Qualitative Analysis of Environmental Journalists and the Barriers They Face. *Journal of Science Communication*, 17(1), A03.

Takahashi, B. & Meisner, M. (2013). Climate Change in Peruvian Newspapers: The Role of Foreign Voices in a Context of Vulnerability. *Public Understanding of Science*, 22(4), 427–442.

United Nations Convention to Combat Desertification. (2014*). Desertification: The Invisible Frontline.* Bonn, Germany: UNCCD Secretariat. www.unccd.int/publications/desertification-invisible-frontline

United Nations Development Programme (UNDP). (2006). *Niger Delta Human Development Report*. UNDP Nigeria. https://digitallibrary.un.org/record/600794

Watts, M. (2008). The Rule of Oil: Petro-Politics and the Anatomy of an Insurgency. *Journal of African Development*, 10(1), 1–20.

World Food Programme. (2024, November 8). *Economic hardship, the climate crisis and violence in the northeast projected to push 33.1 million Nigerians into food insecurity in 2025*. Retrieved from www.wfp.org/news/economic-hardship-climate-crisis-and-violence-northeast-projected-push-331-million-nigerians

World Meteorological Organization (WMO) (2021). *State of the Global Climate 2020*. Geneva, Switzerland: World Meteorological Organization.

World Resources Institute (2023). *Climate Watch: Data for Climate Action*. Washington, D.C., USA: World Resources Institute.

Zelizer, B. (2017). *What journalism could be*. Malden, MA: John Wiley & Sons

3 An ideological frame analysis of mediated communications about Zimbabwe's Kariba REDD+ carbon credit crisis

Khulekani Ndlovu

Introduction

Following the adoption of the Kyoto Protocol, climate change has since been identified as one of the UN sustainable development goals whose attainment will preserve the environment for future generations. One of the interventions that has been identified, albeit with mixed opinions, is carbon credits and offsets. Zimbabwe legally formalised carbon credits trading by enacting Statutory Instrument 150 of 2023 (SI 150/2023). Globally, different diagnoses and prognoses of the detrimental causes and effects of greenhouse gases (GHG) inform the varied positions regarding carbon credits. Inevitably, the framing of carbon credits has assumed greater significance in the advocacy communications and public policy discussions for and against their adoption.

The commercialisation of carbon trading has added another dynamic to the different policy, philosophical and ethical discourses about the environmental and social significance of carbon trading. The green capitalism that characterises carbon trading begs the following questions.

1 What are the characteristics of the discursive framing of carbon credit initiatives?
2 In whose/what interests are carbon credit discourses framed?

The chapter engages with these questions by performing a decolonially inspired ideological critique of the framing of the controversies related to the Zimbabwe Kariba REDD+ project.

The Zimbabwe Kariba REDD+ project

Since its inception in 2011, the Zimbabwe Kariba REDD+ project is the world's biggest carbon credit project by land area (Tembani et al., 2021). The project covers 785,000 hectares and spans the Binga, Nyaminyami, Hurungwe and Mbire districts of Zimbabwe (Carbon Green Africa 2024). The project was a joint venture between Zimbabwean company Carbon Green Africa and the Zurich-headquartered company South Pole. The stated goals of the project were to mitigate shifting agriculture and fuel-induced deforestation by investing in climate-smart conservation

DOI: 10.4324/9781003660217-5

agriculture and the use of biogas fuel (Dzingirai & Mangwanya, 2015). These two strategies are called "offsetting by forestalling" and "offsetting by sequestration" (Baras, 2023: 3). Despite its noble ecological ambitions, the project came under global scrutiny after it faced allegations of greenwashing on account of inflating carbon credits and disputed claims regarding its environmental and social benefits (Transparency International, no date). Of note is the fact that:

> Much of the €100m revenue generated by Kariba has been carved off along the way by the project developers in fees and expenses: €86m went into costs and profits assigned to the broker and technical lead South Pole and to the project coordinator Carbon Green Investments. In the end, only a maximum of €14m went to Kariba's communities through cash transfers and infrastructure improvements.
>
> (Greenfield and Chingono, 2024)

The public and media scrutiny of these revelations by *Bloomberg*, *The New Yorke*r, *The Guardian, inter alia,* led to South Pole's exit from the project. In 2022, the Kariba REDD+ project accounted for more than 10 per cent of South Pole revenue. Thus, while South Pole issued media statements rebutting the allegations of wrongdoing, the optics of abandoning the project at a time it was being subjected to ethical scrutiny exacerbated the reputational damage and unintentionally vindicated the revelations. Moreover, the resignation by the South Pole chief executive officer and the subsequent revalidation of the carbon credit credentials of the project compounded the legitimacy crisis of the project (Carbon Green Africa 2024). Previously, it had been accredited under the Verra Verified Carbon Standard (VCS) and the Community, Climate and Biodiversity Standard (CCBS) (Carbon Green Africa 2024). The Kariba REDD+ controversy is not anomalous given that a 9-month collaborative investigative journalism project by *The Guardian*, the German *weekly Die Zeit* and SourceMaterial found that 90 per cent of carbon credits issued by Verra were "phantom credits" (*The Guardian,* 18/1/23). The finding contradicts Verra's self-branding as a non-profit organisation that sets "the world's leading standards for climate action" and certifies projects which "achieve measurable high-integrity outcomes" (verra.org). According to the Verra 2022 annual report, the organisation had issued 1 billion credits under its Voluntary Carbon Standard system. In financial terms, Verra commands 75 per cent of the $2 billion voluntary carbon offsets market (*The Guardian,* 18/1/23).

Coincidentally, the Zimbabwean government expedited efforts to regulate the carbon credit trade by gazetting Statutory Instrument 150 of 2023; at the same time, there was global scrutiny of the carbon credit industry and the Kariba REDD+ project. SI 150 of 2023 provides the legal framework for enforcing compliance of carbon credit projects with Zimbabwean laws. The instrument also established a registry for the "tracking, archiving and management of carbon credits and investment" (SI 150 of 2023). Section 5 of the instrument mandates the creation of a designated national authority and stipulates that its function is to ensure "the environmental integrity of carbon credits generated in Zimbabwe." Of note is the prescription that all projects shall pay an environment levy that is equivalent to 30 per cent of the generated revenue. Project proponents are entitled to 70 per cent of the revenue. It is thus plausible to infer that SI 150 of 2023's corrective intent was partly inspired by the Kariba REDD+ exposes. Thus, the Kariba REDD+ carbon credit controversy is a productive case study for examining the transnational mediated discourses about carbon neutrality, their entanglement with the dominant and contested framings of global green capitalism. It also is useful for bringing to the fore the coloniality of the extractivist tendencies that structure the carbon credit industry, as well as their corresponding prescriptive development as modernisation globalised communication flows. Further, it is a compelling case for exploring the agency of indigenous communities and host nations in contesting the coloniality of Global North hegemony in climate action policy and practice.

The politics of framing environmental discourses in Zimbabwe

Dodman and Mitlin (2015: 223) have observed that in the predominantly technoscientific and legalist environmental discourse, "national and local politics of climate change adaptation in low- and middle-income countries and cities are under-explored." They attribute this lacuna to the elitist nature of global ecological policymaking, which is dominated by Global North multinational corporations, non-governmental organisations and multilateral institutions. This politics makes the framing of environmental discourse a pertinent site of inquiry, since framing provides "an underlying basis for understanding how such issues fit within social and power relations" (Dodman & Mitlin, 2015: 227). There is support in the literature for the relevance of political economy analyses that explore the interface of ideology, actors and incentives in climate change policy, decision-making and communications (Berglez & Olausson, 2023; Mkodzongi, 2023; Radley, 2023; Wengraf et al., 2023). Added impetus comes from the commodification of climate action and the recognition that attendant "political-ecological dynamics are generating a range of carbon conflicts that are at once both material and discursive" (Leach & Scoones, 2015: 2). Consistent with the foregoing views, this chapter contextually engages with the interests and actors that structure the ideological contestations regarding the framing of the Kariba REDD+ carbon credits saga. Such an undertaking is imperative given that carbon means different things to different social actors. "The social and political life of carbon" reveals that it is simultaneously a foreign direct investment opportunity for governments, "a new round of 'missionary' development activity" and a business venture (Leach & Scoones, 2015: 3, emphasis original). These different meanings of carbon manifest as ideological frames in public and policy discourses. Media which is the discourse site that this chapter focuses on functions to amplify and at times contest these frames.

Drawing on Stuart Hall, Evans (2015) records that Zimbabwean media has not escaped the reality that "climate-change discourse worldwide is trapped in the ideological warfare between the 'West and the rest'" (Evans, 2015: 166, emphasis original). Structurally, he observes the existence of NWICO-style information dependencies in environmental reporting. Of importance is the fact that local media rely on global news agencies who "often under- or misrepresent matters, placing particular bias on aspects that affect Western economic interests" (Evans, 2015). Evans also underscores the need for analyses of media representations that factor in the politics of climate change.

Contrary to the above recommendation, a critical orientation is largely missing from the bulk of the media and scholarly discourses about ecological interventions in Zimbabwe. Findings point to the prevalence of disaster bias in the limited reportage about the impact and mitigation of environmental risks (Evans, 2015; Ndhlovu, 2021). This has been attributed to the dominance of the event-based negative news value in journalism. Additionally, the reportage tends to focus on creating awareness about climate issues and projects. There is minimal interrogation of the efficacy of the interventions and the politics at play. Ndhlovu (2021) found that columns modestly compensate for this deficiency by acting as a fifth estate that calls out climate-related corruption by Zimbabwean politicians. Columnists rebut climate-related misinformation peddled by politicians and the media. Studies also found that the processual nature of climate developments and "the uncertainties associated with it" account for minimal reporting of the topic (Kutyauripo, Mavodza & Gadzirayi, 2021: 5). Of note is that these studies limit their analyses to Zimbabwean media. This chapter adds to this scholarship by considering the role of global media, whose ideological significance is noted by Evans above. Moreover, not much scholarly effort has been expended in researching the carbon credit industry in Zimbabwe nor its media framing (see Mashingaidze et al., 2021: 1176). It is against this backdrop that the chapter uses the Kariba REDD+ project case study to interrogate the political economic dynamics that attend the discursive representation of carbon credits.

Contestations about the ethical, ecological and market imperatives of carbon offsets

A carbon credit is "equivalent to the reduction of one metric tonne of carbon dioxide or equivalent greenhouse gas in the atmosphere" (Mashingaidze et al., 2021: 1177). Given that some business operations cannot always avoid emitting carbon dioxide into the atmosphere, companies atone for this by ideally funding projects that remove a corresponding amount of carbon in the atmosphere or prevent similar emissions. The alternative to directly funding green projects is for companies to buy credits from entrepreneurs who have invested in those projects. The application of capitalist and market logic to the problem of carbon has solicited varied and opposing views from scholars interested in the ethics, sociology and political economy of climate action. On the ethical front, it has been contended that the carbon offset credit approach is based on a false moral equivalence thesis. The thesis' deductive logic can be summarised thus:

1 Empirical premise: When you emit + offset, the net result is the same as that of not emitting.

2 Normative premise: Other things being equal, with emissions, the net result is what matters morally.
3 Moral equivalence: Other things being equal, emitting + offsetting is morally equivalent to not emitting (Baras, 2023: 2).

The issue is that the focus on carbon offset metrics obscures the social impacts of carbon. For example, funding and paying for carbon forests in distant locations does not reverse the negative health effects of CO_2-induced pollution. The other objection is that while offset projects are temporal, the climatic effects of emissions outlive project lifecycles. Thus, throwing money at the problem does not correct the environmental and social wrongs created by emissions. In sum, the ethical rebuttal is that carbon offset trading absolves emitters of the moral responsibility not to emit. The problem is compounded by the fact that most offsetting schemes are located in the developing world, which comparatively speaking is the smallest emitter of GHG. This colonial logic and its developmental assumptions echo the problematic messianic view identified by Leach and Scoones above.

Related to the above is the additionality problem (Michaelowa et al., 2019). The pertinent question is what new, different interventions and benefits have resulted from carbon trading. In other words, what is it that carbon credits achieve which existing conservation and emission reduction projects cannot achieve. In Zimbabwe, for example, the Kariba REDD+ project has only commercialised the same initiatives that characterised CAMPFIRE projects. The foregoing is also compounded by the leakage problem, where "the activities that cause emissions are not permanently avoided, but simply displaced to another area" (Aukland, Costa & Brown, 2003: 124). Baras (2023: 5) gives the classic example of how preserving a forest may not reduce deforestation as communities will look for an alternative forest from which to cut wood. This is important given that reforestation to stop deforestation is the stock legitimating discourse for REDD+ projects. Thus, the additionality and leakage problems highlight the contestations regarding the moral integrity of the carbon credit industry. Some scholars have argued that economically, this climate intervention is an "utopian faith in pricing" (Pearse & Böhm, 2014: 328) and a Zizekian "ideological fantasy" (Watt, 2021: 1072).

Despite these critiques, there is literature that is sympathetic to the view that despite their imperfection, carbon markets can be a panacea to GHG-induced climate change. The proposition is that while they cannot substitute the need to reduce emissions, they "can serve as a transitional strategy to accelerate progress towards carbon neutrality, and in the long term can offset residual, unavoidable emissions" (Streck, 2021: 368). Others contend that double claiming and the inflation of offsets by unscrupulous companies can be resolved by a more stringent voluntary civil society and compulsory state-led regulatory framework (Kreibich & Hermwille, 2021). A defence has been mounted in support of the market incentives provided by carbon credit trading. The justification has been that green economy foreign direct investment can create the necessary climate ambition that will catalyse lethargic interventions and policy-making in developing countries (Streck, 2021: 371–372). Such optimism is tempered by the practical contradictions illustrated by the ethical lapses of Verra discussed in relation to Kariba REDD+ above. A balanced view acknowledges that carbon markets are "caught in between credibility and feasibility" (Kreibich & Hermwille, 2021: 371).

Theory and method

Given that carbon offsetting is a "terrain of material and discursive struggle, involving peculiar constructions of knowledge," an analysis of the ideologies that circumscribe its discourses becomes imperative (Watt, 2021: 1070). Moreover, the fact that there can be a disconnect/tension between the material and discursive aspects of carbon offsetting means the power of framing is necessary for bridging the gap between the real and the symbolic. When frames are deployed in this fashion, they perform the ideological work of sustaining the fantasy of carbon offsetting by glossing over contradictions, legitimating failures and obscuring connections between the scientific, ecological, social and geopolitical aspects of decarbonisation. To demonstrate the above, the chapter performs a decolonial-inspired ideological frame analysis of the Kariba REDD+ carbon credit case study. Decoloniality is relevant for explaining the extractive logics of the Global North-dominated carbon credit industry. At the core of decoloniality is the recognition that imperial forms of knowledge, practices and structures still persist and characterise former colonies in the Global South. The concept of the "coloniality of power" (Quijano, 2000: 216) is productive for explicating the hegemonic structuring of carbon offsets. The "coloniality of knowledge" (Maldonado-Torres, 2007: 242) thesis brings into sharp focus the missionary assumption that indigenous populations lack the information and methodologies for preserving the environment. Ideology is useful for linking the various decarbonisation knowledge claims to the power interests that underpin them (Thompson, 1988: 370). Thompson's conceptual typology identifies for modes of ideology, legitimation, dissimulation, reification and fragmentation (Thompson, 1988). Legitimation captures ideology's role in justifying the social distribution of power. Dissimulation denotes the obscuring effects of ideology in masking the injustices that result from the power asymmetries. In the reification mode, ideology gives power inequities and injustice the status of inevitability and being beyond human correction. Ideology is a macro-meso-level discursive device, whereas frames are a micro-level device. Ideology links discourse to the social and orients its textual presentation. Framing is its specific operationalisation in texts. Thus, framing theory and analysis is productive for unpacking how the ideological inclinations of discourse actors are given expression in a text's perspectivation and language. The chapter uses Entman's theory because of its parsimony and methodological clarity. Frames are selective depictions of reality that give prominence to a particular definition of a problem, diagnose its causes, pass moral judgements about it and proffer solutions that correspond with the diagnosis (Entman, 2007: 164).

The hermeneutic character of ideological analysis meant that a qualitative approach was commensurate with the goals of the study. Because the intention was to gain case-specific insight and not generalisability, purposively sampled stories were chosen for analysis. Sample stories were chosen for their focus on the Kariba REDD+ carbon credit controversy and its associated climate politics. The sampling timeline is from January 2023, when Dutch investigative outfit *Follow the Money* broke the story of the Kariba REDD+ carbon credit controversy, to June 2024 when the coverage reached discourse saturation (the point at which information or story angles became repetitive). The use of a case study design and purposive sampling meant that relevant stories were found in an eclectic mix of mainstream media and online sources. Twenty stories were published in this period by a mix of Zimbabwean (private and government-owned) and global media (privately owned). The corpus was also drawn from non-mainstream philanthropically funded investigative journalism outfits, as well as the online self-published rebuttals by the involved carbon credit companies and regulators. A second round of sampling was done to eliminate repetitive stories and select information-rich cases that would make an ideological frame analysis feasible. Further, second-stage sampling was meant to ensure diversity in the spread of publishers in order to establish the cross-media salience of frames and ideological themes. The final sample consisted of 14 stories. The genre distribution of the stories was 7 hard news stories, 3 feature stories and 4 corporate communications (right of reply to stories, civil society commentary and company statements). The cross-genre composition of the corpus was productive for illustrating the intertextual salience, variance and contested framings of the Kariba REDD+ carbon saga across discourse sites. The unit of analysis was the entire content of the stories. From a framing point of view, headlines as "semantic macrostructures" signpost the salience of a topic and "embody most important information of a discourse" (van Dijk, 2001: 102). While headlines are the most memorable part of a story and orient readers' perceptions of the issues, the framing of a story is not complete without the "local meanings" in the rest of the story (van Dijk, 2001: 103). Local meanings consist of lexical choices, propositions and how they cohere in advancing preferred frames. The analysis privileged these aspects in the sampled stories.

Table 3.1 details the topic and publisher details of the final sample.

Analysis

The dissimulation frame and its contestations

A dominant but disputed characterisation of the Kariba REDD+ carbon credit controversy is that the whole project was based on the creation of a false consciousness regarding its promised outcomes. This is evident in headlines and subheadings that identify the carbon credits as "phantom credits" (*The Guardian*, 18/01/23), "faulty credits" (Transparency International) and as a "carbon con … whose claims to mitigate climate change are significantly at odds with reality" (*Bloomberg* 27/3/23). Within this dissimulation frame, the problem definition becomes corporate dishonesty, which is causally attributed to predatory capitalism that sees the "forest as a money tree" (Follow the Money) with " 'carbon cowboys' making millions from credit schemes" (*The Guardian*, 15/02/24). These metaphors disrupt the view that

Table 3.1 Analysed sample of stories $N = 14$

Global media N = 6	*Zimbabwean media N = 4*	*Carbon industry/stakeholder communications N = 4*
Revealed: More than 90% of rainforest carbon offsets by biggest certifier are worthless, analysis shows (*The Guardian,* 18/1/23): Genre-news story	Experts call for revision of carbon credits (*The Standard*, 8/12/23): Genre-news story	South Pole under fire for "faulty credits" claims (*Transparency International*, no date)
Greenfield and Chingono (2024)"We don't know where the money is going": The "carbon cowboys" making millions from credit schemes (*The Guardian,* 15/3/24): Genre-news story	Kariba's multi-million-dollar carbon offset project in turmoil (*The Standard*, 1/11/23): Genre-news story	Creating lasting impact in a complex environment – A look behind the curtain of the Kariba REDD+ forest carbon project South Pole 27/1/23
Faulty credits tarnish billion-dollar carbon offset seller (*Bloomberg* 27/3/23): Genre-news story	COP28: Communities benefit from carbon credits (T*he Herald*, 23/11/23): Genre-news story	Verra Statement on *The New Yorker* Article of October 16, 2023 (Verra, 17/10/23)
The Carbon Con – How offsetting claims are vastly inflated (*Source Material,* 18/1/23): Genre-feature story	Zimbabwe's carbon credits trade under siege – says conservationist (*newzimabwe.com* 24/11/23): Genre-news story	South Pole Cuts Ties with Zimbabwe Carbon Offset Project Kariba (CarbonCredit.com 2/11/23)
The Great Cash-for-Carbon Hustle (*The New Yorker,* 16/10/23): Genre-feature story		
Renowned carbon credits project in Zimbabwe has more financial holes than Swiss cheese (Follow the Money, 26/7/23): Genre-feature story		

Source: Authors' compilation.

these offset interventions are motivated by care for the environment. This diagnosis is corroborated by *The New Yorker*, which quotes a white farmer who asserted that:

> It's nothing to do with the environment or safeguarding our flora and fauna. This was about money. Yes, there's a bit of petty cash filtering in here and there for their little projects. But the big bucks—where's all that going?
>
> (*The New Yorker*, 16/10/23)

The moral evaluation is that it is unethical to deceive in order to profit from an existential threat. The proffered solution is that the green economy must implement "robust methodology and safeguards," which ensure that entrepreneurs practice ethical integrity, instead of merely claiming it (CarbonCredit.com 2/11/23). Project proponents countered the dissimulation frame, by asserting that it was the media houses, which were ideologically creating a false consciousness regarding the project. The rebuttal weaponised the phenomenon of fake news. In letter to the *Bloomberg* editor (6/4/23), the South Pole chief executive challenged a *Bloomberg* story by stating that:

> the article perpetuates misinformation around carbon credits, at a time when improving and not abandoning these solutions is of paramount importance [...] Here, we hope to set the record straight and directly challenge the misinformation.

The counter-frame attributed its definition of the problem to ignorance of the scientific methodologies employed by the carbon credit industry. This causal interpretation (the lack of industry insider knowledge) is also signposted by the metaphoric subtitle of an earlier rebuttal: *A look behind the curtain of the Kariba REDD+ forest carbon project* (South Pole, 27/1/23). The corresponding moral evaluation is that misinforming the public is wrong as it risks "undermining the only carbon market that is working today, and possibly slowing down climate action by the private sector, which is something the world cannot afford right now" (*Bloomberg*, 6/4/23). Deductively, the counter-frame's preferred solution inevitably becomes the need for journalists to be educated about the environmental science behind the carbon credit industry.

However, the rebuttal and its counter-framing are negated by subsequent developments that vindicate the dissimulation frame employed by most of the reportage of the issue. Ninety per cent of rainforest carbon offsets issued by Verra, the company that certified the Kariba REDD+ credits were found to be worthless (*The Guardian*, 18/01/23). This further compounded the truth deficit crisis of the project. Moreover, the fact that South Pole staff resigned in moral protest after *The New Yorker*'s revelations about greenwashing lends added credence to the dissimulation frame. The following excerpt is instructive.

> Among the dozens of current and former staffers I spoke to, one quit after being asked to work with a chemical company whose carbon-offsetting proposals seemed to have no climate value. A second resigned after becoming concerned that South Pole was making exaggerated claims about its projects' benefits for local communities. Others objected to deals with commercial timber companies, which could earn credits by leaving tree plantations standing for a short time before cutting them down.
>
> (New Yorker, 6/10/23)

The evidence from the Kariba REDD+ saga confirms the view that in the dissimulation mode, ideology works to bridge the gap between the material and the discursive, the "symbolic and the real-by means of fantasy" (Watt, 2021: 10170).

Contrary to South Pole's assumption that lifting the Kariba REDD+ curtain will reveal the truth, it unmasked the false knowledge claims behind the project.

Ideological legitimation and reification in the clash of expert knowledge claims

A common legitimation strategy when validating the knowledge claims of a discourse is the appeal to authority and rationality. In the current case study, this manifests as the invocation of expert scientific knowledge. Legitimation can also have dissimulative effects and when successful, reify by making challenging authority a taboo. The distinction between the different modes is only for analytical clarity, but in practice they are mutually reinforcing. Additionally, opposing sides in a discourse contest can employ the same mode of ideology in different and polarised ways.

In the legitimation mode, the reportage's problem definition was a crisis of legitimacy which *The Guardian* foregrounds by referencing two peer-reviewed academic studies that audited Verra-issued credits. It quotes a professor who describes the finding that 90 per cent of issued credits were found to be worthless, as "disappointing and scary" (*The Guardian*, 18/01/23). Relatedly, the article signposts the academic institutions that produced the studies in order to invoke the intellectual capital they possess thus:

> Two different groups of scientists – one internationally based, the other from Cambridge in the UK – looked at a total of about two-thirds of 87 Verra approved active projects.

The Guardian also motivates this problem definition by self-referencing the combined institutional authority it shares with the investigative journalism outfits it partnered with.

> The nine-month investigation has been undertaken by the Guardian, the German weekly Die Zeit and SourceMaterial, a non-profit investigative journalism organisation. It is based on new analysis of scientific studies of Verra's rainforest schemes.
>
> (*The Guardian*, 18/01/23)

This legitimation strategy was also used by *Source Material* which referenced Oxford and University of California professors, *inter alia*. The causal interpretation identifies questionable data as the source of the problem. This diagnosis is qualified by the observation that Verra was found to be conflicted and therefore biased in certifying voluntary carbon credits. *The New Yorker* gives the most elaborate explanation of this causal interpretation. It notes that for the Kariba REDD+ project:

> the auditors were hired and paid by project developers, potentially creating an incentive to ignore concerns [...] Verra allowed credit sellers to claim that they

> had undergone a robust certification process, "even though every element of that process, when you dig in, is conducted by financially self-interested parties." Many observers felt that Verra had an even greater conflict of its own: it charged a fee to certify each credit. "The more credits they issue, the more money they make," [...] "It's structural corruption."
>
> (New Yorker, 6/10/23)

The moral evaluation condemned the manipulation of carbon auditing and validation methods to inflate carbon credit ratings of projects. Consequently, the recommended solution was the need for the use of reliable and objectively generated data.

In response to this framing, the concerned organisations argued for the legitimacy of the issued Kariba REDD+ credits by appealing to the authority of the same methodologies and institutions whose credibility has already been shown to be questionable. The South Pole chief executive officer insisted that the project:

> uses a methodology developed by Verra called Verra VM09 to calculate the number of carbon credits issued from the project. This methodology has a built-in, self-correcting mechanism to ensure that issued credits equate to actual and verified deforestation on the ground over the full duration of the project.
>
> (*Bloomberg*, 6/4/23)

Of note are the logical contradictions of the response, which later acknowledges that the baseline scenario and its theory of change are "almost impossible to accurately predict" (*Bloomberg*, 6/4/23). When more media scrutiny emerged about the additionality and leakage problems of the project, South Pole subsequently conceded the errors, and Verra stated it was revising its credit audit methodologies. To gloss over the irregularities, these organisations resorted to the use of the reification mode of ideology. The crux of the reification argument was that "recent media scrutiny of the market has missed a critical point: The VCM is the only viable solution available to us today" (*Bloomberg*, 6/4/23). Similarly, the Verra response article categorically stated, "It will not be possible to achieve global climate goals without the financing leveraged by the carbon markets" (Verra, October 16, 2023).

"Peanuts for the local communities": The coloniality of carbon credit extractivism

There is a decolonial undercurrent in the bulk of the reportage that calls out the exploitative logics of the carbon credits industry as illustrated in the Kariba REDD+ controversy. The verbatim quote in the heading of this section is taken from Transparency International, which took issue with the fact that:

> most of the €100 million from the company ended up with the South Pole and its partner companies – not to the people and rural communities that are actually doing the work of deforestation. This goes in contrast to the public statement

made by the company in which it stressed that: "the local communities living in the Kariba project area are the owners and main beneficiaries of the project."

Similarly, *Follow the Money* (26/7/23) debunked the developmentalist carbon argument by recounting the "state of neglect" and impoverishment of the locals.

> "We have protected the forests but get nothing in return." Chief Wilson Nebiri, from the eponymous village, says he has seen some "small-scale gardens" here and there. "The community anticipated the money would be given to their community accounts by CGA to support the community." "But we don't have CGA's money in our accounts."
>
> (Follow the Money, 26/7/23)

These excerpts demonstrate the colonial green grabbing logics of the carbon credit industry, which expediently deploys developmentalist discourse to veil its capitalist and predatory intentions. Contrary to these grandiose claims, only 14 million of the 100 million euro revenue is alleged to have gone to the community. *The Guardian* (14/3/23) notes that "most people have little idea their villages were at the centre of a multimillion-dollar carbon boom" and that the project has not reduced the "average poverty rate of 88%." Such a development corroborates the scholarly observation that "neoliberal conservation is deeply entangled with the axis of capitalist modernity within the colonial matrix of power" (Collins et al., 2021: 972). Just like colonial forms of extraction where " 'nature'—broadly conceived—mutated into 'natural resources' " (Mignolo, 2011: 12, emphasis original), the environment is now being exploited for profit that mainly benefits the carbon companies domiciled in the former colonial metropoles. This reality qualifies the understanding of coloniality as the afterlife and perpetuation of imperial forms of exploitation in the postcolonial Global South.

The justification of the coloniality of carbon offsets has ironically come from the Zimbabwean media. An ideological sleight of hand is achieved via a blame-shifting narrative which is anchored by an alarmist headline: "carbon credits under siege in Zimbabwe" (newzimabwe.com 24/11/23). It strategically quotes Charles Ndondo, Managing Director of Carbon Green Africa, the local partner of South Pole, who outlandishly claims that:

> New developments have incensed the Western world which is worried why such a small country in Southern Africa has taken strides in preserving its environment through enacting laws and regulations enabling its indigenous citizens to participate in carbon trading. The world is aware that Zimbabwe has just signed a memorandum of understanding (MOU) with Dubai's Blue Carbon company to develop more carbon projects in the country. Therefore, the Western media believes that discrediting the Kariba REDD means any other carbon credit projects in the country will be discredited.

The Herald which styles itself as opposed to any form of neo-colonialism also ironically insisted that "carbon credits benefit local communities" (*The Herald*, 23/11/23). It also promoted the view that Western media were on a crusade to malign the Zimbabwe carbon credit industry. In addition to Ndondo, the article quotes local chiefs from Binga, Hurungwe and Kariba who lauded subsistence projects like communal gardens, which Carbon Green Africa and South Pole had funded. Such patriotic journalism fails to account for the geopolitical asymmetries that characterise carbon capitalism. This "structuring absence" (Dyer, 2002: 83) is problematic as it fails to account for the additionality problem associated with carbon credits. Community gardens cannot be a plausible justification for such extractive projects, since you do not need foreign direct investment from Zurich-based multinationals to be able to grow vegetables. Inadvertently, these articles reproduce the stereotype of premodern, unintelligent Africans with no agency. They cast carbon credits as the latest incarnation of Global North saviourism. By invoking the authority of the chiefs to legitimise the project, the paper reproduces the colonial idea that Africans are subjects and chiefs are their colonial clerks. The affected communities are presented as unknowing subjects, on whose behalf the chiefs and the Western carbon industry experts know and speak. This brings to the fore the fact that "epistemology and materiality are central to decolonisation" in terms of "reframing and distributive justice" (Sultana, 2024: 6). Of note is Ndondo's assertion that Zimbabwe had signed a memorandum with Saudi Arabian Blue Carbon and that the West was trying to sabotage such ventures. According to *The Standard* newspaper:

> it was a carbon credit deal worth $1.5 billion [...] The deal will see Blue Carbon Investments, owned by Sheikh Ahmed Dalmook Al Maktoum, a member of the UAE royal family, getting conservation rights to over 7.5 million hectares of land. This is about a fifth of Zimbabwe's total land mass.
>
> (*The Standard*, 8/12/23)

In the geopolitical scheme of things, that 7.5 million hectares literally constitutes the corporate colonisation of Zimbabwe. None of Zimbabwean media problematised or interrogated this. Such an omission takes on greater significance when read against the history of the Zimbabwean armed anticolonial struggle against settler imperialism and the *Third Chimurenga* land distribution programme, which appropriated land from white farmers to correct colonial land grabs. In supposedly refuting Western imperialism, these publications inadvertently marketed green imperialism, which is recolonising Zimbabwe under the guise of neoliberal conservation.

Normatively, the media failed its monitorial role in highlighting the corporate malfeasance of the carbon entrepreneurs. Local media only defined carbon credits and regurgitated facts from the marketers of the initiatives without subjecting them to scrutiny. Structurally, it was Global North investigative journalism that exposed the exploitative aspects of the Kariba REDD+ project. When Africans have to rely on foreign journalism to comprehend their social realities, the coloniality of dependency is not only material but discursive as well.

Discussion and conclusion

This chapter analysed the ideological framing of the carbon credit industry in Zimbabwe using the case study of the Kariba REDD+ credibility crisis that was exposed by the collaborative investigative journalism of *The Guardian, De Zeit and Follow the Money.* The analysis confirms the literature that argues for studies that examine the political ecologies of green capitalism, as well as its discursive legitimation. Further, the ideological critique of the framing confirms that the discourses mobilised to promote carbon credits offsets function to sustain a carbon offset fantasy that promises development to poor communities in the south but only succeeds in exploiting those communities. Importantly, findings show that in the dissimulation and reification mode carbon offset discourses become the new missionary discourse that presents carbon credits as the only salvation for the environment and Global South populations. Inevitably, they secure the hegemony of a new colonialism, that of green land grabs. Local media become complicit in this by virtue of its journalism of omission, which creates an information dependency on Western media in postcolonial Africa. This study contributes to a burgeoning body of critical scholarship that engages with the coloniality of decarbonisation initiatives. By exploring the ideological contestations that attend the varied framings of the Kariba REDD+, the study unravels the material and discursive politics of carbon credits. As has been shown by other scholars, climate action is not an apolitical phenomenon, as insisted by technoscientific solutionism. By centring power dynamics and their ideological legitimation, it highlights the need for acknowledging that awareness-raising climate communications and decarbonisation policies and interventions are not neutral, value-free and interest-free undertakings (Collins et al., 2021; Sultana, 2024). The study complements climate communication research that has hitherto predominantly focused on how media translate the abstract scientific idea of climate change into a communicable object and socially meaningful discourse (see Faimau, Sello & Nkukhu-Orlando, 2023). The chapter lends credence to the view that climate action manifests the logic of coloniality, where Global South countries become the carbon frontier for Global North extractive capital. The study also points to the need for future research to explore the extent to which other climate initiatives like the "just energy transition" are ideologically legitimated in spite of the green imperialism and climate injustices that they perpetuate.

References

Aukland, L., Costa, P.M. & Brown, S. 2003. A conceptual framework and its application for addressing leakage: The case of avoided deforestation. *Climate Policy*. 3(2): 123–136. DOI: 10.3763/cpol.2003.0316.

Baras, D. 2023. Carbon offsetting. *Ethics, Policy & Environment*. 27(3):1–18. DOI: 10.1080/21550085.2023.2223805.

Berglez, P. & Olausson, U. 2023. Climate irresponsibility on social media. A critical approach to "high-carbon visibility discourse." *Social Semiotics*. 33(5): 1011–1025. DOI: 10.1080/10350330.2021.1976053.

Carbon Green Africa (2024). Kariba REDD+ Project https://carbongreenafrica.net/kariba-redd-project/ (retrieved 10 August 2024)Carbon Green Africa (2024). Carbon Green Investments Approval to withdraw from VERRA's VCS Program. https://carbongreenafrica.net/carbon-green-investments-approval-to-withdraw-from-verras-vcs-program/ (Retrieved 27 September 2024).

Collins, Y.A., Maguire-Rajpaul, V., Krauss, J.E., Asiyanbi, A., Jiminez, A., Mabele, M.B. & Alexander-Owen, M. 2021. Plotting the coloniality of conservation. *Journal of Political Ecology*. 28(1): 968–989. DOI: 10.2458/jpe.4683.

Dodman, D. & Mitlin, D. 2015. The national and local politics of climate change adaptation in Zimbabwe. *Climate and Development*. 7(3): 223–234. DOI: 10.1080/17565529.2014.934777.

Dyer, R. 2002. *The Matter of Images: Essays on Representation*. London: Routledge.

Dzingirai, V. & Mangwanya, L. 2015. Struggles over carbon in the Zambezi Valley: The case of Kariba REDD in Hurungwe, Zimbabwe. In *Carbon Conflicts and Forest Landscapes in Africa*. 1st ed. M. Leach & I. Scoones, Eds. Routledge. 142–162. DOI: 10.4324/9781315740416.

Entman, R.M. 2007. Framing bias: Media in the distribution of power. *Journal of Communication*. 57(1): 163–173. DOI: 10.1111/j.1460-2466.2006.00336.x.

Evans, H.-C. 2015. Chapter 8 – Climate-change awareness and online media in Zimbabwe: Opportunities lost? In *Knowledge for a sustainable world: A southern African–Nordic contribution*. T. Halvorsen, H. Ibsen & V.R. M'kumbuzi, Eds. Cape Town: African Minds Publishers.

Faimau, G., Sello, N. & Nkukhu-Orlando, E. 2023. Print media coverage and the socio-contextual representation of climate change in Botswana. In *Climate Change Epistemologies in Southern Africa: Social and Cultural Dimensions*. 1st ed. J. Ahrens & E. Halbmayer, Eds. London: Routledge. DOI: 10.4324/9781003180814.

Greenfield, P., & Chingono, N. 2024. "We don't know where the money is going": The cowboys making millions from credit schemes. *The Guardian*, 15 March 2024. Retrieved from www.theguardian.com/environment/2024/mar/15/money-carbon-credits-zimbabwe-conservation-aoe. Accessed 16/08/2024

Kreibich, N. & Hermwille, L. 2021. Caught in between: Credibility and feasibility of the voluntary carbon market post-2020. *Climate Policy*. 21(7): 939–957. DOI: 10.1080/14693062.2021.1948384.

Kutyauripo, I., Mavodza, N.P. & Gadzirayi, C.T. 2021. Media coverage on food security and climate-smart agriculture: A case study of newspapers in Zimbabwe. *Cogent Food & Agriculture*. 7(1): 1927561. DOI: 10.1080/23311932.2021.1927561.

Leach, M. & Scoones, I. Eds. 2015. Political ecologies of carbon in Africa. In *Carbon Conflicts and Forest Landscapes in Africa*. 1st ed. London: Routledge. 1–42. DOI: 10.4324/9781315740416.

Maldonado-Torres, N. 2007. On the coloniality of being: Contributions to the development of a concept. *Cultural Studies*. 21(2–3): 240–270. DOI: 10.1080/09502380601162548.

Mashingaidze, N., Chirisa, I., Mutambisi, T. & Matamanda, A.R. 2021. Paradoxes surrounding carbon credits and local area development: The case of Mbire District, Zimbabwe. *Local Environment*. 26(10): 1175–1185. DOI: 10.1080/13549839.2021.1967899.

Michaelowa, A., Hermwille, L., Obergassel, W. & Butzengeiger, S. 2019. Additionality revisited: Guarding the integrity of market mechanisms under the Paris Agreement. *Climate Policy*. 19(10): 1211–1224. DOI: 10.1080/14693062.2019.1628695.

Mignolo, W. D. 2011. *The Darker Side of Western Modernity: Global Futures, Decolonial Options*. 1st edition ed. Durham & London: Duke University Press Books.

Mkodzongi, G. 2023. The political economy of the climate crisis in Southern Africa. *Review of African Political Economy*. 50(177–178): 374–387. DOI: 10.1080/03056244.2023.2289748.

Ndhlovu, M.P. 2021. The Fifth Estate: Analyzing Climate Change Punditry in the Zimbabwean Newspaper Columns. *Environmental Communication*. 15(3): 418–429. DOI: 10.1080/17524032.2020.1863826.

Pearse, R. & Böhm, S. 2014. Ten reasons why carbon markets will not bring about radical emissions reduction. *Carbon Management*. 5(4): 325–337. DOI: 10.1080/17583004.2014.990679.

Quijano, A. 2000. Coloniality of power and eurocentrism in Latin America. *International Sociology*. 15(2): 215–232. DOI: 10.1177/0268580900015002005.

Radley, B. 2023. Green imperialism, sovereignty, and the quest for national development in the Congo. *Review of African Political Economy*. 50(177–178): 322–339. DOI: 10.1080/03056244.2023.2277616.

Streck, C. 2021. How voluntary carbon markets can drive climate ambition. *Journal of Energy & Natural Resources Law*. 39(3): 367–374. DOI: 10.1080/02646811.2021.1881275.

Sultana, F. 2024. Urgency, complexities, and strategies to confront climate coloniality and decolonize pathways for climate justice. In *Confronting Climate Coloniality: Decolonizing Pathways for Climate Justice*. F. Sultana, Ed. London: Routledge. 1–28. DOI: 10.4324/9781003465973.

Tembani, M., Mujuru, L., Mureva, A., Mutete, P., Gotore, T., Muchawona, A., Makumbe, P. & Murepa, R. 2021. Institutional arrangements and collective action: Evidence from forest management in Zimbabwe. *Forests, Trees and Livelihoods*. 30(4): 258–274. DOI: 10.1080/14728028.2021.1985625.

Thompson, J.B. 1988. Mass communication and modern culture: Contribution to a critical theory of ideology. *Sociology*. 22(3): 359–383.

van Dijk, T.A. 2001. Multidisciplinary CDA: A plea for diversity. In *Methods of critical discourse analysis*. R. Wodak & M. Meyer, Eds. (Introducing qualitative methods). London; Thousand Oaks [Calif.]: SAGE. 95–11

Watt, R. 2021. The fantasy of carbon offsetting. *Environmental Politics*. 30(7): 1069–1088. DOI: 10.1080/09644016.2021.1877063.

Wengraf, L., Bujra, J., Mfula, C., Greco, E. & Bush, R. 2023. The climate emergency in Africa. *Review of African Political Economy*. 50(177–178): 295–301. DOI: 10.1080/03056244.2023.2293419.

Media sources

Revealed: More than 90% of rainforest carbon offsets by biggest certifier are worthless, analysis shows (*The Guardian*, 18/1/23).

"We don't know where the money is going": The "carbon cowboys" making millions from credit schemes (*The Guardian*, 15/3/23).

Faulty credits tarnish billion-dollar carbon offset seller (*Bloomberg* 27/3/23).

The Carbon Con – How offsetting claims are vastly inflated (Source Material, 18/1/23).

The Great Cash-for Carbon Hustle (*The New Yorker*, 16/10/23).

Renowned carbon credits project in Zimbabwe has more financial holes than Swiss cheese (*Follow the Money*, 26/7/23).

Experts call for revision of carbon credits the (*The Standard*, 8/12/23).

Kariba's multi-million-dollar carbon offset project in turmoil (*The Standard*, 1/11/23).

COP28: Communities benefit from carbon credits (*The Herald*, 23/11/23).
Zimbabwe's carbon credits trade under siege – says conservationist (newzimabwe.com 24/11/23).
South Pole under fire for "faulty credits" claims (Transparency International, no date).
Creating lasting impact in a complex environment – A look behind the curtain of the Kariba REDD+ forest carbon project South Pole 27/1/23.
Verra Statement on *The New Yorker's* article of October 16, 2023 (Verra, 17/10/23).
South Pole Cuts Ties with Zimbabwe Carbon Offset Project Kariba (CarbonCredit.com 2/11/23).

4 Climate change newsmaking practices in Botswana's newsrooms

Journalists' perspectives

Tumelo Koboyatau, Lyton Ncube, and Allen Munoriyarwa

Introduction

At a global level, there is growing consensus amongst media experts, climate change scientists/experts and activists, and even journalists themselves that the mainstream media has largely failed society on climate change news reporting (Rusbridger, 2009; Hertsgaard and Pope, 2021; Schäfer and Painter, 2021; Hagen, 2023; and many others). This criticism of the mainstream has been a global assessment of their performance, which has yielded a global damnation of a kind. The media, as the criticism goes, have not paid much attention to climate change news reporting, choosing to focus on other news beats at the expense of climate change. Rusbridger[1] once called for a reinvigorated conversation on climate change journalism. Hagen (2023) was more damning of their criticism of the mainstream media climate change news reportage, noting thus,

> We wouldn't be in as deep of a hole on climate change right now if it weren't for the mainstream media. Their job is to *inform the public*. They're supposed to cover the most important things happening and effectively communicate what's going on and why so people can make well-informed decisions. But the mainstream media has been failing to do this *for decades* when it comes to the planetary emergency. The climate crisis and ecosystem collapse are the stories of our time and the media is nowhere near where it needs to be on both the quality and quantity of their climate coverage. They've failed to effectively communicate that this is an emergency, what is causing it, how it connects to everything we care about, that we already have most of the solutions that we need to address this, who is working to prevent those solutions from being deployed, and that immediate and systemic action is required to address this … [2]

The criticism has hinged on several factors that range from an inability to inform the public (Hagen, 2023) of the quantity of climate change (Tagbo, 2010) and the inability of the mainstream to influence public discourses on climate change through their climate change news practices (Hertsgaard and Pope, 2021). In this chapter, we acknowledge these criticisms of the mainstream media's performance

DOI: 10.4324/9781003660217-6

ultimate climate change news coverage. We, however, argue that collective condemnation of mainstream media's performance blurs context-specific factors and nuances that explain the current state of climate change newsification in the mainstream media. We use the case of Botswana media to demonstrate that to understand media failings in climate change news reporting, we should explore each context to understand the causal factors. Using our Botswana case, we answer the question: What newsroom factors influence news coverage of climate change? Presently, there is no known research in Botswana that focuses on how journalists perceive climate change coverage in the mainstream media.

This dearth in literature happens at a time when consensus is growing amongst climate change researchers that public and mainstream journalism is central to public messaging of climate change. Furthermore, there is a growing consensus among researchers and mainstream journalists and journalism scholars that climate change journalism is important and essential to fostering global climate accountability and enhancing investigative journalism on climate change. However, the specific newsroom conditions essential to enabling climate change communication have not been interrogated sufficiently. Mechanisms that enable the growth of these networks have not been thoroughly researched. These newsroom factors remain unresolved and are still a critical area for further academic inquiry on climate change communication. This chapter highlights potential obstacles to climate change communication news production within mainstream media by analysing newsrooms. The chapter, therefore, uncovers underlying hindrances to effective climate change news production which could impact the success of future climate change news production.

This chapter draws on in-depth interviews with Botswana journalists. Conducting in-depth interviews for climate change framing research involved several key steps. Firstly, it involved identifying potential participants such as journalists, editors, or media professionals with relevant knowledge and experience in reporting on climate change. Furthermore, it involved developing a set of open-ended questions that explore their perspectives, values, and decision-making processes when it comes to climate change coverage. This is best achieved using a semi-structured format to allow for flexibility and in-depth exploration of participants' responses. For this chapter, interviewees were sampled from reporters from two mainstream newspapers in Botswana – *The Daily News* and *The Weekend Post*. The former is government-owned, while the latter is privately owned.

Purposive sampling of respondents is utilised. It was employed to target specific individuals or groups who are influential in shaping public opinion or policy decisions related to climate change, in this case, journalists and editors. By selectively sampling these individuals, we were able to gather in-depth insights into their perceptions, the underlying motivations, and the potential impact on public perception and action on climate change. The sample size is eight human participants. This is broken down into one editor from each newspaper (making them two), and six journalists (three from each newspaper) who have the most by-lines in terms of climate reporting. The criteria for selecting journalists focused on those directly connected to the research area of climate journalism. Such included if they are titled climate/environmental/science reporters or correspondents and regularly work with reporting climate change.

There is not much of a climate change newsbeat within Botswanan media. However, this is characteristic of most African media institutions. It is only in recent times that climate change news is slowly becoming recognised as an important newsbeat in African mainstream media. The interviews sought to understand, amongst other issues, factors that hinder climate change reporting, climate change knowledge amongst journalists in Botswana, editorial understanding of climate change, financial support for climate change-linked investigative reporting, and climate change news sources in Botswana. These were some of the issues that were discussed during the interviews with journalists. For data analysis, we draw on thematic analysis (Braun and Clark, 2013), which allows us to build prominent themes for analysis from the raw data gathered during interviews.

Climate change news and journalism in Africa

Emerging scholarship on journalists' coverage of climate change shows that African journalists still adopt an episodic approach to climate change news reporting, rather than a thematic approach (Sithole, 2023). In other words, journalists in South Africa, in particular, have been unable to cultivate climate change as an enduring newsbeat. Instead, they tend to focus on climate change as a response to major conferences like COP events or in relation to major climate change-induced disasters (Sithole's (2023).

This view is not only true of Africa's media climate change coverage. It has been noted even in newsrooms beyond Africa. For instance, Hertsgaard and Pope (2021) noted that in the United Kingdom, just like in Africa, mainstream news coverage of climate change has been dispiriting. However, this lack of coverage has manifested itself in several ways. Firstly, there have been a few stories – that is, the coverage fails at a quantitative level (Tagbo, 2010). Tagbo notes very few news stories (in a 2010 sample) on climate change, and the trend gets far worse when Africa is compared with other continents. This is true, especially if we take the example of Zimbabwe and South Africa. We argue that the Durban floods of 2022, brought by the storm Issa,[3] ignited climate change news coverage. In Zimbabwe, it was the devastating 2019 Cyclone Idai[4] that ignited massive climate change news coverage with newsrooms.

Scholars have noted that if climate change is not prioritised in the media, society would face devastating consequences (Tagbo, 2010; Sithole, 2023). Sithole (2023) argues that this lack of coverage means that climate change on the continent would lack a "a human experience" (Sithole, 2023, n.p.). Tagbo (2010) further adds that the quantity of mainstream news being disproportionate clouds public understanding of climate change, and issues around who is responsible for it. What this scholarship fully agrees on is that in Africa, the urgency of climate change has yet to be fully conveyed to the public. How did we arrive here? Ighobor (2022) argues that

African newsrooms find themselves in this situation because the journalists and these media institutions lack training and have not generally taken risks, like travelling, and, perhaps, investigative climate change reporting. What is lacking in all this literature is an interrogation of newsroom-specific factors that impact climate change reportage. This present chapter seeks to add to this literature by examining these newsroom factors using a less-researched context of Botswana media.

Climate change communication and the sociology of news

Sociology of news production theories widely acknowledge that news production is a product of several factors that coalesce at different levels (Tuchman, 2003: Schudson, 2005). As such, news production is either enabled or constrained by both organisational and occupational routines (Benson, 2004). Political and newsroom power can determine or control journalists' experiences with news production (Hout and Jacobs, 2008). For example, how journalists obtain sources – be they institutional political elite or low-level whistleblowers – determines the ultimate news product society interacts with (Munoriyarwa and Chambwera, 2020). From this perspective, anthropological approaches to news production do acknowledge the importance of news generation that links news reporters and other representatives of news bureaucracies, including government officials. In order to understand climate news production, for instance, we need to examine the interface of a host of institutional, technological, newsroom, financial, and any other factors to colour climate change communication within newsrooms. Furthermore, as Tuchman (2003) noted earlier on, newsroom traditions signal journalism as a social practice with roles as specific as news production (e.g., what we see in the news on climate change) and gatekeeping (what we do not see in the news on the same subject). What this means is that there are different people and factors that exert different levels of control on the news production process.

Journalists, on the other hand, are obliged to absorb news production policies and abide by these, even in instances where they are not clearly defined (Benson, 2004; Munoriyarwa, 2018; Ncube Mare and Muzondo, 2023). This often breeds tension between journalists' own motivations and those of news owners and other forces vested ultimately in the news product (Vos, 2013). For instance, the investigative climate change reporter might find themselves at loggerheads, and in conflict with an industrial polluter, or a huge coal-producing conglomerate backed by the state as a job creator. This creates tension between the journalist's obligation to produce news and the conglomerate's interests not to be framed in public discourses as an undesirable agent of climate change. Journalists often see themselves as professionals who publish news in the interest of society that they are part of (Vos, 2013).

There is a normative and supposedly professional view permeating within newsrooms that news is out there, waiting to be discovered and reported (Tunstall, 1971). But this, in the contemporary age, may not be true. News may not be out there but had to be dug and be discovered and disseminated. For instance, corporations that manufacture cars may hide independent assessments of their cars' pollution levels from environmental and climate change journalists. The latter, therefore, may learn that news, after all, is not out there. They had to hunt for news and expose these through whistleblowers, data-driven driven approaches, and other methods (Munoriyarwa and Chiumbu, 2024; Motsaathebe, Munoriyarwa and Chiumbu, 2025). Thus, the sociology of newswork approaches calls for an understanding of broader factors of power, control, structure, institutions, class, and community as influential in news production (Waisbord, 2014). At the same time, the agency of the journalist – as an agenda setter, (news) source identifier, a professional, and a writer (Peterson, 2003) – interacts with these factors in news production. In this chapter, we draw on this theory to understand how these factors provide opportunities or barriers to climate news production in a less-researched context of Botswana.

Findings

From the interviews with purposively sampled journalists from Botswana's two national newspapers – *The Daily News* and *The Weekend Post* – we established that most journalists struggle to competently report climate change discourses due to a plethora of challenges. Such challenges include limited knowledge of climate change language and practices. In addition, our study also notes that some editors have not fully embraced climate change as an important beat which deserves respect and resourcing, just like other genres, such as politics and business. Furthermore, some interviewees expressed that economic pressures and commercial interests also hinder effective climate change reporting in Botswana.

Inadequate knowledge of climate change jargon and practices

Most interviewees noted that Botswana journalists covering climate change struggle due to inadequate preparedness. Several lack familiarity with climate change jargon, routines, and cultures. This is largely due to gaps in journalism education and limited on-job training. Some journalists admitted difficulty distinguishing scientific terms like "greenhouse gas emissions" and "global warming," often using them interchangeably. When faced with complex concepts, they sometimes omit such information fearing misinformation or due to tight deadlines. Interviewee 1 (Journalist) said:

> Since practising, I have learnt that it is a better choice to omit information I do not understand, especially if it is science and numbers. It puts me in a better space of not having to write a rebuttal because of some complex science I could have easily avoided.
>
> (Interviewed, 02 June 2024, Gaborone)

Due to the lack of preparedness of most of the journalists in climate change reporting, audiences end up consuming half-baked stories, which are less informative. As a result of the cautious approach taken in climate change reporting, some of the stories are also less engaging and are largely event-driven than critical. Comparatively, journalists expressed that they feel at home when reporting traditional beats such as business, politics, and sports. Climate change is domiciled in the area of "science" making it difficult for all journalists to easily access, process, and share its jargon. Interviewee 2 (Journalist) said:

> The reason I ended up as journalist is because I have never performed well in maths and science related subjects. Now having to deal with those as a practitioner is quite stressful, the science and numbers are complicated. I think that is also why most stories across all beats do not have compelling data. Yes, there is data here and there in business stories, but I think we end up just regurgitating it to readers not knowing what it means.
>
> (Interviewed, 02 June 2024, Gaborone)

The above perspective is also corroborated by Interviewee 3 who expressed that due to limited knowledge on climate change discourses, journalists find it difficult to participate from an informed perspective in conversations around local and global impacts of climate change unless something bizarre happens. Interviewee 3 (Journalist) said:

> In my view, most of the local articles are event-led. For example, if there is a heat wave or floods. We need to do better and go further by reporting on other angles, such as how climate change affects vulnerable communities and global food security.
>
> (Interviewee 3, 02 June 2024, Gaborone)

From the above responses, lack of appreciation of basic science and mathematics contributes to some of the struggles encountered by Botswana journalists mandated to cover climate change. Traditionally, mathematics and science subjects are not compulsory for students enrolling for Journalism or Media Studies programmes in Botswana and most African tertiary institutions. However, due to the salience of climate change reporting in the current journalism menu, there is a need for educators to revisit their student's admission criteria.

The issue of limited climate change sources is also challenge in Botswana. The only important tool of the reporter is his news sources and how he uses them (Hess 1981). Some interviewees pointed out that they get climate change stories from reports from various institutions such as government agencies and non-governmental organisations. For instance, some interviewees from *The Weekend Post* expressed that they largely rely on reports from Bretton Woods institutions, citing that it saves them the time of having to "sift" between several sources such as industry-sponsored reports and peer-reviewed studies. Moreover, at times, authorities do not play ball, making it difficult for the journalists to work on their stories. Interviewee 4 (Journalist) remarked:

> Government officials and experts may not always provide timely or transparent information on climate change-related issues, which can hinder reporting.
>
> (Interviewed, 6 June 2024, Gaborone)

Scholars assert that how journalists obtain sources – be they institutional political elite or low-level whistleblowers – determines the ultimate news product society interacts with (Munoriyarwa and Chambwera, 2020). In essence, key to journalistic success is access to sources (Hellmueller 2014; Strömbäck et al. 2013). We note that journalists from the government-owned *The Daily News* enjoy relative access to government officials or state-aligned agencies compared to *The Weekend Post* counterparts. The bureaucrat provides a reliable and steady source of news (Schudson 1989). Therefore, the process of news generation is the link between the reporter and official, the interaction of the representatives of news bureaucracies, and the government bureaucracies (Schudson 1989). However, due to trust issues and adversarial relations which generally exist between journalists from the private press and government in polarised contexts, journalists from the privately owned press are at times denied access to information by government officials.

We also gathered that effective climate change news coverage is also hampered by lack of stand-alone climate change news desk in newsrooms. Interviewee 5 (Journalist) said:

> It would be easy to cover such a topic if we had a research desk, solely dedicated to climate change, we would also be interested in specializing as journalists instead of all of us being general news reporters.
>
> (Interviewed, 7 June 2024, Gaborone)

The study also notes that, at times, local media houses tend to rely on Western news media agencies such as AFP, Reuters, and BBC, for climate change news in Africa. Therefore, some of the interviewees expressed the need to capacitate local media houses and journalists so that they can also authoritatively participate in the public sphere on climate change discourses in Botswana and Africa. Moreover, some of the interviewees underscored that in-house research desks will reduce reliance on news wire services or external sources, ultimately leading to original and localised reporting of topics. Interviewee 6 (Journalist) said:

> Media coverage of climate change in Botswana is quite limited, this is due to lack of expertise: Journalists may not have the necessary expertise or training to report accurately on complex climate-related issues.
>
> (Interviewed, 03 June 2024, Gaborone)

We also established that journalists have a limited appreciation of international policies and instruments on climate change. Interviewee 7 (Editor) said:

> We are aware of the existence of the Paris Agreement, COP events and other policies on climate change but they are of no use to us as we have no understanding

> of the nitty gritty of the instruments, but I recognise they have an implication on various sectors that drive our economy such as mining and agriculture.
>
> (Interviewed, 08 June 2024, Gaborone)

The journalists said grasping the policies and instruments on climate change would help supplement other beats like business by introducing the economic costs of climate change action or inaction on the economy. Geopolitics and international politics play a major role in the political framework of climate change journalism. International climate discussions, like the COP conferences, are important occasions that receive a lot of media attention from *The Daily News* and *The Weekend Post*. The data shows that journalists cover the dynamics of international cooperation, the promises and deeds of many nations, and the conflicts between industrialised and poor countries. Journalists from both publications acknowledged that rich information for in-depth reporting is available on the geopolitical ramifications of climate change, including concerns about migration, security, and climate justice. Emphasising how global power dynamics, geopolitical stability, and already-existing inequality are all impacted by climate change.

Editors' attitudes on climate change within newsrooms in Botswana

We also note that the attitudes of most editors in Botswana newsrooms also hinder progress in the area of climate change reporting. In fact, editors' choices result in the marginalisation of climate change stories in newspapers. Editors prefer to devote much space to beats which arouse reader interests. Interviewee 8 from *The Weekend Post* revealed that since they are a weekly publication, they have a limited amount of space in the paper and prioritise political and economic stories which have a huge readership. Of course, the issue of space somehow relatively differs at a daily newspaper – *The Daily News* – and a weekly paper – *The Weekend Post*.

> Stories compete on a weekly basis for space. The editorial process focuses on content that is immediate and has an impact. It is easy to decide to run a story if it is political, economic or has major social impacts on Batswana. These types of stories often take precedence over climate change news.
>
> (Interviewed, 5 June 2024, Gaborone)

Interviewee 7 also expressed interest in generating high readership, emphasising the role of audience preferences.

> Climate change-related stories do not generate high levels of readership. Politics. Sports and Business news are more appealing to readers.
>
> (Interviewed, 02 June 2024, Gaborone)

It is also worth noting that a newsroom is a fast-paced environment that favours news with dramatic impacts and a shorter shelf life. In other words, sensational or tabloidised content precedes "serious" forms of journalism. Most editors are of the view that climate change is a slow and incremental topic, which makes it challenging to dramatise without causing alarm. In addition, some of the interviewees revealed that oftentimes they decide to run a climate change-related story if it is a significant policy development or if a natural disaster occurs. In fact, some of the editors said natural disasters and meteorological warnings such as cold fronts and heat waves make climate change reporting easy. They consider such stories as "human-interest news value". Moreover, some of the editors described such events as "newsworthy," "drive the story," and "strong hook," underscoring that these events simplify the work for journalists and allow them to frame the stories in a compelling way.

The issue of inadequate budgets or resources in particular was also cited as a constraint to effective climate change reporting in Botswana, especially in the privately owned media sector. For example, an editor from *The Daily News* revealed that covering climate change comprehensively requires significant resources, explaining that it is best to cover it as an investigative piece. Furthermore, interviewee 8 (Editor) remarked:

> Investigative journalism is resource intensive, not just financially, it requires time, and the journalist has to be well equipped to execute the assignment. We prioritise cost-effective measures.
>
> (Interviewed, 02 June 2024, Gaborone)

The cost-effective stories require less-intensive budgets to work on them, thus allowing constrained finances to be channelled towards stories that can generate more income for the newspapers. Additionally, geographic location also plays a role. According to the editors, it does not make business sense for them to cover climate change-related news in countries that are not near their readers. They indicated that the proximity news value ensures that they focus on stories with direct and immediate relevance to their readers, thus relegating climate change stories to the back burner unless there is a local angle. The editors also highlighted that there is also a risk of "redundancy of topic," "repetition," and "audience fatigue." Newspapers often struggle to find new angles of reporting climate change in Botswana, contributing to the marginalisation of climate change in everyday news discourses.

Some interviewees revealed that climate change reporting affects avenues of economic interests and political decisions. From a critical political economy perspective, the media are tied to business and political interests. One of the reasons why editors lack interest in climate change news is primarily because the beat does not "bleed". Most commercial media rely on advertising revenue for survival. In essence, advertisers have "acquired a de facto licensing authority status since, without them, newspapers cease to be economically viable" (Herman and Chomsky, 1988:14). Some of the interviewees conceded that as private media, they rely on advertising revenue, and their potential advertisers are from industries such as oil and gas. Therefore, such advertisers might be reluctant to place advertisements alongside content that could be seen as antagonistic to their interests. Interviewee

7 recalled, "As private media, we have received several blows in the past from advertisers who pulled out their adverts because of conflicting views." (Interviewee 7 (Editor), 08 June 2024, Gaborone).

This dependency on advertisers for financial sustenance contributes to the "muzzling" of climate change discourses in some of the newspapers in Botswana. Epstein (1973) argues that in spite of journalists having some shared professional values in relation to what constitutes news, these values have to be subservient to the needs and logic of the organisation they work for. Of course, this creates tension between the journalist's obligation to produce news and the conglomerate's interests not to be framed in public discourses as an undesirable agent of climate change. Journalists often see themselves as professionals who publish news in the interest of society that they are part of (Vos, 2013).

Climate change news sourcing dilemmas within newsrooms

Successful mainstream journalism depends on the ability of journalists and their newsrooms to cultivate and develop reliable sources. In fact, the ability of journalists to develop news sources is part of the journalistic capital that most journalists aim to achieve. In climate change news reporting, just like any many other news beats, a journalist is as good as their sources. Journalists in Botswana still need to develop news sources that relate to climate change. As a result, journalists' reportage of climate change remains limited due to the absence of a wide array of climate change news sources from which to draw news. For example, *The Daily News* and *The Weekend Post* were over-reliant on government experts. The danger is that government sources may actually provide government climate change propaganda, which only foregrounds government efforts to mitigate climate change, and overhyped successes. One journalist said,

> Government officials and experts may not always provide timely or transparent information on climate change-related issues, which can hinder reporting.
> (Interviewed, 4 June 2024, Gaborone)

Furthermore, journalists noted that some of their sources were institutions like COP reports and World Bank reports. This is where they often drew stories on climate change. While this points to a diversification of news sources, a move away from government sources, these are still elite news sources. Therefore, news source diversification is lacking in climate change communication. A lack of diversification in climate change news reporting has the overall effect of backgrounding the ordinary people's experience with climate change.

Another existential challenge with government and organisation-driven sources is that they often bring statistical data. For journalists who already complain of not having enough knowledge with statistical analysis, this may present a serious challenge. In fact, it is a double dilemma. It is double in the sense that they now need to deal with statistical data on climate change, which as they testified, they need more understanding of. Moreover, these statistics anchor a news beat (climate change), which they are not used to. One journalist summed up this lack of knowledge and how it intertwines with other factors, thus,

> In my view, most of the local articles are event-led. For example, if there is a heat wave or floods. **We need to do better and go further by reporting on other angles such as how climate change affects vulnerable communities and global food security**.
>
> (Interviewee 3, 02 June 2024, Gaborone) [emphasis ours]

Therefore, widening news sources on climate change is imperative for diversity in climate change news reporting. However, this is only possible with more training for journalists and much more awareness of climate change issues within newsrooms. The Botswana context raises several other dilemmas. Firstly, there is very little training offered by journalism schools within the country. This raises the question of whether it would be possible for journalists to obtain this kind of training. Secondly, there is the absence of in-house training for journalists. Newsrooms are already struggling with revenue flows, and training for climate change would be difficult from a sustainability perspective.

Discussion

Climate change is a central topic in mainstream journalism in Botswana and beyond. However, most journalists and African media institutions are inadequately prepared for climate change coverage due to a number of factors. Climate change reporting in Botswana newspapers largely takes an event-driven approach and is hardly analytical and critical. Most news stories on climate change focus on climate change workshops, reports, conferences and summits, as well as natural disasters. We follow up on Nixon's (2011) observation that some studies tend to just focus on media framing of specific issues while neglecting structural factors which shape and structure the coverage of such issues. Admittedly, "journalists work within a range of constraints and influences; structural factors that affect their output" (McQuail 2000: 244). Our study used the case of Botswana media to examine causal factors of media failings in climate change news reporting. For feasibility purposes, we focus on two publications – *The Daily News* and *The Weekend Post* newspapers. In addition, researching both a weekly newspaper and a daily newspaper provides valuable insights into climate change perspectives across diverse types of news outlets. Weekly newspapers often have longer article formats, allowing for more in-depth analysis and coverage of complex topics like climate change. This can provide a deeper understanding of the context, nuances, and background of climate change issues. On the other hand, daily newspapers have the advantage of providing more frequent updates and timely news coverage, reflecting the immediate impact and evolving nature of climate change. Researching both types of newspapers enabled a comprehensive examination of climate change framing, considering both long-term trends and current events, and helped to identify similarities and differences in their coverage strategies and audience reach.

Our study benefited from the rich theoretical insights of the sociology of news approach. This was important in exploring the interplay between various organisational factors that shape the production and dissemination of climate change news or lack of it in Botswana. Sociology of news production theories aver that news production is a product of several factors that coalesce at different levels (Tuchman, 2003; Schudson, 2005). As such, news production is either enabled or constrained by both organisational and or occupational routines (Benson, 2004). Critically, news is the result of the methods news workers employ (Fishman, 1980:14). The idea of news values, or the standards by which editors and journalists judge whether a topic is worthy of being covered in the media, is central to the sociology of news theory (Gans, 1979). Such qualities encompass, among other things, human interest, conflict, prominence, timeliness, and closeness. We established that key factors, including limited knowledge of climate change jargon and practices, impede journalists from competently covering the subject. In addition, sourcing routines and practices also limit adequate climate change coverage in Botswana. Critically, newsroom hierarchies and editorial policies influence climate change news coverage. Our study notes that editors' attitudes on climate change within newsrooms in Botswana have a significant role in how climate change news is generally treated.

Our findings corroborate the assertion that news organisations prioritise stories that fit their established values and routines (Gans, 1979). Importantly, when an event occurs, journalists choose either to select or ignore aspects of the event based on the values of their news organisation, rather than their own deep-seated beliefs or ideologies (Munoriyarwa and Chambwera 2020). As profit-driven entities, media institutions generally do not valorise climate change news beats since it does not "bleed, therefore does not lead". The inverse of it, being "if it bleeds, it leads" in journalism, highlights the tendency of news media to prioritise stories involving violence, conflict, or tragedy, as these are believed to attract more audience attention. McManus (1994) asserts climate change is perceived as a low-interest topic among the public; hence, it is less prioritised by news organisations. This approach is driven by the idea that sensational, emotionally charged content captures viewership more effectively, boosting ratings and advertising revenue, hence leading to profitability. Common examples include high-profile crimes, accidents, and natural disasters. Interviewed journalists noted that news with emotionally charged content often emerges from political or business stories. They described climate change news as a "slow burn" which hardly increases newspaper sales compared to other beats. Very few traditional audiences understand and have interest in climate change news. We also noted that climate change becomes "newsworthy" when the president and other world leaders attend a conference or summit. The story is covered on the basis of prominence rather than its relevance.

Both newsrooms' internal structure and resource allocation significantly affect the quality and extent of climate change news coverage. *The Daily News* and *The Weekend Post* have no independent desks dedicated to climate change reporting which could provide a comprehensive reporting on complex climate change issues. Additionally, financial and time constraints limit the ability of news organisations to conduct thorough investigations, yet climate change news requires such investigations. Schudson (2005) argues that investigative journalism is essential for in-depth climate change reporting, although it requires substantial resources that several newsrooms may not possess. Such resource constraints negatively impact on the depth and quality of climate news coverage in Botswana newsrooms.

The negative impact of structure cascades to news sourcing routines that we explained in the last section of our findings. As we noted, there is an overt reliance on elite news sources. We conclude this discussion on a cautionary note. It should be borne in mind that climate change news sourcing routines have an impact on how climate change stories are represented for the entire society

We caution that the problem with elite news sources, like the Civil Society Organisation's reports, think tank reports, government press statements, and presidential communique on climate change, is that these sources generate news that is largely for the elite. These sources if overtly relied on to report climate change within newsrooms may lead to climate change communication that focuses on (pro-elite) news policy debates, pro-government climate change propaganda, and general climate change reporting that focuses on the activities of the elite and government. The danger is that climate change can easily devolve into "calendar journalism". When it devolves into calendar journalism, it means it becomes one of those recurring (news) events that are routinely planned and that often serve as an excuse to elevate something to the status of news. Think tank reports, government-induced press conferences, and news-driven from other elite institutions do not often answer questions like: How are rural communities in Africa dealing with the devastating effects of climate change? How is indigenous knowledge being harnessed to mitigate climate change effects? How are conglomerates engaged in mining, etc., impacting on climate change in rural communities? To answer these questions and many more, climate change, journalists need to move beyond narrow news sourcing approaches. As we have noted, this requires a commitment to climate change reporting within newsrooms and further training of journalists for climate change reporting. In its current form, climate change reportage in Botswana and its overreliance on these elite news sources fail to pinpoint to us what is not happening, especially in rural and other peripheral communities of the country. This leads to a systemic and systematic invisibilisation of these peripheral communities. This adds to our scepticism of elite sources in news reporting and deepens distrust of media that do not question these sources.

Beyond organisational constraints, we gathered that journalists face individual constrictions in covering climate change news in Botswana. Importantly, climate change is a "science" discourse. Some of the journalists admitted that they have limited knowledge in science and mathematics, hence making it difficult for them to comprehend climate change discourse. From our analysis, challenges are compounded by the fact that media and journalism training institutions in Botswana and South Africa do not insist on mathematics and science when enrolling students. We therefore argue that there is a serious need to revisit this model. Moreover, though not the main focus of this study, from our analysis as well as reflexive experiences as journalism educators in African universities, we note that climate change journalism has yet to receive significant attention. As a matter of fact, climate change journalism is not part of the mainstream menu in most journalism curricula in Botswana and Africa. This requires urgent revision. Critically, we also established that a significant number of journalists in Botswana's newsroom lack formal journalism training. In other instances, some of these journalists have never been to college/university or studied disciplines completely divorced from journalism. Such journalists are driven by passion, experience, and on-the-job training. However, such journalists lack the basic research skills required to write good informative stories. Such research skills help them to go through peer-reviewed, credible scientific journals from the field. It is important for newsrooms to find ways of capacitating such journalists, or individual effort is required so that they can competently report on critical matters such as climate change.

Conclusion

Our study contributes to growing literature on journalism and climate change in the global South. Using the Botswana case, we examined newsroom factors influencing news coverage of climate change. Molotch and Lester (1974: 54) contend that newspapers reflect not a "world out there" but the "practices of those who have the power to determine the experiences of others." From the interviews with purposively sampled journalists from Botswana's two newspapers – *The Daily News* and *The Weekend Post* – we demonstrate that most journalists struggle to report climate change discourses due to challenges, including limited knowledge of climate change language and practices. Furthermore, we also gathered that economic pressures and commercial interests also hinder effective climate change reporting in Botswana. Importantly, our study also notes that some editors have not fully embraced climate change as an important beat which deserves respect and resourcing just like other genres such as politics and business. Some of the challenges faced by journalists in climate change reporting are traceable to lack of or inadequate college/university training. Our study also calls for journalism educators in Botswana and Africa to rethink/retool their training model and ensure that their curricula respond to the ever-changing environment. As journalism educators, we note that most journalism curricula in Botswana and the African region do not prioritise climate change reporting or science journalism in general. Where it exists on the menu, it is offered as an optional course. Our study has examined structural factors constraining effective climate change news coverage in Botswana. Future studies could take a comparative approach and examine other African contexts that are equally resource constrained. Moreover, interactive digital social media platforms, including TikTok, Facebook, and YouTube, have also become central in climate change conversations. Future research could also examine such platforms' climate change communication cultures and practices. Furthermore, future research can focus on existing on climate change news sourcing routines across selected newsrooms in selected African countries.

Notes

1 Follow Rusbridger's contributions here: www.iied.org/pubs/pdfs/G02512.pdf.
2 The full article is available here: www.linkedin.com/pulse/how-media-continues-fail-us-climate-what-needs-change-ryan-hagen/
3 The story of the floods can be followed here: www.downtoearth.org.in/africa/2022-durban-floods-were-most-catastrophic-natural-disaster-yet-recorded-in-south-african-province-89047#:~:text=these%203%20countries-,Parts%20of%20KZN%20coastal%20zone%2C%20including%20the%20greater%20Durban%20area,over%2040%2C000%20people%20became%20homeless.
4 The devastation of the cyclone and all its stories can be followed here: www.worldvision.org/disaster-relief-news-stories/2019-cyclone-idai-facts.

References

Benson, R. (2004). Bringing the sociology of media back in. *Political Communication*, 21(3), 275–292.

Clarke, V., & Braun, V. (2013). Teaching thematic analysis: Overcoming challenges and developing strategies for effective learning. *The Psychologist*, 26(2), 120–123.

Epstein, E. J. (1973). *News from Nowhere: Television and the News*. New York: Random House

Fishman, M. 1980. *Manufacturing the News*. Austin: University of Texas Press.

Gans, H. J. (1979). *Deciding What's News: A Study of CBS Evening News, NBC Nightly News, Newsweek and Time*. London: Constable

Hagen, R. (2023). How the media continues to fail us on climate & what needs to change. Accessible at: www.linkedin.com/pulse/how-media-continues-fail-us-climate-what-needs-change-ryan-hagen/

Hellmueller, L. (2014). The Washington, DC Media Corps in the 21st Century: The Source-Correspondent Relationship. New York: Springer. https://doi.org/10.1057/9781137389022

Herman, E. S., & Chomsky, N. (1988). *Manufacturing Consent: The Political Economy of the Mass Media*. New York: Pantheon Books.

Hertsgaard, M., & Pope, K. (2021). The media is still mostly failing to convey the urgency of the climate crisis. The Guardian 21 October. Available at: www.theguardian.com/commentisfree/2021/jun/03/media-climate-change-crisis-emergency

Hess, S. (1981). *The Washington Reporters*. Washington DC: The Brookings Institution.

Ighobor, K. (2022). African journalists: More training and resources will boost climate change coverage Environment reporting is expensive; it needs a lot of traveling and risk-taking. Accessible at: www.un.org/africarenewal/news/african-journalists-more-training-and-resources-will-boost-climate-change-coverage.

McManus, J. H. (1994). *Market-Driven Journalism: Let the Citizen Beware?* Newbury Park, CA: Sage.

McQuail, D. (2000). *Mass Communication Theory*. London: Sage Publications.

Molotch, H, & Lester, M. (1974). News as purposive behaviour: On the strategic use of routine events, accidents and scandals. *American Sociological Review*, 39, 101–12.

Motsaathebe,G., Munoriyarwa, A. & Chiumbu, S.H. (2025). Data journalism appropriation in African newsrooms: A comparative study of Botswana and Namibia, *Journalism Practice*, 1–20, DOI: 10.1080/17512786.2024.2447805

Munoriyarwa, A. (2018). Have they got news for us? The decline of investigative reporting in Zimbabwe's print media, *Communication*, 44(1), 71–88, DOI: 10.1080/02500167.2018.1441888

Munoriyarwa, A., & Chiumbu, S. H. (2019). Big Brother is Watching: Surveillance Regulation and its Effects on Journalistic Practices in Zimbabwe. *African Journalism Studies*, 40(3), 26–41, DOI: 10.1080/23743670.2020.1729831

Munoriyarwa, A and Chiumbu, S. (2024). Data journalism and investigative news reporting practices during the pandemic. In Tong, J. (2024). *Data Journalism and the COVID-19 Disruption*. London: Routledge.

Munoriyarwa, A., & Chambwera, C. (2020). Who are the arbiters of truth? Mainstream journalists' responses to fake news during the 2017 Zimbabwe coup. *Communication: South African Journal of Communication Theory and Research*, 46(4), 75–95.

Ncube, L., Mare, A., & Muzondo, I. (2023). Social Media and Sports Journalism in Zimbabwe. *Journalism Practice*, 19(7): 1637–1657. DOI: 10.1080/17512786.2023.2279334

Nixon, R. 2011. *Slow Violence And The Environmentalism Of The Poor*. Cambridge, MA: Havard University Press.

Peterson, M, A (2003) *Anthropology & Mass Communication. Media and myth in the new millennium.* New York: Berghahn Books.

Rusbridger, A. (2009). I've seen the future and it's mutual. *British Journalism Review*, 20(3), 19–26.

Schäfer, M. S., and J. Painter. 2021. "Climate Journalism in a Changing Media Ecosystem: Assessing the Production of Climate Change-Related News Around the World." *Wiley Interdisciplinary Reviews: Climate Change* 12(1): 1–20.

Schudson, M. (1989). The Sociology of News Production. Media, Culture & Society 11(3):263–282.

Schudson, M. (2005). News as stories. *Media anthropology*, 121–128.

Sithole, E. (2023). "Climate change journalism in South Africa misses the mark by ignoring people's daily experiences" — https://theconversation.com/climate-change-journalism-in-south-africa-misses-the-mark-by-ignoring-peoples-daily-experiences-207229

Strömbäck, J., R. Negrine, D. N. Hopmann, C. Jalali, R. Berganza, G. U. Seeber, and M. Belluati. (2013). "Sourcing the News: Comparing Source Use and Media Framing of the 2009 European Parliamentary Elections." *Journal of Political Marketing* 12(1): 29–52. https://doi.org/10.1080/15377857.2013.752227

Tagbo, E. (2010). Media coverage of climate change in Africa: a case study of Nigeria and South Africa. Reuters Institute for the Study of Journalism & University of Oxford, London. Retrieved from: https://reutersinstitute.politics.ox.ac.uk/our-research/media-coverage-climate-change-africa-case-study-nigeria-and-south-africa

Tuchman, G. (2003). The production of news. In Klaus Bruhn Jensen (ed.), *A Handbook of Media and Communication Research: Qualitative and Quantitative Methodologies*. London: Routledge, pp. 78–90.

Tunstall, J. (ed.) (1971) *Journalists at Work.* London: Constable.

Van Hout, T., & Jacobs, G. (2008). News production theory and practice: Fieldwork notes on power, interaction and agency. *Pragmatics. Quarterly Publication of the International Pragmatics Association (IPrA)*, 18(1), 59–85.

Vos, T. P. (2013). Historical mechanisms and journalistic change. *American Journalism*, 30(1), 36–43.

Waisbord, S. (2014). *Media Sociology: A Reappraisal*. London: Polity.

5 Bringing Climate Change Back Down to Earth

Intersectional Climate Justice Communication Strategies in South Africa and Beyond

Alex Lenferna

Introduction – Reframing Climate Action

Many stories told about the climate crisis emphasise distant and remote effects that are far-off both in time and space. Meanwhile, those with vested interests who oppose climate action, like coal, oil, and gas corporations, have capitalised on telling stories that emphasise the immediate pain of any action we would take to solve the climate crisis (see Lamb et al., 2020). You can imagine a formula: *climate action = real pain now + distant uncertain benefits far away in time and in space.* In a world where many are struggling to get by in the present, it's hard to sell people on swallowing a bitter pill now to pass on distant and uncertain benefits in the future. However, such a framing of climate change is just that, a framing. It is a framing that chooses to highlight certain benefits and costs, while ignoring, exaggerating, or downplaying others, which is why I call it the Climate Action as Sacrifice or Sacrifice Framing for short.

This chapter explores how through elevating different stories, framings, and narratives on climate change, activists are challenging that framing and building more support for climate action and stronger movements that drive forward climate justice. By showing how climate action can bring about major, meaningful, and multiple benefits for communities, they are working to expand support and movements for climate action beyond just those who care more narrowly about climate change. By challenging the narratives of vested interests and intersectionally connecting action on climate change to the lived struggles of people, they are better motivating collective action on climate change. One can imagine replacing the previous formula with this one: *climate justice = leaving behind our current rapacious, destructive, and expensive fossil-fuelled economic system + transformative action to build a better, more just world for both present and future generations, with an incredible array of short-, medium- and long-term benefits.*

This second framing is what I call the Climate Justice as an Urgent and Awesome Opportunity or Opportunity Framing for short. This piece argues that this framing is not only more powerful as a motivating framing. Additionally, it is a more accurate than the Sacrifice Framing, as will be shown by citing relevant empirical literature demonstrating the major, meaningful, and multiple benefits of acting on

DOI: 10.4324/9781003660217-7

the climate crisis, especially if climate action centres justice, intersectionality, and transformation.

Shifting climate discussions to an Opportunity Framing becomes particularly important in the context of countries like South Africa, where people's perceptions of climate change vary widely across communities, with many seeing other social problems as more urgent (Mahl et al., 2020). In this context, showing how addressing the climate crisis can help tackle other interconnected and pressing social and environmental crises can help grow stronger public support for climate action. For example, a 2023 survey showed that climate change was third on South Africa's priorities, behind poverty and the rising cost of goods and services (Leigh-Chetty, 2023). If narratives can draw on the body of evidence that shows that transformative climate action can help address poverty and the rising costs of goods and services, then climate justice can potentially become a more urgent priority and opportunity for the people of South Africa. This can be coupled with narratives that demonstrate the urgency of the climate crisis to also help climate action rise as a public priority. Doing so can tap into and build on existing public support for climate action, as surveys show that South Africans do largely believe that the climate is an emergency that requires urgent action (UNDP, 2021).

To demonstrate how movements are pioneering the application of the Opportunity Framing, the piece draws on and analyses activist and scholar discourse from across the world. It focuses largely on the South African context and draws heavily on the work of the South African Climate Justice Coalition (CJC) – a coalition of over 70 community-based, trade union and civil society organisations working to advance a transformative climate justice agenda. The author served in the elected leadership of the coalition and will draw heavily on the discourse of the coalition and its members to demonstrate the prevalence and efficacy of the Opportunity Framing in how the coalition discusses the climate crisis. Among the sources of analysis will be articles from coalition members, campaign documents, and conversations from members, including from the CJC's own podcast, Just Us and the Climate. Demonstrating the centrality of the Opportunity Framing, that podcast's tagline is about the need to "bring climate change back down to earth and show how it's not only a crisis, but an opportunity to build a better, more just world" (CJC, 2023a).

This article is written from a scholar activist perspective, and the research methodology is heavily based in participatory action research. Drawing on Cornish et al. (2023), "participatory action research (PAR) is a scholar–activist research approach that brings together community members, activists and scholars to co-create knowledge and social change in tandem". The aim of the research is to produce emancipatory scholarship, driven by the desire to tackle injustices and build futures supportive of human thriving. Reflecting this, the author works alongside, is connected to, and supports activists and groups at the forefront of climate and environmental justice struggles.

For the sake of focus and prioritisation, this piece focuses primarily on the benefits and burdens of transitioning away from fossil fuels, as they are the main driver of climate change. Fossil fuels for energy and industrial processes constitute about 80% of all global GHG emissions (Le Quere et al., 2016). Thus, the central (but not only) task of tackling the climate crisis is to end the fossil fuel era – the era in which fossil fuels dominate our energy and industrial systems and the fossil fuel industry dominates and corrupts our political and economic systems. While fossil fuels are this article's focus, similar dynamics also apply to a lesser or greater extent to other forms of climate action too.

Four False Sacrifice Narratives

To show why the Sacrifice Framing is inaccurate and problematic, this chapter attempts to classify four central narratives which are deployed or promoted to support the Sacrifice Framing, especially but not only in propaganda propagated by polluting interests opposed to climate action. Those narratives are also prevalent in the works of academics, writers, and reporters who may unwittingly elevate a disempowering narrative that serves vested interests. This chapter shows how movements are pushing back against such narratives by deploying the Opportunity Framing. The Four False Sacrifice Narratives the piece focuses on are:

1. **Intergenerational Sacrifice**: This narrative paints climate action as a sacrifice that the current generation takes on to protect future generations. It is based on an outdated or distorted view of the science that fails to recognise how the climate crisis is already here and that acting on climate change is very much in the interests of current generations.
2. **Greenhouse Gas (GHG) Parochialism:** This narrative paints the costs of climate action as outweighing their benefits. This is based on a view of the costs of climate inaction, which typically focuses solely or predominantly on GHG emissions, ignoring or downplaying the broader harms of fossil fuels and the broader benefits of climate action.
3. **Solution Bashing**: This narrative falsely sees climate solutions, such as renewable energy, as prohibitively expensive and a driver of energy poverty and inflation. Such a narrative is typically inaccurate and partial or misleadingly relies on outdated data from before the rapid cost reductions and advances that have occurred in the renewable energy sector.
4. **The Neo-Liberal Imaginary**: The pervasiveness of neoliberal thinking has led to limited conceptions of what is possible and appropriate in response to climate change. It has driven analyses which distort the costs and benefits of action by promoting an individualistic view which fails to understand the underlying structures that shape the costs and benefits of fossil fuels and renewable energy.

Intergenerational Sacrifice

An oft-repeated narrative is that climate change is something we do for future generations, even a sacrifice taken on for them. While it is true that the most harmful effects of climate change will be visited upon future generations, where the Sacrifice Framing goes wrong is by claiming that current generations will not benefit from climate action. To see why, we can turn to the Intergovernmental Panel on Climate Change (IPCC) Special Report on 1.5°C (IPCC, 2018). As their 2018 report warned, unless we dramatically reduced our emissions, then by 2040 the world could push passed the Paris Climate Agreement's more ambitious target of keeping the global average temperature increase to 1.5C. What's notable about 2040 is that the majority of people alive today will still be alive. Millennials will on average be middle-aged, and many babies born in 2020 will have just finished high school and entered adulthood into a world experiencing the ravages of warming above 1.5°C. What's more, due to inaction since the report's publication and worrying acceleration of warming trends, the world has already at least temporarily passed 1.5°C and may go over the long-term 1.5°C average target on current emissions trajectory around 2030, if it has not done so already (Lamboll et al., 2023).

The IPCC estimates the economic damage will be $54 trillion if we breach the 1.5°C average mark. That massive but dry number does not quite convey the immensity of suffering and harm that will be unleashed. In the words of David Wallace-Wells,

> hundreds of millions of lives would be at stake, nearly all coral reefs would die out, wildfires and heat waves would sweep across the planet annually, and the interplay between drought and flooding and temperature would mean that the world's food supply would become dramatically less secure.
>
> (Wallace-Wells, 2018)

Clearly, these are devastating and widespread climate impacts within the lifetimes of the majority of people alive today, if we do not take urgent and sweeping action. So, contrary to Intergenerational Sacrifice Narrative, both future *and* current generations will enjoy the benefits of climate action and suffer the consequences of inaction.

Furthermore, increasingly catastrophic climate change will not simply start or end at 1.5°C. Rather every additional amount of warming that we add into the atmosphere compounds the dangerous climate impacts that we are already facing. As advisers to the Least Developed Countries group, Achala Abeysinghe and renowned late Bangladeshi climate scientist Saleemul Huq reported in significant detail nearly a decade ago: the effects of climate change are already here and bringing more "frequent and extreme climate- and weather-related disasters such as floods, cyclones, tornadoes, landslides, droughts, heatwaves, and malaria outbreaks" (Abeysinghe and Huq, 2016, p. 9). Likewise, Indian Fridays for Future youth activist Disha Ravi (2021) wrote about how unlike for many in the global north who are relatively shielded from the impacts of climate change, for many in the global south it is an immediate crisis already upon them – a central point in how young global south activists are framing the issue.

On the negative side, within the next few years or decades, we have the potential to push the climate into devastating tipping points that threaten to spiral the climate system and human responses out of control (Carrington, 2023). On the positive side, there is also increasing evidence that shows that current generations can be the ones to halt the worsening of climate change. Evidence from economic models shows that within the next few decades, we can bring total or net GHG emissions down to zero globally and do so in ways that better achieve sustainable development, poverty alleviation, and equality than not acting on climate change (IPCC, 2018). If we are able to reach the point of net zero emissions, then as climate scientist Zeke Hausfather highlights, the "best available evidence shows that… warming is likely to more or less stop once carbon dioxide (CO_2) emissions reach zero" (Hausfather, 2021). What's more, if we are able thereafter to move to negative emissions, we could begin to reduce the levels of warming.

Thus, climate action is very much something that benefits both the present and future generations. Indeed, the current generation can be the generation that stops the worsening of the climate crisis in their lifetimes, as well as significantly reduces the scale of the climate impacts that they and future generations face. That is truly an awesome opportunity, if it can be seized fast enough to prevent climate change from spiralling out of control. What's more, as the following sections will show, the non-climate benefits of doing so will be massive, especially if done in a just and equitable manner. However, if we don't act fast enough, it is current generation that risks pushing the world into irreversible tipping points, within the next few years or decades, while keeping the world locked into the harms of an extractive, fossil-fuelled, economic system. Thanks to decades of climate inaction, obstruction, and delay, it is current generations that are faced with that staggering and urgent planetary responsibility.

Greenhouse Gas Parochialism

One of the most problematic framings in climate discussions has been the tendency to limit analysis of what is harmful about fossil fuels and climate inaction to include only the effects of GHG emissions – what I term GHG parochialism and others have termed "carbon tunnel vision" (Achakulwisut, Almeida and Arond, 2022). GHG parochialism, as I am defining it, is a framing which focuses on just the harms caused by the warming effect of GHGs, rather than considering the wider context and the broader harms of fossil fuels and pollution. Examples of GHG parochialism abound in the climate literature. Indeed, as researchers highlighted in a literature review, a shift in analysis to take into account the full lifecycle impacts of fossil fuels is largely missing from the energy justice literature (Healy and Barry, 2017). The lack of such a framing brackets out the broader harms of fossil fuels and the immense non-climate benefits of shifting away from fossil fuel dependency. If we broaden our analysis and framing to include the non-climate-related harms of fossil fuels, then climate action is even more in our own interests.

While some climate advocates may have not fully realised the harmful impacts of GHG parochialism, those who sought to block action did and worked to deploy it. Consider Saudi Arabia, for example, who has consistently sought to undermine climate action to protect its vast oil interests. In one example of such behaviour, they attempted to weaken the 2018 IPCC report. More specifically, they pushed back against including the broader benefits of climate action within the IPCC's analysis (Darby, 2018). They were pushing for a framing akin to the Sacrifice Framing, which emphasised the costs of climate action and downplayed the broader sustainable development benefits. Saudi Arabia was shrewd in doing so, for the broader harms of fossil fuels are both numerous and immense, as I will highlight in the following sections.

Environmental Injustice, Racism, and Classism

As a report by climate activist organisation 350.org highlights, oil, gas, and coal companies are directly or indirectly responsible for some of the worst human rights abuses committed by corporations around the world in the last three decades (Savaresi and McVey, 2020). For indigenous peoples especially, the fossil fuel industry's extractive processes have long served to violate their rights, undermine their communities and the ecosystems they rely on, and directly harm their people through violence, murder, and intimidation. For instance, in the Niger Delta, the indigenous Ogoni people were subject to significant violence and intimidation from Shell who worked with the state military and police forces to quell resistance to Shell from communities. The experience of the Ogoni in the Niger Delta echoes that of many indigenous communities.

As a result of the violence and intimidation often visited upon indigenous communities, indigenous peoples' analyses of the climate problem often differ significantly from those who employ GHG parochialism. As indigenous scholar and organiser Kyle Powys Whyte details, indigenous peoples often perceive climate change not through the lens of the far-off and remote effects of GHG, but rather through "their experiences of already having been deeply harmed by the economic, industrial, and military drivers behind anthropogenic (human-caused) climate change" (Whyte, 2017, p. 2). Whyte catalogues how settler colonialists across the globe have long "displaced, terrorized and polluted Indigenous communities for the sake of profiting from coal oil and gas". It is for this reason that Whyte argues that indigenous climate justice movements are distinct "in putting their resistance to the nexus of colonialism, capitalism and industrialization at the vanguard of their work" (Whyte, 2017).

Echoing the disproportionate impacts on indigenous communities, study after study has long shown that thanks to a range of structural factors, the burdens and harms of pollution also fall disproportionately on low-income, Black, and/or communities of colour (Schlosberg and Collins, 2014). Meanwhile, opposition to clean energy, like wind farms, is driven disproportionately by rich white men (Larsen, 2023). Similarly, it is also rich, conservative white men, who are most likely to deny climate change, given that they benefit most from the fossil-fuelled status quo and are least impacted by climate and other environmental impacts (Malakoff, 2011). What the evidence above illustrates is that environmental racism and class exploitation are at the core of the climate crisis.

Communication and campaign strategies which show and tackle the intersection between environmental justice, class struggle, and anti-racism can help the climate movement connect to broader struggles for racial, economic, and social justice. It can also help the movement shed its image as being about the concerns of white middle-class environmentalists. That is an image which prevails in South Africa, partly due to legacies of colonial, racist, and exclusionary environmentalism and partly due to propaganda efforts which attempt to falsely paint climate action as the preserve of just white middle-class environmentalists (Rawoot, 2023). That image prevails, despite the evidence showing that the climate crisis is built on both environmental racism and exploitation of the poor, and that many of the movements pushing for climate justice and on the frontlines of climate injustice are indigenous, black, and working class.

Deadly Air

Keeping in mind that the costs and harms of fossil fuels fall disproportionately on communities of colour, indigenous peoples, and low-income communities, let us analyse the range of benefits that can come from reducing our dependence on fossil fuels. We can start by considering air pollution, arguably one of the most important reasons to transition away from fossil fuels. As research shows, in 2015 alone over 8 million people are estimated to have died as a result of air pollution – that is as many people as died from smoking, AIDs, and all forms of violence including wars combined (Lelieveld et al., 2023). Reducing fossil fuel use, the biggest cause of air pollution, thus provides one of the great lifesaving measures around – even if you take climate change out of the picture.

As multiple studies show, the benefits of reducing air pollution are so great that they are more than enough to pay for the costs of investing in climate action. The Lancet Commission on Health and Climate Change, for example, is a body set up to map out the impacts of climate change on health and make recommendations to improve health standards worldwide. According to their analysis, curbing air pollution, rapidly phasing out coal, and providing access to renewable energy would have substantial and immediate gains for human health (Watts et al., 2015). In the words of their report, "tackling climate change could be the greatest global health opportunity of the 21st century". Additionally, most of the needed actions "are 'no-regret' options, which lead to direct reductions in the burden of ill-health, enhance community resilience, alleviate poverty, and address global inequity" (Watts et al., 2015, 1861).

What becomes clear from these results is that excluding air pollution from our analysis of why we need to end the fossil fuel era leaves out one of the great moral and pragmatic motivators for why we would want to act – a motivator that is more tangible, immediate, and proximate than GHGs. Air pollution impacts such as asthma, bronchitis, ischaemic heart disease, strokes, lung cancer, acute lower respiratory infection, death, and not being able to go outside for fear of breathing pollution. These and other impacts get glossed over by GHG parochialism. By focusing only on GHGs, we can obscure how the same fossil fuels that are causing the climate crisis are also creating air pollution emergencies which are taking the lives of tens of millions of people and diminishing the lives of billions more every single year. In South Africa, climate and environmental justice campaigners are recognising the centrality of air pollution and have taken the government to court in their Deadly Air Case. They have made fighting deadly air pollution central to their fight against coal plants, both existing and new (Ncube, 2022).

Water Is Life

Air pollution, furthermore, is just one of the many deadly impacts of fossil fuels. Another devastating impact of fossil fuels is water pollution. Thanks in large part to indigenous-led resistance to pipelines such as the Dakota Access Pipeline (DAPL) and the East African Crude Oil Pipeline (EACOP), the public is becoming increasingly aware of the connection between water pollution and fossil fuel extraction. Many advocating against DAPL, for instance, referred to themselves as Water Protectors and resisted the pipeline under the mantra "Mni Wiconi", Lakota for Water is Life (Whyte, 2016). A significant focus for both the Water Protectors and those resisting EACOP was the threat of pipeline spills contaminating vital water supplies. Furthermore, beyond pipeline spills and leaks, the potential for fossil fuels to pollute water also manifests in many other ways, such as oil tanker and rig spills, acid mine water pollution from coal mines, toxic metal pollution from coal power plants, and toxic pollution from fracking leaking into water supplies during drilling.

Pollution, however, is not the only impact that fossil fuels have on water. In addition to water pollution, fossil fuels are also incredibly water intensive in terms of how much water is used in their extraction, refining, and combustion processes. According to the International Energy Agency (2012), switching to renewable energy can help free up the 15% of global water use that is estimated to be used in fossil-fuel-powered production. Examining the water–energy nexus, we see the possibility of a virtuous or vicious cycle depending on which energy pathway the world chooses. If we stay reliant on fossil fuels, we will drive further water scarcity through climate change. We will also be locked into a water-intensive mode of energy production which competes for increasingly scarce water supplies and which also pollutes many remaining sources of water. Alternatively, if we switch to renewable energy, we will reduce water pollution, reduce water scarcity driven by climate change, and reduce the amount of water needed for energy production by several orders of magnitude.

Conserving water through clean energy is vital for societal functioning given how central the availability of clean water is to health, ecosystems, food production, and communities in general. The massive impacts that fossil fuels have on using and polluting water are profound and have been a central motivation for resistance to fossil fuel extraction, refining, storage, burning, and transportation projects the world over. In South Africa, it has been at the heart of struggles against the water-intensive and polluting process of fracking for fossil gas in the water-scarce Karoo region (Toit, 2013). It has also been central to the fight against the

Musina Makhado Special Economic Zone – a proposed industrial zone targeted for water-scarce Limpopo that would be heavily powered by coal. The industrial zone would require a massive water scheme that transfers water from neighbouring Zimbabwe (CJC 2023b). Even Eskom, South Africa's national energy utility, which is the biggest polluter on the African continent, recognised the major water benefits of transitioning away from coal. In a presentation, they highlighted that accelerating the transition from coal to renewable energy could save over 16 billion litres of water per year by 2030 (Ruyter, 2022). The importance of such water savings has been a key point in the Climate Justice Coalition's campaign for a Green New Eskom (CJC 2022).

Air pollution and water pollution are just some of the harmful impacts of fossil fuels that are left out by GHG parochialism. If we look across the full life cycle of fossil fuels, we see a litany of violence, pollution, and costs that are left out when we focus merely on the GHG emissions, gasses which are invisible to humans and play out in complex and hard-to-track ways. By shedding GHG parochialism, movements are showing the much more tangible harms and impacts of fossil fuels and better motivating for solutions to move away from them. The case for shifting away from fossil fuels becomes even more robust when we consider how expensive they are relative to clean energy solutions, which are rapidly becoming much cheaper, more accessible, and technologically superior.

Solution Bashing

Thanks to rapid advances in technologies, renewable energy technologies such as wind, solar, and storage now provide the cheapest form of energy across most of the world. Yet, many continue to use outdated analyses to argue that a rapid transition away from fossil fuels will be prohibitively expensive and impose significant costs. Companies like Shell, Exxon, and BP exaggerate the costs of renewable energy and downplay its potential to try persuading the public of the continued need for fossil fuels (Muttitt, 2017). Fossil fuel industry propaganda against renewables helps to control energy narratives and the policies and public understanding surrounding them. Their narratives make a fossil-fuelled future seem inevitable and a renewable energy future seem like unrealistic whimsy, even though the opposite is becoming increasingly true.

One study by researchers at Stanford University, for example, provided renewable roadmaps for 143 countries, representing 99.7% of world's CO_2 emissions (Jacobson et al., 2019). The roadmaps call for countries to move all energy to 100% clean, renewable wind–water–solar (WWS) energy, efficiency, and storage no later than 2050, with at least 80% by 2030. Their study found that worldwide renewable future would have massive benefits, including reducing energy costs from $17.7 to $6.8 trillion/year (61%); lowering social (private plus health plus climate) costs from $76.1 to $6.8 trillion/year (91%); creating 28.6 million more long-term, full-time jobs than are lost in fossil fuels; and requiring only 0.17% and 0.48% of land for footprint and space. This study is just one of many in a growing and overwhelming body of research that demonstrates that a transition to a renewable energy future is not only possible, but more safer, more prosperous, and affordable on numerous fronts too.

Another major talking point that defenders of fossil fuels try to use is that fossil fuels are key to ending energy poverty, especially in Africa and Asia (Lamb et al., 2020). It forms the basis of a major fossil fuel industry public relations campaign. However, claims that fossil fuels are the key to tackling energy poverty are also undermined by the overwhelming amount of evidence that renewable energy is the cheapest, most effective, and cleanest way to tackle energy poverty. For example, one study found that unlike fossil fuels, investing in distributed renewable energy systems could actually end energy poverty in Africa and Asia by 2030 – where energy poverty is most concentrated (Curtin, 2021). In addition, it would also reduce billions of tonnes of climate pollution and create 25 million direct energy jobs and hundreds of millions more jobs in other industries. That's 30 times the number of energy jobs that would be created by a comparable investment in fossil fuels – thus also showing the falsehood of another sacrifice narrative that fossil fuels create jobs, whereas renewables do not.

Turning the fossil fuel industry propaganda on its head, climate justice activists are increasingly driving home the message that it is fossil fuels, not renewable energy, that are driving up energy prices, costing jobs, and causing economic pain in the present. They are demonstrating how price spikes in fossil fuels are a key driver behind inflation and the cost-of-living crisis that so many face. This dynamic was heightened during the Ukraine–Russia war, where war profiteering by oil and gas corporations sent oil and gas prices soaring, driving the cost-of-living crisis across the world. For example, on one estimate, fossil fuel price increases drove about 40 per cent of inflation in the United States (Leber, 2022). Renewable energy, on the other hand, is increasingly being advocated for and used to shield economies from fossil-fuelled inflation, volatility, and energy insecurity.

In heavily coal-dependent South Africa, activists are connecting skyrocketing energy prices to corruption and cost increases associated with the coal industry. For example, township activists in Soweto have fought against energy price increases on the grounds that they refuse to pay for the corrupt coal deals which have been central to driving up energy costs (Bond, 2022). They argue instead that renewable energy is key to tackling and ending energy poverty. Along a similar vein, they are also connecting the dots between corrupt coal deals and austerity. They argue that loans taken out to pay for corrupt coal deals have driven up South Africa's debt burden and put increasing strain on the government's fiscus. This has helped justify some of South Africa's harshest austerity measures and the cutting back of vital public services, such as social grants, medical care, and education. Activists are further showing how fossil fuels, by driving climate change, then contribute to climate-fuelled inflation for essential items like food, water, goods, and services.

Neoliberal Imaginary

The fourth and final problematic narrative that supports the Sacrifice Framing is what I term the neoliberal imaginary. I refer to this as the problem of neoliberal imaginary for several different reasons. Firstly, it is because neoliberal policies are partly, but not wholly, responsible for inhibiting climate action – where a neoliberal policy is one that is intended to promote laissez-faire economic liberalism through policies such as privatisation, austerity, deregulation, free trade, and reduction in government spending (Birch, 2017). By shrinking the role of the state, neoliberal thinking has obstructed some of the transformative, large-scale, and structural solutions that could drive major climate action that provides widespread benefits to many, such as investments in mass public transit; large-scale government renewable energy and energy efficiency programmes; affordable, green housing programmes, and urban design that addresses spatial injustice; and mass job programmes that work to restore the environment and deliver low-carbon care jobs.

Instead of focusing on such structural and political solutions, the neoliberal imaginary permeates how people often understand solutions to the climate crisis (Lenferna, 2025). In the popular discourse around climate change, calls to action on climate change often translate into exhortations to reduce one's personal emissions to buy green, drive less, use less plastic, be a green consumer, and tax people so they can change their behaviour, etc. This emphasises potentially costly individual action, rather than the opportunities for structural transformation that can create a more socially, economically, and ecologically just and prosperous future.

For many, finding out how they can do their own small personal bit to reduce their impacts comes from a place of good intention. However, the prominence of such individualistic thinking can also be traced to a decades-long fossil fuel industry public relations campaign (Westervelt, 2018). The industry tried to ensure that the problem of climate change was framed as a question of individual responsibility, rather than as a structural problem driven by industry lobbying, misinformation, and pushback against climate legislation and structural transformation. By shifting the focus onto individual action, the fossil fuel industry and polluting industries would be able to avoid responsibility for climate change. They would paint themselves as just meeting demand, rather than as active players stifling climate action as much as possible. They would also paint climate action as a costly individual responsibility, rather than an opportunity to build a better world.

The fossil fuel industry's attempts to shift blame to consumers bring me to the reason why I refer to the problem as the neoliberal imaginary, rather than simply neoliberalism (Lenferna, 2020). That's because defenders of the fossil fuel industry often wield neoliberal free market ideas to defend the fossil fuel industry, as being defenders of the free market who are simply meeting demand from consumers in a free market. However, far from being the competitive winners in some elusive free market, the fossil fuel industry relies on rigging economies, capital markets, and infrastructure in their favour, while often (directly and indirectly) employing militaries and police to secure their interests in violent, neocolonial, and extractive ways. The industry has also had many decades' worth of government support and funding, with subsidies amounting to trillions of dollars a year.

Indeed, the fossil fuel industry's dominance is largely the result of the heavy and often bloody hand of the government working to serve their interests. As scholars and activists have pointed out, trillions of dollars have been spent on wars, police, and military interventions to help secure fossil fuels (Shrader-Frechette, 2017, p. 392). In Southern Africa, for example, the South African National Defence Force has played a significant role in advancing the interests of extractive oil and gas corporations in conflicts in Mozambique and the broader region. Those wars are also enabled in significant part by the erosion of democracy. Fossil fuel corporations, through their centralisations of power and capital, have played a major role in corrupting, undemocratising, and undermining governments across the world (Mitchell, 2011). In response, activists often argue that a renewable energy future can help liberate people from oppressive fossil-fuelled regimes and provide a potential path to ensure a more democratic and decentralised energy future.

Climate justice activists in South Africa are working to push back against the neoliberal imaginary. They argue that if we reclaim the power of the state from captured fossil fuel interests and extractive mining and energy corporations, we could invest in a transformative and just transition to a more socially and ecologically just future. Along those lines, several trade union and movement leaders have been at the forefront of calling for a transformative state-led programme that creates millions of new jobs aimed at tackling South Africa's interconnected ecological, economic, and social crises simultaneously (Cry of the Xcluded, 2020). In the words of General Secretary of the South African Federation of Trade Unions, Zwelinzima Vavi (2021), acting on climate change and a just transition should be the basis of a "profound socio-economic transformation – this is the only way to get to achieve a zero-carbon world".

Conclusion

As this chapter has demonstrated, the Climate Action as Sacrifice Framing is doing significant damage to understandings around climate change. The framing is not only demotivating, but it is also largely inaccurate. Rather, the Climate Justice as an Urgent and Awesome Opportunity Framing is both more accurate and a much more powerful way of motivating action on climate change. That framing shows that climate change is not only a crisis. It is also an opportunity to build a better, more just world for both current and future generations. If we examine activist discourses in South Africa and beyond, we see that this Opportunity Framing is already being widely deployed in their campaigning and communications.

The challenge facing many activists, though, is that they are up against a very sophisticated fossil fuel industry public relations machinery. Climate advocates face one of the most powerful and wealthiest industries on earth that works with some of the world's biggest public relations firms. Their propaganda has worked to elevate the Sacrifice Framing and to capture, dissuade, and limit the public imaginary as to what solutions are possible to the climate crisis. Here, there is a vital responsibility for the media, scholars, and all who communicate around climate change to think critically about which narratives and framings they are elevating and whether they might unwittingly be echoing disempowering propaganda. In this context, climate communication is clearly not a neutral activity, especially in the face of the grave injustices perpetuated by inaction. Worse, if communicators are elevating the industry's climate obstructionist narratives, then they may, even if unwittingly, be helping perpetuate the vast harms caused by climate inaction.

References

Abeysinghe, A. and Huq, S. (2016) 'Climate justice for LDCs through global decisions', in Roser, D. and Heyward, C. (eds.) *Climate Justice in a Non-Ideal World*. Oxford University Press, pp. 189–207.

Achakulwisut, P., Almeida, P. C. and Arond, E. (2022) 'It's time to move beyond "carbon tunnel vision"', *Stockholm Environment Institute*, 28 March. Available at: www.sei.org/perspectives/move-beyond-carbon-tunnel-vision/.

Birch, K. (2017) 'What exactly is neoliberalism?', *The Conversation*, 2 November. Available at: https://theconversation.com/what-exactly-is-neoliberalism-84755.

Bond, P. (2022) 'In South Africa, resistance rises to the World Bank's climate-killing mega-projects', *Committee for the Abolition of Illegitimate Debt*, 16 October. Available at: www.cadtm.org/In-South-Africa-resistance-rises-to-the-World-Bank-s-climate-killing-mega.

Carrington, D. (2023) 'Earth close to "risk tipping points" that will damage our ability to deal with climate crisis, warns UN', *The Guardian*, 25 October. Available at: www.theguardian.com/environment/2023/oct/25/climate-crisis-threatens-tipping-point-of-uninsurable-homes-says-un

Climate Justice Coalition (2022) *Green New Eskom*. Available at: GreenNewEskom.org (accessed: 2 July 2023).

Climate Justice Coalition (2023a) *Just Us and the Climate*. Available at: http://justusandtheclimate.org/

Climate Justice Coalition (2023b) *The Public Versus the Musina-Makhado Special Economic Zone, Just Us and the Climate*. Available at: https://pod.link/cjc/episode/c814cdefa02db94a71ef7a10f8386db1

Cornish, F. et al. (2023) 'Participatory action research', *Nature Reviews Methods Primers*, 3(1). doi: 10.1038/s43586-023-00214-1.

Cry of the Xcluded (2020) 'We want a radical new deal that provides three million jobs', *Daily Maverick*, 13 May. Available at: www.dailymaverick.co.za/article/2020-05-13-the-cry-of-the-xcluded-we-want-a-radical-new-deal-that-provides-three-million-jobs/.

Curtin, J. (2021) *Transforming a Billion Lives: The Job Creation Potential from a Green Power Transition in the Energy Poor World*. Available at: www.rockefellerfoundation.org/report/transforming-a-billion-lives-the-job-creation-potential-from-a-green-power-transition-in-the-energy-poor-world/.

Darby, M. (2018) 'Climate weekly: Saudis push back as IPCC talks head into overtime', *Climate Home News*, 5 October. Available at: www.climatechangenews.com/2018/10/05/climate-weekly-saudis-push-back-ipcc-talks-head-overtime/.

Hausfather, Z. (2021) 'Explainer: Will global warming "stop" as soon as net-zero emissions are reached?', *Carbon Brief*, 29 April. Available at: www.carbonbrief.org/explainer-will-global-warming-stop-as-soon-as-net-zero-emissions-are-reached/.

Healy, N. and Barry, J. (2017) 'Politicizing energy justice and energy system transitions: Fossil fuel divestment and a "just transition"', *Energy Policy*, 108(June), pp. 451–459.

IEA (2012) *Water for Energy: Is Energy Becoming a Thirstier Resource?* Available at: www.iea.org/news/for-world-water-day-iea-shares-in-depth-analysis-of-energy-sectors-use

IPCC (2018) 'IPCC special report on the impacts of global warming of 1.5 °C – Summary for policy makers', (October 2018). Available at: www.ipcc.ch/report/sr15/.

Jacobson, M. Z. et al. (2019) 'Impacts of green new deal energy plans on grid stability, costs, jobs, health, and climate in 143 countries', *One Earth*, 1(4), pp. 449–463. doi: 10.1016/j.oneear.2019.12.003.

Lamb, W. F. et al. (2020) 'Discourses of climate delay', *Global Sustainability*, 3(e17), pp. 1–5. Available at: www.cambridge.org/core/journals/global-sustainability/article/discourses-of-climate-delay/7B11B722E3E3454BB6212378E32985A7.

Lamboll, R. et al. (2023) 'Assessing the size and uncertainty of remaining carbon budgets', *Nature Climate Change*, 13, pp. 1360–1367. doi: 10.1038/s41558-023-01848-5.

Larsen, L. (2023) 'Is race a major factor behind opposition to wind farms?', *Inside Climate News*, 5 October. Available at: https://insideclimatenews.org/news/05102023/inside-clean-energy-wind-projects-opponents-race/.

Leber, R. (2022) 'Fight climate change. End fossilflation. Here's how', *Vox*, 12 August. Available at: www.vox.com/science-and-health/2022/8/12/23290488/fight-climate-change-end-fossil-fuel-inflation.

Leigh-Chetty, R. (2023) 'Climate change ranks among top three concerns for South Africans', *HyperText*, 23 October. Available at: https://htxt.co.za/2023/10/23/climate-change-among-top-three-concerns-for-south-africans/.

Lelieveld, J., Haines, A., Burnett, R., Tonne, C., KlingmÃller, K., MÃzel,T. et al. (2023) 'Air pollution deaths attributable to fossil fuels: Observational and modelling study', *BMJ*, 383, e077784. doi:10.1136/bmj-2023-077784

Lenferna, A. (2020) 'Fossil fuel welfare versus the climate', in G. Wood, K. B. (ed.) *The Palgrave Handbook of Managing Fossil Fuels and Energy Transitions*. Palgrave Macmillan, pp. 551–567. Available at: https://link.springer.com/chapter/10.1007/978-3-030-28076-5_20.

Lenferna, A. (2025) 'The 3 radical Rs of environmentalism', in Barnes, B. and Barnwell, G. (eds.) *Community Psychologies for Climate Justice*. Springer. Available at: https://link.springer.com/book/9783031992223.

Mahl, D. et al. (2020) '"We are a Bit Blind About it": A qualitative analysis of climate change-related perceptions and communication across South African Communities', *Environmental Communication*, 14(6), pp. 802–815. doi: 10.1080/17524032.2020.1736116.

Malakoff, D. (2011) 'Conservative, white men more likely to be climate change sceptics, study shows', *The Guardian*, 27 July. Available at: www.theguardian.com/environment/2011/jul/27/conservative-white-men-climate-sceptic.

Mitchell, T. (2011) *Carbon Democracy: Political Power in the Age of Oil*. Verso.

Muttitt, G. (2017) *Forecasting Failure: Why Investors Should Treat Oil Company Energy Forecasts With Caution*. Available at: http://priceofoil.org/content/uploads/2017/03/forecasting-failure.pdf.

Ncube, V. (2022) 'South African court urges action on deadly air pollution', *Human Rights Watch*, 24 March. Available at: www.hrw.org/news/2022/03/24/south-african-court-urges-action-deadly-air-pollution.

Le Quere, C. et al. (2016) 'Global Carbon Budget 2016', *Earth System Science Data*, 8(2), pp. 605–649. doi: 10.5194/essd-8-605-2016.

Ravi, D. A. (2021) 'The climate crisis is about the Global South's present', Al Jazeera, 12 November. Available at: www.aljazeera.com/opinions/2021/11/12/climate-crisis-is-global-south-today-not-global-norths-tomorrow.

Rawoot, I. (2023) 'Decolonizing the South African climate movement', *Africa is a Country*, 16 October. Available at: https://africasacountry.com/2023/10/decolonizing-the-south-african-climate-movement.

Ruyter, A. de (2022) 'The Minimum Emission Standards and Eskom's 2035 Just Energy Transition strategy', *Presentation given to the Climate Justice Coalition* at the Eskom Offices in Johannesburg on 5th July, 2022.

Savaresi, A. and McVey, M. (2020) *Human Rights Abuses by Fossil Fuel Companies*. Available at: https://350.org/press-release/as-fossil-fuel-companies-keep-driving-the-climate-crisis-the-scale-of-human-rights-violations-is-increasing-alerts-350-org-report/.

Schlosberg, D. and Collins, L. B. (2014) 'From environmental to climate justice: Climate change and the discourse of environmental justice', *WIRES Climate Change*, 5, pp. 359–74.

Shrader-Frechette, K. (2017) 'Ethical energy choices', in Gardiner, S. M. and Thompson, A. (eds.) *The Oxford Handbook of Environmental Ethics*. Oxford University Press, pp. 391–403.

Toit, J. Du (2013) 'Fracking: Groundwater threats to Karoo towns', *Karoo Space*, 28 August. Available at: https://karoospace.co.za/fracking-groundwater-threats-to-karoo-towns/.

UNDP (2021) *The Peoples' Climate Vote*. Available at: www.undp.org/publications/peoples-climate-vote.

Vavi, Z. (2021) 'Their just transition, and ours', in *Remarks on the Just Transition to the School of Labour and Urban Studies*. Available at: https://drive.google.com/file/d/1Fl-iwhes7i_BDrGX35YzQGvgDBZiTaos/view?pli=1.

Wallace-Wells, D. (2018) 'UN says climate genocide is coming. It's actually worse than that', *New York Magazine*, 10 October. Available at: https://nymag.com/intelligencer/2018/10/un-says-climate-genocide-coming-but-its-worse-than-that.html.

Watts, N. et al. (2015) 'Health and climate change: Policy responses to protect public health', *The Lancet*, 386(10006), pp. 1861–1914. doi: 10.1016/S0140-6736(15)60854-6.

Westervelt, A. (2018) 'Exploiting scientists kyryponite', *Drilled*, 17 November. Available at: https://podtail.com/podcast/drilled/exploiting-scientists-kryptonite-certainty/.

Whyte, K. (2017) 'Way beyond the lifeboat: An indigenous allegory of climate justice', in Munshi, D. et al. (eds.) *Climate Futures: Reimagining Global Climate Justice*. Zed Books, pp. 11–20.

Whyte, K. P. (2016) 'Why the Native American pipeline resistance in North Dakota is about climate justice', *The Conversation*, 16 September. Available at: https://theconversation.com/why-the-native-american-pipeline-resistance-in-north-dakota-is-about-climate-justice-64714.

6 Climate Change Journalistic Lenses

An East African Perspective

Shitemi Khamadi

Introduction

The research utilized climate change stories submitted to the Aga Khan University Graduate School of Media and Communication (AKU-GSMC) during fellowship programs aimed at enhancing climate reporting in the region. The fellowship programs run by the school are designed to provide learning forums from experts in the subject as well as grants to support storytelling. To merit in the programs, journalists are invited to apply through an open call and thereafter applications are reviewed by a selection committee to determine the stories that best resonate with the theme in the call and would be impactful. Successful applicants published their stories on various media platforms. The fellowships, held from January to June 2022, included two programs in seven East African countries: Kenya, Uganda, Tanzania, Rwanda, Burundi, South Sudan, and the Democratic Republic of Congo (DRC). The first cohort received 143 applications, selecting 22 fellows, while the second had 233 applications, selecting 31 fellows.

Introduction

Climate change has emerged as a critical topic of global discourse, affecting lives and livelihoods across various contexts. Despite the presence of climate change deniers and conspiracy theorists, the media is compelled to cover the multifaceted impacts of climate change due to its pervasive nature (Moyers, 2019). Media coverage not only informs the public but also stimulates debate and influences policy discourse surrounding climate issues (Broadbent et al., 2016). Research indicates that such coverage is often driven by significant events, press releases, and environmental disasters, particularly in the Global South (Schäfer & Painter, 2020). High-profile gatherings like the annual Conference of Parties (COP) underscore the media's role in amplifying discussions about climate change globally.

The complexities inherent in climate change journalism necessitate a deep understanding of the subject matter for journalists to produce credible and authoritative content (Gess, 2012). This challenge is exacerbated by the politicization and commercialization of environmental issues, as well as the polarized nature of the media itself, raising questions about the legitimacy of climate change coverage

DOI: 10.4324/9781003660217-8

(Gutsche & Pinto, 2022). As Kunelius (2019) notes, the intricacies of climate change compel journalists to seek collaborative partnerships, which can undermine their independence. This need for collaboration is particularly pronounced given that most journalists lack formal scientific training, making it essential for them to navigate various partisan interests carefully before disseminating information.

Despite a scientific consensus on global warming and its anthropogenic causes (Evans, 2016; Nuccitelli, 2020), significant divisions persist, particularly in the Global North, often exacerbated by media representation (Boykoff, 2008). This centers the media's role in shaping public perception and engagement with climate change. This is because its framing, prioritization, and issue selection can enhance or diminish public concern (Appelgren & Jönsson, 2021). This paper seeks to build on existing research by exploring the key issues that journalists in the Global South prioritize in their climate change coverage. Focusing on East Africa—specifically the DRC, South Sudan, Burundi, Rwanda, Uganda, Kenya, and Tanzania—this study aims to contextualize the narratives, sourcing practices, and funding impacts that shape climate change journalism in the region.

Literature Review

Media Coverage of Climate Change: News Narratives and Journalistic Norms

Narratives play a crucial role in shaping public understanding of events and phenomena, particularly in the context of climate action (Hinkel et al., 2020; Naguimbing-Manlulu, 2021; Nordensvard & Ketola, 2021). Research suggests that narratives often hold more sway over public perception than scientific facts themselves (Bohm et al., 2019). The dominant narrative surrounding climate change tends to be one of doom and gloom, focusing on problems and adverse impacts (Meyer et al., 2021). Conversely, transformative narratives aim to provide a more positive outlook by emphasizing solutions rather than merely highlighting issues (Hinkel et al., 2020).

Narratives can vary significantly depending on the context—be it international, local, or hyperlocal. These variations may reflect the polarized political landscape surrounding climate issues and how the media chooses to frame them (Christensen et al., 2028). For instance, international discussions may prioritize climate policy and adherence to treaties, while local narratives might focus on climate action and protection (Böhm et al., 2019). Other relevant narratives at national and international levels include energy transition, the urgency of climate action, the declaration of climate emergencies, and the need to listen to science (Nagel & Schäfer, 2023). An emerging narrative is that of gender equality, which highlights the disproportionate impacts of climate change on women (Gaard, 2014). Gaard argues that women are 14 times more likely to be negatively affected by natural disasters due to various factors, such as household responsibilities and a lack of access to critical information.

Globally, journalism is navigating the complexities of covering climate issues that significantly affect lives and livelihoods (Boykoff & Yulsman, 2013).

Recent shifts in terminology, such as *The Guardian* and Norwegian newspaper *Aftenposten* referring to "climate crisis" instead of "climate change," illustrate a growing acknowledgment of the urgent nature of the issue (Elgesem, Knudsen, & Fløttum, 2024). This shift aligns journalism with the scientific consensus on the climate emergency. Historically, mass media coverage of climate change has faced criticism for its focus on scientific uncertainty and the equal representation of dissenting views (Hiles & Hinnant, 2014; Evans, 2016; Schäfer & Painter, 2020). However, contemporary coverage has evolved, with more balanced content emerging, particularly outside of partisan outlets.

The effectiveness of climate action is closely linked to how media reports on climate change and the sources it utilizes. Yet, media organizations face challenges such as digital transformation, disinvestment, and ownership issues tied to partisan interests, which have led to budget cuts and reduced specialist reporting (Boykoff & Yulsman, 2013; Gibson, Harper, & Alpert, 2016). This has made climate reporting a daunting task; journalists often encounter significant backlash, including online harassment and skepticism from editors and sources (Hiles & Hinnant, 2014). Some solutions to these challenges involve integrating climate stories through various lenses and relying on individual efforts for funding and reporting (Gibson, Harper, & Alpert, 2016).

Approaches to climate change coverage depend on how journalists and editors construct the narrative at the publication stage. Elgesem, Knudsen, and Fløttum (2024) argue that climate journalism is undergoing a transformative process aimed at fostering societal sustainability and adopting more constructive approaches. They suggest that while complete objectivity may be unattainable, journalists must remain aware of their subjectivity to represent climate issues fairly. Studies indicate a stark contrast in climate change coverage between the Global North and Global South, though data on the Global South remains limited (Boykoff & Yulsman, 2013; Schäfer & Painter, 2020). Comparative studies of media coverage in the US and UK reveal that a conflict-laden approach to anthropogenic climate change hinders progress toward international climate policy cooperation (Boykoff, 2007). Nevertheless, there is optimism that media reporting can catalyze public pressure for decisive climate action.

Climate Change Frames

Framing theory posits that media emphasizes certain aspects of reality while downplaying others, influencing public interpretation (Entman, 1993). Frame building involves selecting, omitting, and expanding information, thereby providing context, causal responsibility, and recommendations (Moernaut, Mast, & Pauwels, 2018). This process leads to frame production, wherein journalists develop specific frames around climate change (Guenther, Gaertner, & Zeitz, 2020). Common climate change frames include the human rights frame, environmental justice frame, cycles of nature frame, economic challenge frame, and inscrutable nature frame, each with its sub-frames.

Broadbent et al. (2016) assert that frames lend meaning to issues and events, guiding public interpretation. For instance, climate change may be framed as a "real" or "illusionary" phenomenon and discussed in various contexts, such as economic impacts or energy technology. Globally, one prominent frame is the moral imperative to reduce or eliminate carbon emissions. The scientific frame often carries a risk narrative, warning of catastrophic outcomes if climate change remains unchecked. Ultimately, the framing of climate change can significantly influence public perception and policy responses. With this in mind, the study asks:

***RQ1**: What narratives emanate from journalists' coverage of climate change?*
***RQ2**: How do journalists go about sourcing for climate change stories?*
***RQ3**: How do grants influence their climate change storytelling?*

Theoretical Framework

Narrative theory, also called narratology, is a framework that examines the structures and functions of storytelling across various media and cultural forms. It is rooted in literary theory but now spans disciplines such as psychology, sociology, media studies, and communication (Cebik, 1986; Cutting, 2016). It posits that storytelling is not merely a method of entertainment or artistic expression but a fundamental mode through which humans make sense of the world (Abbott, 2008).

The theory goes to the core of storytelling that individuals understand and interpret their experiences and the experiences of others through narrative form (Cebik, 1986). It posits that human cognition is inherently narrative in nature; we structure our memories, construct our identities, and communicate meaning through stories. This concept, often termed "narrativity," suggests that storytelling is a primary cognitive tool, not a secondary aesthetic device (Bruner, 1991).

A foundational distinction within narrative theory is that between story and discourse, or *fabula* and *sjuzhet* (Cutting, 2016). The fabula refers to the chronological sequence of events. In contrast, the *sjuzhet* refers to the manner in which these events are organized and presented to the audience (Chatman, 1978; Genette, 1980; Cutting, 2016). For example, a narrative may begin by revealing the past through flashbacks, or employ nonlinear sequencing to challenge or manipulate the audience's understanding. This distinction allows scholars to analyze not only what a narrative conveys, but how it conveys it.

Another critical aspect of narrative theory involves focalization and perspective. The narrative is always delivered through a particular point of view, which can significantly shape the audience's perception of characters and events. Gérard Genette's (1980) work on narrative discourse distinguished between the narrator (who speaks) and the focalizer (who perceives), highlighting the layered nature of narrative voice and subjectivity.

In addition, narratives also manipulate time in complex ways (Cebik, 1986). They can compress, stretch, or reorder time to influence emotional engagement or reveal information selectively. Such temporal strategies include analepsis (flashback), prolepsis (flashforward), and ellipsis (gaps in the timeline), each serving to build suspense, create surprise, or deepen psychological insight (Genette, 1980).

Several hypotheses within narrative theory explore its persuasive and psychological power. Walter Fisher's (1984) narrative paradigm, for instance, argues that people are more persuaded by stories than by logical arguments. According to Fisher, audiences assess the credibility of stories based on narrative coherence (internal consistency) and narrative fidelity (resonance with personal or cultural truths). Similarly, psychologists such as Jerome Bruner (1991) and philosopher Paul Ricoeur (1984) suggest that individuals construct identity through narrative. One's sense of self is shaped by internalized life stories, which are continually revised in light of new experiences and social contexts.

Narrative also plays a crucial role in persuasion and behavioral change. Research in communication studies, particularly in the area of narrative transportation (as theorized by Green & Brock, 2000), shows that when individuals become immersed in a story, they are more susceptible to its messages. Immersion fosters emotional engagement and identification with characters, reducing counter-arguing and making narratives a powerful medium for influencing attitudes and beliefs.

Narrative theory provides a rich, multifaceted approach to understanding how stories are constructed, how they operate, and why they matter (Cutting, 2016). Whether analyzed through the lens of structure, perspective, psychology, or ideology, narratives remain central to how we process the world, make meaning from our experiences, and engage with others. Their power lies not just in their form, but in their function as vessels of human understanding and as tools for social and personal transformation (Cebik, 1986). Narrative theory is central to this paper because it explains how societies use storytelling to debate, evaluate, and make decisions about responding to complex social issues (Harcourt et al., 2020). Narratives are essential in the construction, formation, and maintenance of collectives, like in this case, climate change, in which stories help to make sense of issues and shape ideals and values (Shenhav et al. 2014). It is especially relevant as theorized by Green and Brock (2000) on the power of narratives in persuasion and behavioral change.

Methodology

The study was investigated qualitatively using interviews and qualitative content analysis. First, the researcher undertook content analysis of the stories that the journalists had done as part of the Excellence in Journalism programs run by the AKU-GSMC. This was done to better understand the data, distill it into content-related categories, and classify it. This was meant to build broad descriptions of the phenomenon (Elo & Kyngäs, 2008) of climate change storytelling and use this to guide the interview process, including the questions to ask. This was done by open coding, creating categories, and abstractions.

The researcher then conducted seven in-depth interviews with science and environmental editors from across East Africa to get their insights. Creswell and Creswell (2019) state that 5–25 interviews are sufficient for qualitative studies.

The saturation point also determines when to stop interviewing participants. These editors not only commission stories but also determine which ones will be published. Hence, getting their insights was critical to the study. The editors interviewed were both those who had published stories from the fellows as well as some who hadn't but focus on environmental stories.

The researcher also generated data through interviews with journalists who had participated in two Excellence in Journalism programs run by the AKU-GSMC. The fellowship program had 53 fellows, and 35 shared their insights through interviews. The researcher was the project coordinator of the fellowship program and sought and was granted consent from all the fellows. The fellows are part of WhatsApp groups where the researcher is the administrator and has access to their phone numbers and emails. The WhatsApp groups provide a forum where relevant climate change information, including grants, can be shared for their consideration.

Data was analyzed thematically. Thematic analysis is a widely used method for identifying and interpreting patterns in qualitative data (Braun & Clarke, 2006). The approach involves starting with familiarization, where researchers immerse themselves in the data through reading and re-reading transcripts or notes. This helps to generate initial codes which are essentially labels that capture important features of the data (Nowell et al., 2017). These codes are then sorted into potential themes that reflect broader meanings and insights derived from the data. This thematic framework allows researchers to organize and interpret the data coherently, revealing underlying patterns, relationships, and nuances.

The researcher discusses the results with direct quotes that exemplify the themes. This provides contextualized insights and helps to ground the analysis in the participants' voices (Elliott, 2018). Yardley (2000) emphasizes that effective presentation of thematic analysis communicates the findings but also emphasizes their significance. This demonstrates how they contribute to the existing body of knowledge and offer implications for practice or further research.

Results and Analysis

Dominant Narratives and Themes in Climate Change in the Media in East Africa

The study found that the most dominant frames are transformative frames. Transformative frames in research refer to theoretical lenses that exemplify the potential for social change and empowerment within subject matters (Schulz et al., 2016). These frames focus on understanding how individuals and groups navigate their experiences and challenges. They often highlight the interaction between personal narratives and broader sociopolitical contexts (Schulz et al., 2016). By employing transformative frames, researchers can analyze how narratives contribute to social justice by amplifying the voices and experiences of those who are often overlooked in traditional academic discourses (Tilly, 2006). Ultimately, transformative frames serve to reshape the understanding of social issues. They foster a more inclusive and equitable research landscape.

Journalists and media houses in East Africa are driven by climate change stories as they relate to lives and livelihoods in their surroundings. This indicates an inclination to transformative narratives that Hinkel et al. (2020) write about in describing what these frames do. Many economies in the region are agriculturally based, and this is the primary driver of what they cover. Agriculture is an important socio-economic activity for people in the region, both for subsistence and for commercial purposes, including exports.

> I am particularly interested in covering the impact on agriculture and food security by examining how climate change affects crop yields, food production, and the livelihoods of farmers, especially in vulnerable regions. I am also interested in covering climate change adaptation and resilience, specifically exploring how communities are adapting to the changing climate, including infrastructure development, disaster preparedness, and resilience-building strategies.
>
> (FL 6)

Journalists in the region are also interested in telling stories that affect not only their communities but also specific groups within them like women and indigenous communities. They are also interested in specific climate polluters like plastics as well as telling stories that have solutions angles, like resilience approaches that farmers are using. The human–wildlife conflict that has been exacerbated by droughts that push wild animals into community neighborhoods also informs what journalists want to cover. Other topics include food waste, biodiversity, conservation initiatives, energy and renewable energy, and other sustainable practices that impact lives and livelihoods.

In retrospect, communities that are driven by other economic and cultural activities like mining and wildlife conservation also see journalists interested in covering these issues. Furthermore, natural disasters exacerbated by climate change become natural stories for journalists, a view also shared by editors from the region.

> As an editor, Uganda media often cover things around food security, because we get quite a bit of, extremes every, so often in the year, with, famine and hunger, some landslides, that cause a bit of flooding and, displacement, for areas that have, sort of most disaster attached to, flooding, droughts, migration, mostly, rural, and peri-urban.
>
> (KI 2)

An examination of the articles published indicates that transformative narratives are the dominant focus of journalists. There were tangible efforts to highlight and explain solutions to the community issues. The stories also discussed the challenges of the solutions. In this approach, they like telling stories on climate change response, scientific research on climate change, and human-interest stories on climate change. It also includes innovations in small-scale farming, particularly agroecology farming, to bring to frame positive changes among local farmers. This solution journalism angle was also echoed by an editor who stated that increasingly, there is a paradigm shift away from doom and gloom storytelling. It needs to consider possible solutions. This approach is cutting across climate change stories in agriculture, energy, and clean energy, among others.

> At ibihe.org [a Rwandan publication], we are interested in covering a wide range of issues related to the environment and climate change. Our focus includes the impacts of climate change on local communities, sustainable agricultural practices, biodiversity conservation, and renewable energy projects. We are also keen on reporting on governmental policies, community resilience initiatives, and innovative solutions to environmental challenges. Additionally, we strive to highlight stories that educate and raise awareness about the importance of environmental protection and climate action.
>
> (KI 4)

As seen above, the response from KI 4 demonstrates the relationship between theoretical arguments (see, e.g., Harcourt et al., 2020) and the findings.

Community interests heavily determine the climate change stories journalists will cover. These community interests and issues affect the lives and livelihoods of people, including women (Gaard, 2014) and children. The issues are agriculture, mining, lumbering, and natural disasters like floods and droughts. This is partly because by targeting community interests, they can find audiences.

> We prioritize stories with a direct impact on the daily lives of Burundians, emphasizing credibility and timeliness. We aim for a balanced coverage that educates and raises awareness, focusing on both challenges and solutions. Audience interest and feedback also play a crucial role in shaping our editorial choices.
>
> (KI 4)

There seems to be an interest, across East Africa, to move away from only covering problems or climate change as catastrophes. These stories look to the future by sharing information on resilience, sustainability, and biodiversity, among others.

> Airing stories about community resilience initiatives, sustainable agricultural practices, and renewable energy projects in Burundi by highlighting these success stories can inspire other communities and showcase practical solutions to climate challenges. Additionally, I am keen on covering the impacts of climate change on local biodiversity and ecosystems, as well as efforts to preserve them.
>
> (FL 3)

The above sentiment alludes to a solutions approach to climate change stories. This is also evidenced in the stories reviewed since there was a significant inclination toward solutions to the concerns the locals have. These solutions include sustainable agricultural practices, renewable energy options, and biodiversity conservation approaches.

Sourcing of Climate Change Stories in Media in East Africa

Sourcing can be a science (Smith, 2005), in the sense that some may want and actively influence stories in general and a highly controversial subject like climate change. However, there is a hierarchy of credibility among sources (Anderson, 2017). The majority of the sources used by journalists are government, local, national, or intergovernmental institutions. Either they were cited by seeking expert opinion, or government data, laws, and regulations were referenced. Where possible, affected communities or victims help tell the stories. There is, however, a dearth of universities and independent experts in the region.

The stories journalists published suggest that there are fewer local experts, especially organizations with rich knowledge to reference in the stories. An overwhelming majority of the stories had global data and organizations as their sources. However, some had local non-state organizations. Even fewer were local universities and think tanks. This could explain the phenomenon where most journalists resort to using press releases and attending press conferences to file their stories. One way in which this manifests itself is through stories that are derived largely from press releases and conferences. A respondent noted below,

> Dependence on press releases and conferences is half the truth since the vast majority who haven't accessed these forums and meetings do not have a way around getting these data. This can be improved on by enhancing journalists' collaboration and workshops plus training that involve experts in the space.
>
> (FL 18)

Another respondent, (FL10), was blatant in the use of press releases, as Schäfer and Painter (2020) explained.

> Journalists depend on press releases because up to this moment, journalists only report on climate issues when a certain phenomenon goes awry. What this means is that journalism training has not incorporated training in weather and climate reporting so very few journalists can go out there to look for a climate story.
>
> (Schäfer & Painter, 2020)

One of the sources is the regional intergovernmental organization under the Intergovernmental Authority on Development (IGAD), the Climate Predictions and Applications Center (ICPAC). The organization has weekly forecasts, monthly forecasts and seasonal forecasts. Journalists are part of their newsletter which ensures they receive them frequently and use them for their stories.

> The seasonal forecast is very important because it gives up various storylines on how the season will impact the various sectors. For instance, the long rains

> seasonal forecast also known as the March, April to May forecast allows journalists to break the story on whether the season will be good. In this I mean if the season will be good for agriculture, especially the grain basket regions. We also ask questions such as will there be sufficient rainfall in the regions that produce hydroelectric power? What about the regions that suffer perennial floods?
>
> (FL15)

There is also an overreliance on the same sources for stories, as (Anderson, 2017) asserts that these are often government sources who are considered "authorized" and legitimate speakers. These government sources are equally reliant on the media to maintain their public standing. The study found that most journalists prefer to call the same sources every time to clarify an issue regarding climate change. However, some sources in the climate change space are not readily available and on short notice to assist journalists. In circumstances where there is no rapport between the journalist and the expert, this can further be difficult, because of the perception of being misquoted or failing to capture accurate details in a story, which could land the sources in trouble with their employers.

The sourcing and resources of stories by journalists in the region are diverse. But, it depends on their networks and contacts in the climate change space. Those who are more experienced or seasoned will find it easier to access a variety of sources while those who do not have experience will struggle, especially in getting experts.

> Key resources (for stories) are research papers, and reports on various research conducted in the space. Journalists must read detailed reports from scientists get credible information and attend webinars of scientists or forums majorly from think tanks in climate discourse discussions. But getting most of these resources can be difficult, similarly, to getting access to scientists and researchers in the sector.
>
> (FL 33)

Resources for stories are there, but knowledge of them and access determine whether a journalist will do a rich story or not. They could then revert to press releases and conferences as seen earlier.

Grants in Climate Change Storytelling

There are concerns that without grants, the quality and quantity of climate stories would be minimal. The sentiments are affirmed by both journalists and editors, alluding to travel expenses to the locations where the stories are and the inability of media houses to fund them.

> Grants for climate change storytelling are crucial as they enable in-depth investigation and high-quality reporting, which can be resource intensive. These

> funds help train journalists in environmental reporting techniques and raise public awareness about climate issues. Ultimately, grants support investigative journalism and promote a broader, nuanced understanding of climate change, which is vital for informed public discourse and effective action.
>
> (KI 7)

This argument contextualizes studies (Schäfer & Painter, 2020; Wetts, 2020) that posit that most climate change stories rely on press releases and press conferences, often held in towns. A review of all the stories indicates that they undertook some travels to remote and other locations to file the stories. It lends credence to the argument made in the interviews that grants play a key role in ensuring stories from the communities are done.

Asked what inspired them to apply for the climate change grant, many of the respondents affirmed that they are largely driven by the economic challenge frame and human rights frame (Moernaut, Mast, & Pauwels, 2018). They see the importance of utilizing the grants to tell stories that affect the lives and livelihoods of the community.

> The social impacts of climate change for instance marginalized communities and how climate change affects some of their habits like the ability to feed their livestock, forcing men to migrate to further locations in search for pastures, which sometimes results in them dying or not returning to their families, turning women to breadwinners.
>
> (FL 25)

The grant provided in the program also came with a learning element that was lauded as informative.

> The funding from Aga Khan University has helped expand my knowledge on climate reporting particularly when I attended the online workshop my understanding on science reporting changed leading to more in-depth reporting on various aspects of climate change, including its impact on agriculture, migration, urban planning, the economy, and culture. This increased coverage helps raise awareness and understanding of climate-related challenges globally.
>
> (FL 29)

Funding is considered in the context of the weak financial ability of media houses to support their journalists. But such opportunities are not always available, yet critical. Countries like Burundi rarely see such opportunities, while countries like Uganda that will soon start exporting oil will be feeling more environmental challenges, and they stated that a continuous program would make it easier to sustain climate change content.

> What interested me in the grant was the fact that people living around the EACOP (oil pipeline) in Kasenyi, in the Albertine region of Uganda were experiencing

> floods. Their homes and farmlands were submerged by water run-offs, yet most of them didn't know what was triggering that. I found out that Total Energies, the company that is meant to export the oil, uprooted trees – which help in controlling floods – on a large scale (7,000 hectares) to pave the way for the Central Processing Facility thus exacerbating floods.
>
> (FL 8)

Summary and Conclusions

This study contextualizes the practice of climate change storytelling in East Africa as one that relies on grants to tell transformative stories, with the default being stories from press releases. Transformative narratives are dominant in East Africa, as journalists seek to provide answers to the well-being of communities affected by climate change. While doom and gloom narratives did occur and painted how climate change is affecting communities, they were a minority. This is facilitated by grants, begging the question whether they would still be transformative without grants. In these stories, government sources and data and global environmental organizations are the dominant sources of information for secondary sources. Communities including affected persons or victims have carried the storytelling weight in the stories reviewed as primary sources. But without grants, these stories by the grantees would not have been told. Implicitly, organizations in the climate change space, particularly civil society, universities, and research institutions, need to explore ways of partnering with media to better the sourcing of their stories. The study also extends narrative theory as it plays a persuasion and behavioral change role as theorized by Green and Brock (2000).

This article was limited to a funding program that had a time limit. The resounding response from the climate change fellows is that funding for stories plays a critical role in supporting storytelling on climate change. Future research can consider climate change journalism in East Africa that is not funded or through a grant. There can also be a comparative study on how TV, radio, and television cover climate change in the region. Furthermore, considering that all respondents rallied on the need for funding to support climate change stories, it would be important for future research to consider models within newsrooms that can directly ensure climate change stories are consistently covered.

References

Abbott, H. P. (2008). *The Cambridge Introduction to Narrative* (2nd ed.). Cambridge University Press.

Anderson, A. (2017). Source influence on journalistic decisions and news coverage of climate change. *Climate Change Communication*. https://doi.org/10.1093/acrefore/9780190228620.013.356

Appelgren, E. and Jönsson, A. M. (2021) Engaging citizens for climate change—challenges for journalism. *Digital Journalism*, 9(6), pp. 755–772. https://doi.org/10.1080/21670811.2020.1827965

Böhm, G, Pfister, H, Salway, A and Fløttum, K. (2019). Remembering and communicating climate change narratives – the influence of world views on selective recollection. *Frontiers in Psychology*, Sec. Environmental Psychology, 10, 2019. https://doi.org/10.3389/fpsyg.2019.01026

Boykoff, M. (2007). Flogging a dead norm? Newspaper coverage of anthropogenic climate change in the United States and United Kingdom from 2003 to 2006. *Area,* 39(4), pp. 470–481. https://doi.org/10.1111/j.1475-4762.2007.00769.x

Boykoff, M. T. (2008). Media and scientific communication: A case of climate change. *Communicating Environmental Geoscience, D. G. E. Liverman, C. P. G. Pereira, B. Marker.* Geological Society of London. https://doi.org/10.1144/SP305.3

Boykoff, M. T and Yulsman, T. (2013). Political economy, media, and climate change: Sinews of modern life. *WIRES Climate Change*, 4(5), 359–371. https://doi.org/10.1002/wcc.233

Braun, V., and Clarke, V. (2006). Using thematic analysis in psychology. *Qualitative Research in Psychology*, 3(2), 77–101.

Broadbent, J., Sonnett, J., Botetzagias, I., Carson, M., Carvalho, A., Chien, Y.-J., Edling, C., Fisher, D., Giouzepas, G., Haluza-DeLay, R., Hasegawa, K., Hirschi, C., Horta, A., Ikeda, K., Jin, J., Ku, D., Lahsen, M., Lee, H.-C., Lin, T.-L. A., … Zhengyi, S. (2016). Conflicting climate change frames in a global field of media discourse. *Socius*, 2, 1–17. https://doi.org/10.1177/2378023116670660

Bruner, J. (1991). The narrative construction of reality. *Critical Inquiry, 18*(1), 1–21. https://doi.org/10.1086/448619

Cebik, L. B. (1986). Understanding narrative theory. *History and Theory*, *25*(4), 58–81. https://doi.org/10.2307/2505132

Chatman, S. (1978). *Story and Discourse: Narrative Structure in Fiction and Film*. Cornell University Press.

Christensen, M., Åberg, A., Lidström, S., and Larsen, K. (2018) Environmental themes in popular narratives. *Environmental Communication*, 12(1), pp. 1–6. https://doi.org/10.1080/17524032.2018.1421802

Creswell, J. W. and Creswell, J. D. (2019). *Research Design: Qualitative, Quantitative, and Mixed Methods Approaches* (5th ed.). Sage Publications.

Cutting, J. E. (2016). Narrative theory and the dynamics of popular movies. *Psychonomic Bulletin & Review,* 23, 1713–1743. https://doi.org/10.3758/s13423-016-1051-4

Elgesem, D., Knudsen, E. and Fløttum, K. (2024) The impact of climate change on lifestyle journalism, *Journalism Studies*, 25(4), pp. 337–357. https://doi.org/10.1080/1461670X.2023.2299463

Elliott, J. (2018). *Reflecting on Qualitative Research: A Critical Approach.* Routledge.

Elo, S. and Kyngäs, H. (2008). The qualitative content analysis process. *Journal of Advanced Nursing*, 62(1), 107–115. https://doi.org/10.1111/j.1365-2648.2007.04569.x

Entman, R. M. (1993). Framing: Toward clarification of a fractured paradigm. *Journal of Communication*, 43(4), 51–58. https://doi.org/10.1111/j.1460-2466.1993.tb01304.x

Evans, S. (2016) Journalistic norms, cultural values, and coverage of climate change in the Philippines, *Environmental Communication*, 10(4), pp. 492–507. https://doi.org/10.1080/17524032.2015.1088459

Fisher, W. R. (1984). Narration as a human communication paradigm: The case of public moral argument. *Communication Monographs,* 51(1), 1–22. https://doi.org/10.1080/03637758409390180

Gaard, G. (2024). What's the story? Competing narratives of climate change and climate justice. *Forum for World Literature Studies,* 6(2), 272–291. https://shorturl.at/iw6MH

Genette, G. (1980). *Narrative Discourse: An Essay in Method* (J. E. Lewin, Trans.). Cornell University Press. https://shorturl.at/i15iB

Gess, H. (2012). Climate change and the possibility of 'slow journalism.' *Ecquid Novi: African Journalism Studies*, *33*(1), 54–65. https://doi.org/10.1080/02560054.2011.636828

Gibson, T. A., Craig, R. T., Harper, A. C. and Alpert, J. M. (2016). Covering global warming in dubious times: Environmental reporters in the new media ecosystem. *Journalism*, 17(4), 417–434. https://doi.org/10.1177/1464884914564845

Green, M. C. and Brock, T. C. (2000). The role of transportation in the persuasiveness of public narratives. *Journal of Personality and Social Psychology, 79*(5), 701–721. https://doi.org/10.1037/0022-3514.79.5.701

Guenther, L., Gaertner M. and Zeitz, J. (2020): Framing as a concept for health communication: A systematic review. *Health Communication*. https://doi.org/10.1080/10410236.2020.1723048

Gutsche Jr, R. E. and Pinto, J. (2022) Covering synergistic effects of climate change: Global challenges for journalism. *Journalism Practice*, 16(2–3), 237–243. https://doi.org/10.1080/17512786.2021.2016475

Harcourt, R., Bruine de Bruin, W., Dessai, S., and Taylor, A. (2020) What adaptation stories are UK newspapers telling? A narrative analysis, *Environmental Communication*, 14(8), 1061–1078. https://doi.org/10.1080/17524032.2020.1767672

Hiles, S. S. and Hinnant, A. (2014). Climate change in the newsroom: Journalists' evolving standards of objectivity when covering global warming. *Science Communication,* 36(4), 428–453. https://doi.org/10.1177/1075547014534077

Hinkel, J., Mangalagiu, D., Bisaro, A. and Tàbara, J. D. Transformative narratives for climate action. *Climatic Change* 160, 495–506 (2020). https://doi.org/10.1007/s10584-020-02761-y

Kunelius, R. (2019). A forced opportunity: Climate change and journalism. *Journalism*, 20(1), 218–221. https://doi.org/10.1177/1464884918807596

Meyer, K., Koren, E., McCaffrey, M and Slean, C. (2021). Transforming the stories we tell about climate change: from 'issue' to 'action'. *Environmental Research Letters,* 16(1). https://doi.org/10.1088/1748-9326/abcd5a

Moernaut, R., Mast, J. and Pauwels, L. (2018). Working mechanically or organically? Journalism Practice, 12(10), 1311–1331. https://doi.org/10.1080/17512786.2017.1387070

Moyers, B. (2019). What if reporters covered the climate crisis like murrow covered World War II?. *Nation*, 309(1), pp. 32–34. Available at: https://research-ebsco-com.proxyiub.uits.iu.edu/linkprocessor/plink?id=0d964065-cc3a-3e51-b880-08cff2df67e7

Nagel, M. and Schäfer, M. (2023). Powerful stories of local climate action: Comparing the evolution of narratives using the "narrative rate" index. *Review of Policy Research,* 40(6), 1093–1119. https://doi.org/10.1111/ropr.12545

Naguimbing-Manlulu, M. F. (2021) Climate change narratives in Philippine print news media. *Media Asia*, 48(3), pp. 190–206. https://doi.org/10.1080/01296612.2021.1944541

Nordensvard, J. and Ketola, M. (2021). Populism as an act of storytelling: Analyzing the climate change narratives of Donald Trump and Greta Thunberg as populist truth-tellers. *Environmental Politics*, 31(5), pp. 861–882. https://doi.org/10.1080/09644016.2021.1996818

Nowell, L. S., Norris, J. M., White, D. E. and Moules, N. J. (2017). Thematic analysis: Striving to meet the trustworthiness criteria. *International Journal of Qualitative Methods*, 16(1), 1–13.

Nuccitelli, D. (2020) How we know the Earth is warming and humans are responsible. *Bulletin of the Atomic Scientists*, 76(3), pp. 140–144. https://doi.org/10.1080/00963402.2020.1751969

Ricoeur, P. (1984). *Time and Narrative* (Vol. 1, K. McLaughlin & D. Pellauer, Trans.). University of Chicago Press.

Schäfer, M. S. and Painter, J. (2020). Climate journalism in a changing media ecosystem: Assessing the production of climate change-related news around the world. *Wires Climate Change*, 12, (1). https://doi.org/10.1002/wcc.675

Schulz, A. J., Israel, B. A. and Lantz, P. M. (2016). Empowerment and participatory research in health promotion. In: D. K. Cohnand & C. A. Clarke (eds.), *Advancing Health Literacy: A Guide to Participatory Research and Community Engagement* (pp. 34–50). Springer.

Shenhav, S. R., Oshri, O., Ofek, D., and Sheafer, T. (2014) Story coalitions: Applying narrative theory to the study of coalition formation. *Political Psychology*, 35(5), 661–678. https://onlinelibrary.wiley.com/doi/pdf/10.1111/pops.12072 .

Smith, J. (2005). Dangerous news: Media decision making about climate change risk. *Risk Analysis*, 25(6). https://doi.org/10.1111/j.1539-6924.2005.00693.x

Tilly, C. (2006). Why and how to study contentious politics. In: D. S. Meyer and S. G. Tarrow (eds.), *The Social Movement Society: Contentious Politics for a New Century* (pp. 7–21). Westview Press.

Wetts, R. (2020). In climate news, statements from large businesses and opponents of climate action receive heightened visibility. *Proceedings of the National Academy of Sciences*. 117(32) 19054–19060. https://doi.org/10.1073/pnas.1921526117

Yardley, L. (2000). Dilemmas in qualitative health research. *Psychology & Health*, 15(2), 215–228.

7 Newspaper Framing of Climate Change Responsibilities in Nigeria

Uchenna Oluchi Eze and Victor C. Eze

Introduction

Climate change is one of the most pressing issues of the 21st century. Scientists have agreed that Africa, most especially Nigeria, is one place that is likely going to suffer the grave consequences of climate change (Agwu and Amu, 2015). Nigeria is committed to a net-zero future, just like many other nations, to address climate change risks as evidenced by rampant flooding incidents (Adekola and Lemond, 2018). The commitment is also a legally binding obligation under the Paris Agreement to mitigate and address climate change globally. While addressing the adverse effects of climate change is a priority on Nigeria's development agenda, some problems are involved in achieving this endeavour. Some of these problems include unemployment and economic instability, which disproportionately affect the poor and marginalized groups in the society.

According to the Intergovernmental Panel on Climate Change's report (AR6 Synthesis Report: Climate Change, 2023), "climate change is already affecting every inhabited region across the globe, with human influence contributing to many observed changes in weather and climate extremes". According to Afinota (2022), Nigeria is a major contributor to approximately 65% of the total amount of gas flared globally, flaring 60% of its gas, place it as the country with the worst record of gas flaring in the world. This is a major cause of climate change. Thus, Nigeria is both a victim and a contributor. This affirms the claim that Nigeria is not just a victim but also a contributor to global climate change and its consequences (Agwu, and Amu, 2013). This shows the probability of a bigger climatic change in the future if unchecked. Scientists are more certain than ever that people have had a significant impact on the climate and that further human-caused climate change is on the horizon (IPCC, 2023).

Furthermore, Egbra (2014) warned that global warming increases the global temperature of the atmosphere and oceans, causing climate changes that affect rainfall patterns, arctic ice (and the northern jet stream that drives weather), the frequency and intensity of storms and droughts, growing seasons, humidity, melting polar and glacial ice, and sea level. Pucharis (2007) also predicted that in Africa by 2020, between 75 and 250 million people will be exposed to water stress due to climate change. He added that profits from rain-fed agriculture could be reduced

DOI: 10.4324/9781003660217-9

by up to 30% in some African countries. This is because so many systems are tied to the climate. The implication is that climate change can influence where people, animals, and plants inhabit. It can also affect the production and availability of food, water use, and health risks.

Existing research asserts that understanding how journalists frame climate issues is vital to how the public and policymakers respond to lifestyle changes necessary to mitigate and adapt to climate change (Ukonu, Akpan and Anorue 2012; Agwu and Amu 2015). In the context of controversial topics such as climate change, a difference has been made between causative and treatment responsibilities (Iyengar, 1994). They are significant because they address both the parties accountable for the causes and remedies (Kim, Lee, Lee and Kim, 2010) and the public's knowledge of the concerns (Schlesinger and Lau, 2000). The media offer the space where social actors are likely to blame others to insulate themselves from criticism bothering on responsibilities. According to (Kim et al., 2010), political beliefs and professional routines influence how responsibilities are portrayed in the media. While previous studies have dwelt more on how the media frame climate change issues, this research seeks to address the media attribution of responsibility gap in the climate change literature. It investigates how newspapers attribute climate change responsibility via the framing of the actions of various agents in the face of the scientific problem. *The Sun* and *Daily Trust* newspapers were selected for this study because both dailies have strong national outlook and are among the top 10 major dailies in Nigeria (Nigerian Waves, 2024). While *The Sun* newspaper focuses more on events that happen in the southern part of Nigeria, *Daily Trust* captures more incidents from the northern part of the country. They are both privately owned dailies and have limited government influence in their reportage.

This chapter follows this order: The next section reviews literature that focuses on media framing of climate change; it also critically digs deep into studies that examine how the media hold various agents accountable for the facing of increasing climate change problem. The next section focuses on the theoretical framework and methodology. The findings are discussed, highlighting how the media frame climate change responsibility. The final part of the study captures the conclusion and recommendations.

Media Framing of Climate Change

More people get information about global issues like climate change from the media. The media frame issues in different ways that they deem appropriate for their audience. This framing procedure leads to the selection of certain aspects of perceived reality and giving them more salience (Entman, 1993). Several scholars have investigated how the media frame climate change issues around the world. Some of these studies are reviewed here.

Günay, İşeri, Ersoy and Elega (2021) analysed newspapers in carbon-locked developing nations to find out if their climate action frames tilt towards adaptation or mitigation issues. They conducted a comparative quantitative analysis of climate change news framing in newspapers from two carbon-locked-in developing countries: Nigeria and Turkey. They found more adaptation framing in Nigerian newspapers, whereas adaptation and mitigation frames are more evenly seen in Turkish newspapers. They also found that the use of foreign sources in news with a climate action frame is negatively correlated with the use of an adaptation frame in sampled news. Their findings improve understanding of the factors shaping climate change communication in developing countries.

Ajaero and Anorue (2018) compared the influence of newspaper framing on climate change mitigation in Nigeria and Ghana and evaluated the challenges newspapers face in reporting climate change. They found that climate change was offered less coverage, and the adopted dominant frame in Nigerian newspapers is the action frame, while Ghanaian newspapers employed environmental frame. Flooding was the major climate change issue identified in both nations. Challenges of reporting climate change were hinged on funding, lack of technical reporting skill on climate change, poor documentation, and expertise.

Nwabueze and Egbra (2016) investigated how two newspapers in Nigeria and Ghana framed climate change issues within 7 months. They found that the dominant frame was the environment and action frame, which highlighted the predicted effect of climate change on the landscape and relief in Nigeria and Ghana. Both Nigerian and Ghanaian media source their reports more from international fora.

Asogwa, Ezeorah and Onyekachi (2023) investigated public perception and newspaper framing of environmental challenges in southeast Nigeria. They found that climate change and desertification were the most reported environmental challenges. The newspapers adopted these frames in this order: mitigation, government efforts, and public reaction. There is a strong relationship between attitude to environmental challenges and the framing used by the newspapers.

Jang (2013) investigated how climate change communication can influence public perception about who is responsible for the problem – nature or human actions. He found that participants who were exposed to information about excessive utilization of energy were more likely to attribute climate change to natural rather than human causes. This attribution of climate change to nature was negatively associated with concern for climate change and policy support for climate change mitigation.

Liu and Huang (2023) investigated how *China Daily* (CD) and *The New York Times* (NYT) frame responsibilities for climate change and found that CD largely tilted towards human causes of climate change, while NYT problematized human causes for climate change. CD underscored developed nations' role in causing climate change, while NYT prefers to highlight how developing nations are causing climate change.

How the public understands the cause of climate change influences public attitudes towards global environmental issues (Ferguson and Branscombe, 2010). Hormio (2023) averred that the problem of climate change is collective responsibility, which focuses on the collective entities that bear responsibility for climate change. The implication is that any concrete solution will involve the aggregation of individual actions, which could result in a better outcome. This means that the public a play vital role in combating the crisis.

Alawade and Obun-Andy (2024) noted that the media contributes to shaping public perception and creating awareness about climate change. Anchoring on framing theory while adopting the desk research method, they found that to a great extent social media, online news sites, and blogs influence discourse and public attitude towards climate change and also encourage people to take more action directed at climate change mitigation or adaptation.

Brossard, Shanahan and McComas (2004) compared the newspaper coverage of global warming in France and in the United States from 1987 to 1997 as a way of analysing the effect of culturally bound journalistic practices on media attention cycles. They found that France presented event-based coverage that highlighted international relations and offered more streamlined perspectives on global warming than the United States did. The US coverage hinted more at the conflict between scientists and politicians.

Ukpong (2020) examined the volume of climate change articles published in *The Punch, The Guardian, Vanguard*, and *This Day* newspapers between 2015 and 2019 to determine how much coverage these publications received of climate change issues. His research revealed that *The Punch* published 28.1% of all articles, or 85 articles, while *The Guardian* published 78 stories, or 25.7% of all articles. Comparably, *This Day* newspaper published 67 news items (22.1%), and *Vanguard* newspaper produced 73 articles (24.1%) about climate change, while *This Day* newspaper published the fewest articles (67). The research concluded that climate change articles are rarely given prominence in Nigerian newspapers.

Studies on media framing of climate change reveal that narratives of global responsibility are very salient, suggesting that all countries should take equal responsibility in addressing the climate crisis (Murali, Kuwar and Nagendra, 2021). However, there is limited literature on how the Nigerian media attribute climate change responsibility to Nigeria as a country, government officials, individuals, CSOs, public officials, corporations, international communities and other global actors. This is the gap the present study addresses.

Framing Theory

Frames are closely related to "the processes through which people signal 'what's going on' in a communication", as stated by Erving Goffman in 1974 and further demonstrated by George Lakoff in his 2004 book *Frame Analysis* (Jones, Jaworska and Aslan, 2020, p. 26). Framing, according to Entman (1993), is the process of selecting and emphasizing features of reality to support a specific cause. As a result, the framing process can change people's interpretation and knowledge of a specific issue by describing the problem, its causes, and potential solutions (Entman, 1993). News framing of climate change responsibilities can guide the audience to identify the sources and solutions to the problem (Billet, 2010; Olausson, 2009). For instance, admitting human influence on climate change will add to the people's contribution to addressing climate change (Billet, 2010).

Recent studies have investigated media framing of responsibilities and discovered that media coverage may give less attention to causative responsibilities than it does to treatment responsibilities (Freeman, 2017). The examination of causes focuses mostly on the human effect on climate change. For example, Nigerian newspapers are generally sceptical of human impacts on climate change, government negligence and the impact of manufacturing companies on climate change and will most likely attribute the cause to natural fluctuations (Kuhaa, 2009), which may divert the government's attention and result in ineffective actions. In some nations, such as the United Kingdom, the media ideally presents agreement on human impacts on climate change (Grundmann and Krishnamurthy, 2010; Freeman, 2017).

Other scholarships concentrate on the causal responsibility of both developed and developing countries. According to (Post et al., 2019), most reports use the in-group-protecting strategy, which involves assigning less responsibility to in-group members. For example, Indian media rarely assign causal responsibility for climate change to other developing countries (Billet, 2010). Regardless, several studies have discovered that the media in developed countries will prefer to blame their own or other developed countries for contributing to climate change (Liang et al., 2014; Post et al., 2019). This can be allocated to the international agreement on developed countries' historical responsibilities for climate change (Post et al., 2019). These studies have indicated that the framing of climate change responsibilities is often shaped by national interests. As suggested by (Lee et al., 2002), "media representations of international news events tend to be filtered by the prism of national interests". However, there is little knowledge on whether the Nigerian newspapers hold appropriate actors and entities accountable for climate change.

Media frames can clarify "media reception" and how people interpret media narratives based on their own expectations about how and what stories should be told; Goffman (1974) was more interested in how frames were used in everyday interactions (Jones, Jaworska and Aslan, 2020, pp. 26–27). Frames "may be used to pass down information, giving greater weight to certain considerations and elements over others", as (Nisbet, 2009, p.16) demonstrates. Therefore, framing gives a deeper insight into how the media construct communication. Nisbet (2009) discovered, for example, that in the discourse around climate change in the United States, conservative commentators tended to emphasize frames of scientific doubt and economic repercussions. Conversely, leftist pundits tended to exaggerate the conversation surrounding climate change. Both strategies contribute to the maintenance of political divisions and have an impact on public participation and policy implementation, which emphasizes how crucial frame analysis is to a critical discourse analysis. Similarly, Molek-Kozakowska (2018) discovered that "some frames and narratives are more likely to mobilize diverse audiences" (p. 75) in their study employing discourse analysis with frames, narratives, and news values. They contend that the discussion of danger in climate change must be powerful yet controllable, frightening but not alarmist. This study employs framing theory to explore the frames bothering on attribution of responsibility. These frames reveal who is to be held accountable for climate change mitigation and adaptation or the increase in the climate crisis because of inaction.

Methodology

In assessing how Nigerian newspapers frame climate change responsibility, we adopted qualitative frame analysis of 24 national newspapers in Nigeria. We randomly selected two newspapers (*Daily Trust* and *The Sun*) out of the 24 national newspapers in Nigeria. The newspapers are among the most read newspapers in Nigeria. The timeframe spans from 2015 to 2023, which focuses on the administration of President Muhammadu Buhari, which was active in climate change activities. While both are national dailies, *Daily Trust* gives more prominence to events that happen in the north, while *The Sun* pays more attention to incidents in the southern part of Nigeria. They are both privately owned dailies, and their reports have limited control from the government. Both dailies report their stories in the English language. The unit of analysis for this study was the newspaper articles. We defined an article as one of the following: news stories, editorials, features, interviews and opinion essays. We did not consider the size and prominence of the article.

The data from the two newspapers were gathered through Nexis Uni, an easy-to-use academic search engine that delivers relevant content that makes scholarly research more efficient, using the keyword "climate change in Nigeria". The keyword fits into the primary objective of this work. Combined with the chosen time frame, the keyword and the chosen news source were entered on Nexis Uni and provided the corpus for this analysis. We purposively selected 50 articles each from *Daily Trust* and *The Sun* newspapers for data analysis. The analysis focused on the objective of this research, which investigates how the dailies frame climate change responsibilities. Taking a cue from the study of Liang, Tsai, Mattis, Konieczna and Dunwoody (2014), this chapter arrived at these frames: causal responsibility, treatment responsibility, morality, global justice – climate debt, global justice – vulnerability and national efficacy frames. These frames are analysed in the next section.

Causal Responsibility Frame

This frame mentioned the agents that are responsible for causing climate change or making it hard for climate change policies to be put in place. In a sense, the government of Nigeria has some blame in the face of the mounting climate problem. While the crisis is growing every day, the right steps are not being urgently put in place for mitigation of the climate change effect or adaptation to the crisis. The frame clearly relays the lackadaisical attitude that is common with the government when it comes to attending to certain burning issues that affect the nation and the citizens at large. CSOs try to hold the government accountable by calling out the slow pace of response that characterizes the whole process. In some cases, funds are withheld; in other instances, appropriate policies or bodies are not put in place to effect actions that can bring about the much-needed changes in lifestyle. Instance of CSOs lamenting the attitude of the government was captured thus: "There are concerns from members of civil society organisations (CSOs) that

that despite mounting life-threatening climate change crises in Nigeria, the federal government is yet to inaugurate the National Council on Climate Change" (*Daily Trust*, Wed, 8 Jun 2022). Spokespersons of the CSOs were the primary sources of the causal responsibility frame as they lay the blame for slow climate action on the Nigeria which takes slow paces or almost none on key issues that could offer appropriate solutions. For instance, at one time, it took several months to get the head of Nigeria's climate change department. As reported, , "almost four months after Dr. Samuel Adejare Adejuwon retired from the Ministry of Environment as the Director and Designated to have his services extended by the ministry after his retirement but that proved abortive" (*Daily Trust,* Tue, 16 Feb 2016).

Treatment Responsibility Frame

This frame pointed out agents that proffered solutions to climate change issues. In most cases, other nations, especially developing countries, look up to the advanced nations like the United States and others to take the lead in the effort against climate crisis. President Buhari was the primary source of this frame as he commends the United States and at the same time beckons her to continue to do more for others in fight for improved global climate. "The President, who said the fight to redress the impact is the responsibility of all countries and stakeholders, congratulated President Joe Biden of the United States for renewed interest in the Paris Agreement". He said,

> I would like to express my profound appreciation to the President of the United States, His Excellency, Joe Biden, for convening this Summit at a time the world is experiencing tremendous vulnerability from the COVID-19 pandemic and Climate Change impacts, among several other global challenges.
>
> (*Daily Trust*, 23 Apr 2021)

With the devastating effects of the climate change crisis ravaging. Nigeria, the call for more action appears cogent and urgent. Advanced nations possess more resources at their disposal for engaging in the fight against climate degradation. Deploying episodic frames, the call for urgent actions presented direct instances of the impact of the crisis. Increased emissions of greenhouse gases are causing changes in the way the climate used to be. Even though Africa can be said to have contributed in a small way to the climate problem when judging from the volume of industries, manufacturing firms and other factors that induce the crisis, the impact is felt globally, with overwhelming impact in developing countries. Most of the clamour for solutions by Nigerian officials are calling for collective action piloted by more advanced nations like the United States and other developed nations. Even as pressure mounts in Nigeria to do the right things in face of our increasing climate change menace, President continued to seek foreign help to mitigate the crisis in Nigeria. "President Muhammadu Buhari has solicited the help of the international community in mitigating the effect of climate change in Nigeria. The president disclosed this after he received Letters of Credence from three envoys to Nigeria, yesterday" (*The Sun*, 9 May 2018).

Morality Frame

This frame presents articles that hint at moral messages or social prescriptions with focus on climate change or policy formulation. There were several of these frames deployed in the analysed articles where prescriptions are made on how to combat the crisis. Many instances of the morality frame were portrayed through the thematic write-ups that proffered solutions by striving to call appropriate bodies to act in a particular way. In most cases, the primary sources of this frame are government officials like governors, political appointees and experts on environmental and climate change issues. The call is for intensification of effort to stem the problem of the crisis. One of such calls was captured thus: This article suggests immediate measures these groups can implement to address climate change and protect Africa, especially Nigeria's environmental health. The Federal Capital Territory (FCT) and states such as Anambra, Rivers, Lagos, and Kano should immediately prohibit all single-use non-recyclable disposable polymers (*Daily Trust*, Fri, 25 Aug 2023). The irony is that many of such calls to action are not matched with adequate responses, even when colourful promises are made, policies put in place and money mapped out for them.

On the corporate level, companies are also enjoined to join in the fight against climate crisis in the form of corporate social responsibility (CSR). Corporations are encouraged to innovate sustainable services and goods. They are also enjoined to partner with NGOs and government agencies on projects directed at sustainable development and environmental conservation. Actions such as tree planting in the cities and fighting against desert encroachment are worthwhile climate actions corporate entities can embark on for the good of the society and the nation at large. Corporations showcasing transparent reports that lay bare their effort to make impact on the environment play a major moral role in the effort to mitigate the impact of climate change or encourage appropriate adaptation.

Global Justice – Climate Debt Frame

This brings into perspective the difference in responsibilities between developed and developing nations because of past and current actions. Series of thematically framed articles highlighted the need for developed nations to assist developing ones in the effort to combat climate change. Some of these efforts are captured in the agenda of the United Nations Climate Change Conference, often tagged COP. In one of such conferences, it was articulated that "The United Nations Framework Convention on Climate Change (UNFCCC) needs assessment estimated that developing countries need $6 trillion by 2030 and that Africa will particularly require $2.5 trillion by 2030 to effectively deal with climate change issue" (*Daily Trust*, Wed, 28 Sep 2022).

To underscore the urgency of the call to action, the frame sources are usually from the United Nations agents, country representatives at the UN and other major experts in the field of environmental science. Most of the slant deployed through this frame is pro-developing nations. They support climate change payment to developing nations to aid their effort in the fight.

Global Justice – Vulnerability Frame

This takes cognizance of the slow adaptive capacity and vulnerability of developing nations in terms of climate change effects. Nigeria is vulnerable to the impacts of climate change. This is seen in the constant flooding, increased rise in temperature causing heatwaves, desertification, prolonged drought destroying crops, food crisis, environmental degradation and other climate change effects.

Nigeria's vulnerability to the climate change crisis is such that the number of Nigerians in Internally Displace Peoples camp is soaring. This is due to resource crisis that has led to herders and farmers struggling over who has the right to use the land – for grazing or for agriculture. Climate change has aggravated the lingering herder-farmer crisis with attendant traumatic impacts on victims, according to the Global Internal Displacement data compiled by Dataphyte, a media, research and data analytics organization with the mission to deploy data tools and technology for socio-economic development of Nigeria (*Daily Trust*, Sun, 16 Oct 2022).

Several thematically framed articles portray this negative effect of climate crisis in Nigeria, calling for urgency in seeking a solution. Research shows that while the ravaging effects of climate change are felt across the length and breadth of the nation, the northern part of Nigeria faces more noticeably negative impact of the crisis. This is captured in *Daily Trust* thus: "Extensive research shows that Northern part of Nigeria has already experienced noticeable changes in climate and predicts that more changes will occur soon. Many of these changes are having and will continue to have negative impact on the environment" (*Daily Trust*, 3 Sept 2015). The level of destruction that comes from vulnerability resulting from flooding has led to the loss of a lot of lives and properties worth millions of Naira in Nigeria. Every year, flooding causes death, destruction and displacement of unimaginable proportion. This frame was well captured by one of the ministers of President Buhari. It reads thus: "This is just as the Minister of Humanitarian Affairs, Disaster Management and Social Development, Sadiya Umar-Farouq, said flood has so far killed 612 Nigerians and affected over 3.2 million others this year" (*Daily Trust*, Wed, 26 Oct 2022).

National Efficacy Frame

This frame referred to articles that pointed out that Nigeria or foreign countries could mitigate climate change. The tones of the articles are usually negative, pointing to inadequate national effort directed at solving the problem of climate change either by mitigation or adaptation. Episodic articles were employed to report on how ready Nigeria is when it comes to combating climate change problem. One of such stories was captured thus: The President of the Movement for the Survival of the Ogoni People (MOSOP) Legborsi Saro Pyagbara said so far the response to the issue of climate change by the Nigerian government is poor (*Daily Trust*, Wed, 11 Jul 2018). Commitment to combatting climate change menace is one that comes with government resolve. It moves from resolve to action. Severally, President Buhari promised the international community that Nigeria will continue to strive assiduously in the climate change fight. He affirmed that Nigeria will attain the Paris Agreement on climate change.

Using this rhetorical device, "A Race We Can Win. A Race We Must Win" as a theme of his presentation at the United Nations Climate Action Summit, Buhari announced robust plans and initiatives by his administration to reverse the negative effects of Climate Change in Nigeria (*The Sun*, 23 Sept 2019). He boasts that the plan takes care of water, environment and energy, portraying on global state some level of confidence that the right things will be done. Several times, he gave such reassuring stance on how Nigeria will be effective in fighting the menace of climate change.

"President Muhammadu Buhari assured the international community of Nigeria's commitment to galvanize relevant stakeholders for climate action, and support regional, continental and global multilateral processes for the attainment of the objectives of the Paris Agreement on climate change" (*Daily Trust*, 23 Apr 2021). Despite the show-off on global stage by Nigerian politicians, a lot of policies are yet to be implemented in face of the increasing climate problems. Thus, Nigeria's effectiveness in combatting the crisis is still far from what it should be.

Conclusion and Recommendation

The Attribution of Responsibility Frame points directly at the entities that are responsible for the climate change problem that faces Nigeria and the world at large. Frame analysis identified how the selected newspaper framed the climate change responsibilities to elicit more actionable plan and disposition in climate change mitigation and adaption. The frames of climate change responsibility are deployed to create awareness for increased government intervention in response to the impacts of climate crisis such as farmer–herders clash, drought, flooding, and environmental degradation. The Nigerian newspaper adopts more frames that tend to hold other actors like the international community and the United Nations more accountable than the Nigerian government and officials. This will not serve as a wake-up call to Nigeria. The media should continue to hold the government accountable by asking questions and offering solutions on the issue of climate change.

While this study investigated just two newspapers, this can limit the generalizability of the outcome of this research. Further studies can expand the research by investigating more newspapers in the stable of the Nigerian Press Council (NPC).

Also, further studies can look at how other types of media like the television, radio, and online platforms frame climate change responsibilities.

References

Adekola, O. and Lamond, J., 2018. A media framing analysis of urban flooding in Nigeria: current narratives and implications for policy. *Regional Environmental Change*, 18, pp. 1145–1159.

Afinotan, U. 2022. How serious is Nigeria about climate change mitigation through gas flaring regulation in the Niger Delta? *Environmental Law Review*, 24(4), pp. 288–304. https://doi.org/10.1177/14614529221137142 (Original work published 2022)

Agwu, A.E. and Amu, C.J., 2015. Framing of climate change news in four national daily newspapers in Southern Nigeria. *Agricultural Information Worldwide*, 6, pp. 11–17.

Ajaero, I.D., Anorue, L.I., 2018. Newspaper framing and climate change mitigation in Nigeria and Ghana. *Etude de la Population Africaine*, 32, 4228–4238. https://doi.org/10.11564/32-2-1195

Akpan, C.S., Anorue, L.I. and Ukonu, M.O., 2012. An analysis of the influence of the Nigerian mass media on public understanding of climate change. *Journal of alternative perspectives in the social sciences*, 4(4), pp. 688–710.

Alawade, P., Olayinka, S. and Obun-Andy, M.K., 2024. The role of media in shaping public perception of climate change. *International Journal of Research and Innovation in Social Science*, 8(1), pp. 2442–2448.

Asogwa, F.O., Ezeorah, I.G. and Onyekachi, U.M., 2023. Newspaper framing and public perception of environmental challenges in southeast, Nigeria. *Innovations,* 73, 796–812. https://doi.org/10.54882/7320237316702

Billett, S., 2010. Dividing climate change: Global warming in the Indian mass media. *Climatic Change*, 99, pp. 1–16. https://doi.org/10.1007/s10584-009-9605-3

Brossard, D., Shanahan, J. and McComas, K., 2004. Are issue-cycles culturally constructed? A comparison of French and American coverage of global climate change. *Mass Communication and Society*, 7(3), pp. 359–377. https://doi.org/10.1207/s15327825mcs0703_6.

Egbra, S., 2014. *Newspaper Coverage of Climate Issues in Nigeria and Ghana* (Unpublished MSc dissertation). Department of Mass Communication, Anambra State University, Igbariam Campus, Anambra State, Nigeria.

Entman, R.M., 1993. Framing: Toward clarification of a fractured paradigm. *Journal of Communication*, 43(4), pp. 51–58.

Ferguson, M.A. & Branscombe, N. R (2010). Collective guilt mediates the effect of beliefs about global warming on willingness to engage in mitigation behavior. *Journal of Environmental Psychology*, 30(2), pp. 135–142. https://doi.org/10.1016/j.jenvp.2009.11.010

Freeman, B.C., 2017. Claims, frames, and blame: coverage of climate change in ASEAN's English-language newspapers, 2002-2012. *Sage Open*, 7(1), pp. 1–12. https://doi.org/10.1177/2158244016675199

Goffman, E., 1974. *Frame Analysis: An Essay on the Organization of Experience*. Harper and Row, New York.

Grundmann R and Krishnamurthy R. 2010, The Discourse of Climate Change: A Corpus-based Approach. Critical Approach to Discourse Analysis across Disciplines vol 4(2):113-133

Günay, D., İşeri, E., Ersoy, M., and Elega, A.A., 2021. Media framing of climate change action in carbon locked-in developing countries: Adaptation or mitigation? *Environmental Communication*, 15, 663–677. https://doi.org/10.1080/17524 032.2021.1885 462

Hormio, S. 2023. Collective responsibility for climate change. *WIREs Climate Change*, 14(4), e830. https://doi.org/10.1002/wcc.830

IPCC. 2023. AR6 Synthesis Report: Climate Change 2023. www.ipcc.ch/report/sixth-assessment-report-cycle

Iyengar, S., 1994. *Is Anyone Responsible?: How Television Frames Political Issues*. University of Chicago Press, Chicago.

Jang, S.M. (2013). Framing responsibility in climate change discourse: Ethnocentric attribution bias, perceived causes, and policy attitudes, *Journal of Environmental Psychology*, 36, 27–36, https://doi.org/10.1016/j.jenvp.2013.07.003.

Jones, R.H., Jaworska, S. and Aslan, E., 2020. *Language and Media: A Resource Book for Students*. Routledge, London and New York. https://doi.org/10.4324/9781003084211

Kim, H.R., Lee, M., Lee, H.T. and Kim, N.M., 2010. Corporate social responsibility and employee–company identification. *Journal of Business Ethics*, 95, pp. 557–569.

Kuha, M., 2009. Uncertainty about causes and effects of global warming in US news coverage before and after Bali. *Language and Ecology*, 2(4), pp. 1–18.

Lee, C.C., Pan, Z., So, C.Y.K. and Chan, J.M., 2002. *Global Media Spectacle: News War over Hong Kong*. Hong Kong University Press, Hong Kong.

Liang, X., Tsai, J.-Y., Mattis, K., Konieczna, M., and Dunwoody, S., 2014. Exploring attribution of responsibility in a cross-national study of TV news coverage of the 2009 United Nations Climate Change Conference in Copenhagen. *Journal of Broadcasting and Electronic Media*, 58(2), pp. 253–271.

Liu, M., and Huang, J. 2023. Framing responsibilities for climate change in Chinese and American newspapers: A corpus-assisted discourse study. *Journalism*, 25(8), pp. 1792–1811. https://doi.org/10.1177/14648849231187453

Molek-Kozakowska, K., 2018. Popularity-driven science journalism and climate change: A critical discourse analysis of the unsaid. *Discourse, Context & Media*, 22, pp. 73–81. https://doi.org/10.1016/j.dcm.2017.09.013

Murali, R., Kuwar, A. and Nagendra, H., 2021. Who's responsible for climate change? Untangling threads of media discussions in India, Nigeria, Australia, and the USA. *Climatic Change*, 164, pp. 1–20.

Nigerian Waves, 2024. Top 10 Newspapers in Nigeria and Their Publishers https://nigerianwaves.com/top-10-newspapers-in-nigeria/

Nisbet, M.C., 2009. Communicating climate change: Why frames matter for public engagement. Environment: Science and policy for sustainable development, 51(2), pp. 12–23. https://doi.org/10.3200/ENVT.51.2.12-23.

Nwabueze, C. and Egbra, S., 2016. Newspaper framing of climate change in Nigeria and Ghana. *Applied Environmental Education & Communication*, 15(2), pp. 111–124. https://doi.org/10.1080/1533015X.2016.1164094

Olausson, U., 2009. Global warming—global responsibility? Media frames of collective action and scientific certainty. *Public Understanding of Science*, 18(4), pp. 421–436.

Post, S., Königslöw, K.K. and Schäfer, M.S., 2019. Between guilt and obligation: debating the responsibility for climate change and climate politics in the media. *Environmental Communication*, 13(6), pp. 723–739.

Pucharis, K., 2007. IPCC climate change assessment report. Retrieved November, 12.

Schlesinger, M. and Lau, R.R., 2000. The meaning and measure of policy metaphors. *American Political Science Review*, 94(3), pp. 611–626.

Ukonu, M.O., Akpan, C.S. and Anorue, L.I., 2012. Nigerian newspaper coverage of climate change, 2009-2010. *Change*, 5, pp. 22–37

Ukpong, I.I.U., 2020. *Environmental Journalism in Nigeria: Representation of Climate Change in Nigeria* (Master's thesis), Eastern Mediterranean University (EMU)-Doğu Akdeniz Üniversitesi (DAÜ)), Gazimağusa, North Cyprus.

Part II

Visual Arts, Community Media, and Climate Change Communication

8 The role of visual arts in climate change communication in Africa

Nancy Mtambo, Sehlisiwe Tshuma, Takudzwa Mandiringa, Daniel Maposa, and Munetsi Zvavahera

Introduction

Climate change is a reality whose effects are felt the most by those communities that depend directly on nature for livelihoods, especially the poor communities (Skolczylas, 2020). The ever-increasing incidences of climate-related natural disasters are a cause for concern in Africa and the world over. African countries are often unprepared for extreme weather events such as droughts, flooding, tropical cyclones and heat waves which have a bearing on livelihoods (Sidford et al., 2016). While numerous African countries have tried to mitigate climate change through the United Nations Framework Convention on Climate Change (Muchuru & Nhamo 2017), to achieve effective climate change adaptation strategies, there is a need for clear communication channels that reach the widest audiences. Thomas et al. (2011) argue that there is a need to diversify coping and adaptation strategies to respond to shifts in climate change. Climate change requires comprehensive coverage to disseminate information and updates. This is because climate-linked disasters affect not only the livelihoods and food security but also the economies of African countries. Increasing the understanding of climate change and possible mitigation strategies will enhance people's coping and adaptive capacity.

In Zimbabwe, artists use numerous art forms to convey climate change messages. This chapter mainly focuses on visual arts. Visual arts create lasting impressions of the impacts of climate change first-hand. This chapter examines the potential of visual arts in changing communication about climate change issues. It addresses the following question: How are visual arts deployed to communicate climate change?

The role of art as a tool for understanding climate change

Arts create a conducive platform for climate change communication (Cosgrove, 2022). Gabrys and Yusoff (2012) propose that art–science may not provide the ultimate solution to climate change communication, but it will go a long way in simplifying complex issues. Art facilitates the interaction with climate change, simplifying complex scientific findings to levels that can be understood by all. The collaborative work between art and science can improve the communication

DOI: 10.4324/9781003660217-11

of climate change and resilience projects. Art–science collaboration can simplify complex information to simpler information that can be understood and assimilated by the intended audience (Gough et al., 2016). Visual art is one significant form of climate communication that the Zimbabwean environment activists and arts practitioners have adopted to educate and create a hazard-free environment. The notion of visual art can be explained as motionless pictures that seek to represent a concept, an idea, an emotion or a feeling based on human intellectual capabilities, observation and interaction with nature and the environment. The relationship between visual art and climate is intertwined as artists draw much inspiration from their surroundings, especially the environment (O'Hara, 2014). The central motivation for using visual art in climate change advocacy is that it can create mental images and achieve the anticipated results. Popov (2013) notes that forms of visual art include photography, digital images, paintings, sculpture and architecture created by artists. Visual communication takes various aspects such as colour combinations, size and brightness of symbols. Such visual communication makes it easier to understand climate change information. Calvo et al. (2022) argue that visual art makes climate information more readily available to the user than other forms of communication. Visual art helps to effectively disseminate climate change information (Valli, 2021), using pictures and social media. The images are used to tell a story and change the community's perceptions towards climate change. However, there is a need to ensure that visualisation is as simple as possible to avoid confusion or minimise the chances of being misinterpreted.

Visual arts empower communication between humans and the environment, limiting climate threats. It inculcates critical thinking and promotes critical analysis by reconstructing meanings behind exhibited artefacts, leading to changes in perspective and behaviours. This is an indispensable facet of climate change advocacy as it will lead to a better understanding of the environment and its conservation. Adaptive capacity is achieved when people attend art exhibitions. Therefore, visual arts are significant in creating a safe and healthy climate in Zimbabwe. This chapter looks at how sculptures and paintings are used to communicate climate change.

The contemporary climate change movement requires the involvement of artists, environmentalists, policymakers and the global community to create a healthy environment. There is limited knowledge of how arts help climate change communication. This chapter attempts to fill this gap by examining the role of arts in climate change communication in a less researched context of Zimbabwe.

Conceptualising climate (change) visuality

The chapter deploys the concept of climate visuality (Sheppard, 2012; Chapman et al., 2016) to gain insights into how the arts have been used to communicate climate change. Climate change visuality is an emerging culture of communicating pertinent climate issues. It seeks to explain how visual artworks, sculptures, photographs and other visualisations drive the message of climate change. Advances in technology have led to the availability of visual images that feature climate narratives and discourses (O'Neill & Smith, 2014). O'Neill and Smith (2014, p. 73) further argue that "Visual images are everywhere in the portrayal of climate change. They appear in the media, in scientific reports, and in NGO and governmental campaigns, and artists create them". For other researchers (see, e.g., Brenthel, 2016), 3D and 4D visualisations, in addition to artworks like sculptures and paintings, now feature what used to be largely invisible climate change issues like ice flows melting, earth carbon releases and deep ocean floor pollution (Sheppard, 2012). These forms of communication used to be on the periphery of climate change communication. Now, they are recentring climate change communication. Brenthel (2016) refers to visuality in climate change communication as outreach material. There have been calls for humanities and the arts to increase their contribution to climate change discourses (Hahn & Berkers, 2020). The argument has been that artists can make meaningful contributions to the climate change debate through their power of imagination and storytelling. The ArtCOP21, for instance, included performances, installations and exhibitions to advocate for climate action creatively (Hahn & Berkers, 2020). And, as the same authors note, music, literature, film, visual arts, theatre arts and sculptures, amongst others, have become visible forms of climate change. These forms of communication used to be on the periphery of climate change communication. They are increasingly becoming central to climate change in a world filled with many variegated debates about climate change. In this chapter, we deploy climate change visualisation as a conceptual lens to understand how the arts in Zimbabwe have been deployed. Our broad argument is that the arts are recentring climate change communication from the periphery.

Methodological issues

This study adopted a qualitative case study approach to examine how Zimbabwean visual artists communicate climate change narratives through their creative practice. In-depth interview with purposively selected visual artists was conducted to gain insights into the experiences, creative processes, and interpretations of visual artists about the potential of visual arts in communicating climate change. Four primary interviews were conducted, each lasting approximately 60 minutes, with additional follow-up sessions as needed to clarify or expand upon key points. The interviews took place at the Bulawayo National Gallery during October 2024, coinciding with the anniversary of the notable "Black the New Green" exhibition. This timing provided natural contextual relevance and facilitated richer discussions about the evolving role of art in climate change communication. While the sample size of four primary participants might appear limited by quantitative standards, the depth of data obtained from these carefully selected informants ensured theoretical saturation was achieved. Each participant's extensive experience and reflective capacity yielded rich, nuanced data that provided comprehensive insights into the research problem. The study's focus on quality rather than quantity of data aligned with established qualitative research principles that emphasise information richness over statistical representativeness.

Additional data were collected from content analysis on visual arts to explore the visuals that are used in climate change communication. The art pieces were chosen due to their explicit focus on climate change issues. In relation to this study, the researcher investigated visual materials from the art organisations under scrutiny to further understand their operations in relation to climate change.

Findings and discussion

The relationship between visual arts and climate change is intrinsically intertwined; this research established that nature serves as a source of inspiration. This relationship can be explored in diverse aspects of creativity and communication which entails representation, symbolism, abstraction and safe environment advocacy. Nature is a muse and a message in visual arts as realistic elements are transformed into environmental artworks used for climate maintenance and environmental conservation. Artists interviewed for the purposes of this study portrayed nature by capturing how human activities were detrimental to the environment and natural elements were used in conveying nature preservation and in stirring environmental consciousness. The use of raw materials in creation of artefacts, implementation of colour theory in abstraction and photographing human activities demonstrate how visual arts have been used enormously in climate change advocacy. The artists maximised the use of plastic materials, craft paper and cardstock in creation of artefacts that emphasised the concept of categorising waste for recycling purposes. Out of these materials, they created sculptures, flower vases and pots. Some showcased abstracts deployed the use of the colour theory in communicating the intended message and the colours used include orange, yellow, brown, white and red.

Visual arts for climate change communication

The strategic deployment of visual arts in climate change advocacy stems from its unique capacity to generate powerful mental imagery and engage public consciousness, thereby fostering awareness and driving behavioural change. This premise finds particular relevance in Zimbabwe, where shifting climate patterns have been extensively documented in relation to anthropogenic factors. Peer-reviewed studies, including the Zimbabwe National Climate Change Response Strategy (2019), have systematically linked environmental degradation to unsustainable practices such as deforestation, intensive agricultural production, and fossil fuel-dependent energy generation through mining operations. These activities contribute significantly to greenhouse gas accumulation, creating a compelling need for innovative communication strategies that can translate complex climatological data into accessible public knowledge.

Within this context, visual arts and particularly sculptural practice have emerged as critical tools for both waste management education and practical environmental intervention. Academic research by Nkomo (2024) provides empirical evidence that structured art initiatives have directly influenced waste management behaviours in urban Zimbabwe. The study documents how community art exhibitions in Bulawayo and Harare have not only raised awareness but also led to the establishment of 17 formalised recycling stations at previously unregulated dumping sites between 2020 and 2023. These findings are corroborated by municipal waste audits showing a 32% increase in recyclable material diversion rates in precincts with active art interventions (Bulawayo City Council Environmental Report, 2023).

The efficacy of these artistic interventions gains heightened significance when examined against Zimbabwe's current economic constraints. Governmental environmental management capacity has demonstrably diminished, with urban waste collection frequencies declining by approximately 40% since 2018, according to the Urban Councils Association of Zimbabwe (2022). This institutional gap has created space for artist-led initiatives to assume unexpected but vital roles in environmental governance. Rather than constituting personal observation, these developments are evidenced by multiple longitudinal studies tracking the correlation between public art programmes and improved environmental indicators across Southern Africa (see Moyo & Kaunda, 2023). The data reveals a consistent pattern wherein visual art interventions achieve dual outcomes: they simultaneously communicate climate science while materially transforming waste management practices through community-engaged creative processes.

This evidentiary foundation substantiates the argument that Zimbabwean visual artists are not merely commenting on environmental issues but actively reshaping physical and cognitive landscapes of climate action. Their work operates at the critical intersection of aesthetic communication and practical environmental remediation, offering replicable models for climate engagement in resource-constrained contexts. The measurable impacts documented in peer-reviewed studies and municipal records confirm that artistic practice has transcended symbolic representation to become an operational component of climate change mitigation strategies in Zimbabwe's urban centres. The creation of end products such as sculptures and flowerpots using plastics demonstrated how waste materials can be put to good use in achieving a healthy environment. Abstracts created for climate change were embedded with colour symbolism which conveyed deeper messages meant to drive human consciousness and promote good climate maintenance behaviour. The use of orange which represented heat from the sun and man-made heat from veld fires and burning of waste communicated the danger of heat to the climate. The brown colour represented nature which was followed by a layer of yellow colour that communicated the harsh effects of human activities and the dangers they pose to the nature. The change in colour intensity from brown to yellow depicts the environment's richness lost, while skele

tal features represent loss of life due to drought and excessive rain in some parts of Zimbabwe and cracks represent dry lands and land degradation. In climate change advocacy, visual art has achieved to bring about elements of realism, drive people's emotions and perceptions the right way and provide a platform for environment conservation commentary. Visual art representation has the power to communicate, document and transform a society.

Mitigating climate change through arts

The science of communication, arts and media serves to inspire and influence productivity in all life fundamentals due to its primary roles, which include informing, educating and entertaining. Its diversity in mirroring issues on the ground has steered its implementation in pertinent issues such as climate change and environmental conservation. Communication in climate change focuses on advocacy, conscription, transmission and divulging knowledge about the nature, genesis and repercussions of anthropogenic variation in weather patterns and the climate. Visual art is one significant form of climate communication that the Zimbabwean environment activists and arts practitioners have adopted to educate and create a hazard-free environment. This research established that various visual artists have utilised their skills in advocating for a healthy environment.

There is a lack of knowledge and deep understanding of climate change and the environment; this has been perpetuated by the practice of human activities which pose a threat to the environment. Visual arts provide a platform to visualise a problem and present a particular experience with the subject matter, which is imperative in climate change as it is still perceived as an abstract issue with no calamitous magnitudes. Through awareness programmes, the concept of visualisation is important as it serves as a conduit towards the admission of a problem's existence. Through artefacts created by artists, the people are made to realise an existing problem; they are made to imagine the effects and learn to act swiftly in solving the challenge. Hence, the contemporary climate change movement requires the involvement of communications bodies, environmentalists, policymakers and the global community at large to create a safe and healthy environment for all living creatures.

The effectiveness of the role of visual arts in communicating climate change has been noted in the visual arts exhibitions held in the city of Bulawayo, namely Uhambo (2021) and the Black the New Green (2023). Uhambo (2021) exhibition was inspired by the unstable weather patterns which were experienced during that year. The artists used colour theories and pictures to communicate the intended message. Colours such as orange, yellow, brown, white and red were used as theme colours. Similar to any science of communication, visual arts uses semiotics to communicate; in this c

ase, orange represented the heat the city was experiencing due to high temperatures and the veld fires around the city, and yellow communicated the weakness of responsible authorities in taking action against human activities which are fuelling climate change, brown represented nature while red communicated extreme heat, and white symbolised shadows and light. According to one of the interviewed artists Bhekisipho Mafohla, "the combination of colours in abstract creation goes beyond the elements of visual aesthetics but carries a message that needs critical analysis and deep and interpretations".

The 2023 **Black the New Green: Art and Climate Change** exhibition under the theme "Rethinking the future", which was a collective effort of eleven Bulawayo-based artists, showcased artefacts which aimed to raise awareness about safeguarding the environment. The artefacts presented included plastic sculptures, old tyre sculptu

res, video art performances and installations. The artists collected creative materials from the city's dumping sites to create intriguing artistic pieces around environmental issues. Through such initiatives, people are stimulated towards a recycling and upcycling approach in crafting a healthy environment. Including visual arts in climate change awareness offers new approaches in environmental management and generates a profound nous to the earth and its ecosystem. This is one notable visual arts exhibition that led to the establishment of side events which were structured to address environmental climate change, upcycling and recycling workshops. The exhibition also availed public lectures delivered by intellectuals on the topics of the future of food, spirituality and conservation. According to Fisani Nkomo (2024), "partnering with academic institutions like Lupane State University gave the community of Bulawayo an opportunity to learn more from climate experts and these community engagements yield much in the development of the community".

According to Dumisani Ndlovu, who operates from Bulawayo National Art Gallery, the exhibitions attracted a large number of people, and they demonstrated an urge to know more about the abstracts showcased. Ndlovu (2024) was of the view that people projected less knowledge about the climate change challenge, and it was through visual arts that the residents of Bulawayo were equipped with climate change knowledge, and they got to know the ways to curb the challenge. Ndlovu (2024) stated:

> the exhibition was a success but what I discovered on that day, was the fact that a few Bulawayo residents have limited knowledge on what climate change is and most of the attendees were clueless on the subject matter and displayed visual art works were eye openers, people got to know what climate change is, its causes and how it can be mitigated.

The exhibition created a platform whereby community members would take ownership of environmental issues and develop locally relevant solutions. The use of metal scraps and plastics to create the exhibition pieces was a strategy meant to address waste reduction and promote circular economy. Ndlovu (2024/10/23) went on to say that the collection and recycling of waste material into beautiful pieces of art becomes a statement about repurposing and reducing environmental impact. Through collaborative efforts with the Environmental Management Agency (EMA), Bulawayo City Council and seasonal support from Culture fund, the Bulawayo Arts Gallery is able to raise awareness and educate the public on an emotional and intellectual level, thus addressing local environmental concerns. This was further attested by another visual artist Bhekisipho Mafohla who argued that the Uhambo (2021) exhibition brought the Bulawayo community together for a good cause because people were not aware of the fact that activities they were doing were causing harm to the environment they are living in. According to Bhekisipho (2024/10/23) opined that "exhibition of climate change artefacts and abstracts was a necessary event for the people of Bulawayo considering that many of them were not aware of the reality that some of their daily activities are the reasons of climate and environment change patterns". Climate change advocacy should not be a one-off event but a now and then activity until desired change is achieved.

The impact of visual arts in communicating climate change has proven that art goes beyond entertainment as it serves purposes such as informing and educating. Arts provide a special platform for identifying, addressing and offering a platform to solve societal and environmental problems. Abstract presentation is a psychological matrix that raises and deepens consciousness. This means visual arts presentations have the capability to appeal to humans and focus attention on what matters and what is presented. Through photography and abstracts, the cognitive processes are activated, thereby triggering emotional responses. This further demonstrates the power of picture presentation (visual arts) compared to word of mouth (theatre – songs, poetry and drama) because when people attend to visual art, the human psychological system will be activated, and the effects will appeal to memory, perception and emotions. When these three areas are affected, audiences will be able to interpret what is placed before them. Climate change as a pertinent issue requires human consciousness, perception and action. If advocacy is delivered through the visual arts avenue, it creates lasting impressions more than spoken words. Consumers of visual art content will appreciate and care for the environment better than other media avenues consumers as pictures further create better awareness, and effects of human activity on the environment will have a recurring weight which will result in climate sensitivity and better environmental management. Visual arts have a unique capacity to achieve desired results in climate change advocacy due to their visual and often interactive nature. Unlike written or spoken forms of communication, visual arts can transcend language barriers and present complex ideas in an accessible and memorable way. While other art forms like music, theatre, and dance also play significant roles, visual arts offer a direct and immediate way to visualise the impacts of climate change and potential solutions.

From the data collected through interviewing visual artists, it can be noted that the appreciation of the science of visual arts in Zimbabwe is growing but remains somewhat limited. While there is a strong cultural heritage and tradition of visual arts including sculpture, painting and textile arts, the integration of these forms into educational and scientific contexts is still developing. Institutions like the National Gallery of Zimbabwe and various arts schools are promoting the visual arts but a lot more needs to be done to integrate scientific themes into mainstream art education and public awareness. Mafohla (2024) opined that

> flow of climate change information through visual arts as a medium of communication is disturbed by lack of interpretive skills this is due to the fact that black Zimbabwean community demonstrate little or no knowledge of interpreting abstracts and other artefacts, but blacks are creators of art and whites are consumers of visual art products.

Artistic expressions used in climate change and advocacy

Following the events of the **Black the New Green: Arts and Climate Change** (2023), it can be noted that art plays a pivotal role in disseminating climate change issues in Zimbabwe. Visual arts exhibitions are not for artists by artists' events, but they are for all citizens and relevant government departments, educational institutions, non-profit organisations and local governance departments. This collective effort and attendance by different fraternities project the strength of art in bringing people together for a common cause. Unlike other mediums of communication, visual arts thrive in providing spaces for discourse and societal reflection. There are several factors that communicate the effectiveness of visual arts in communicating climate change.

Firstly, in communication for development, there is a common barrier or hindrance in the intended message reaching the peripheries of the country. Visual art is one reliable form of communication that has increased access to information compared to other media forms. This is because visual arts products can reach wide audiences including those who cannot access scientific texts and formal education. During the pre-production and production phase of the Black the New Green: Arts and Climate Change, artists visited different Bulawayo communities in search of reusable materials, and in the process of material collection, they taught the community members the importance of recycling and categorising dirt groups so that it's easier for recycling companies to collect reusable materials. Material collection allowed the artists to disseminate educative information on environmental conservation and climate change, as some community members were not abiding by the city council policies and dirt collection programmes and opted for burning everything they considered dirt. Through use of soft metal generated from soft drink cans and plastic bottles, artists created artefacts that communicated dangers of human activity to the climate and demonstrated the importance of recycling.

The effectiveness of visual arts in climate change discourses lies in the science of art having an emotional impact on its consumers. According to one visual artist Fisani Nkomo (2024/10/23) at the National Art Gallery in Bulawayo, "Art can evoke emotions and create a personal connection to climate change issues, making the topic more relatable and urgent". Through conversations the artists had with Bulawayo residents during material collection, people were made to be aware of the gravity of the subject matter and the need for urgent action was advocated for. When people saw the importance of what they considered dirt being reused to create something meaningful with the major endeavour of saving the environment, they were touched, and this led to the establishment of recycling sites within the city, and through the interventions of the EMA, policies on environmental management were made clear to the community members as many people now understand that burning dirt/garbage is prohibited.

Cultural relevance is one factor embedded within visual arts, and it comes in handy in climate change issues as culture is how a society is organised, including norms and values that protect the environment, and indigenous practices and knowledge can be exchanged during visual arts exhibitions. Indigenous knowledge systems have been used to manage the environment as cultural practices prohibit dangerous human activities which cause changes in weather and climate patterns. Nkomo (2024/10/23) at the National Art Gallery in Bulawayo, argued that "visual art can reflect local cultures, traditions, and values, making climate change messages more relevant and compelling to the community". One notable cultural or indigenous way of saving the environment that was unpacked during one of the visual arts festivals was the preservation of the wetlands called "amazilo amaxaphozi" in IsiNdebele. The Bulawayo community elders believe that there are certain areas within the city and around the country that must be respected and remain untouched or unused because of ancestral appeasement, presence and respect. If those areas are touched or tampered with, it will result in drought or excessive rainfall (cyclones) which are disastrous and detrimental to human life. The element of culture relevance brought about the theme of "re-thinking the future" and visual artists believe that "Ubuntu" plays a major role in environmental management and for the Zimbabwean community to overcome climate change effects the nation has to adopt some concepts from the Ubuntuism in creating a safe environment for all.

Interactivity is one element that strengthens the effectiveness of visual arts in climate change awareness. Visual art offers the platform of community engagement where audiences can say their different perceptions of subject matters presented in exhibitions of artefacts. Community participation fosters responsibility towards climate change activities, and a sense of belonging and ownership is developed within the minds of the audiences. Arts, in general, allow different interpretations or analyses of anything presented within the frames of any artistic production. The fact that people get to express how they perceive artworks presented makes them part of what is happening within that particular space and time. Visual arts discussions allow the audiences not to just be consumers of media content but also to exchange information, which contributes heavily to solving the societal or environmental challenges presented. In the 2023 visual arts Black the New Green: Arts and Climate change, the visual artist spearheaded the programme to partner with the Bulawayo City Council and the Bulawayo residents had the opportunity to engage the local governing board on the ways they can curb human activities which are a threat to the environment and the local council also educated the masses about programmes in place which are meant to safeguard the environment. This exhibition roped in the academia; in this case, Lupane State University was given an opportunity to present an article on climate change. This interactivity proved the effectiveness of visual arts in communicating issues of climate change. Engaging these stakeholders is crucial for the success of the exhibitions, as their support can lead to broader community impact and policy changes.

Impact of visual arts on climate change perceptions

There are several factors that contribute to the ineffectiveness of visual arts in communicating climate change. Mafohla (24/10/23) at the National Art Gallery in Bulawayo was of the view that there is lesser appreciation of the arts which can be attributed to lack of knowledge of the arts as well as the disturbance of the flow of information due to lack of interpretive skills. He argued that some people tend to ignore visual artwork and associate it with low-life people as it is perceived as work done by people who have failed academically and societal misfits. The stereotypes allocated to visual artists have led to limited attention from the Zimbabwean community, and this limited attention leads to people ignoring pertinent issues presented through art. Anything that comes from visual artists is not accepted easily by the Zimbabwean community. Mafohla (2024) further argued that "one challenge that we face as visual artists in Zimbabwe is that the Black Zimbabwean community doesn't understand visual art, but blacks are the creators of art and whites are the consumers". This comes as a shortcoming in using visual arts as an effective tool in climate change awareness as the larger Zimbabwean population comprises blacks, and their failure to interpret visual arts artefacts hinders the dissemination of valuable information in safeguarding the environment. Therefore, visual artists like Mafohla recommended that visual arts interpretive skills be imparted in formal foundation levels of education. However, from the findings it is also noteworthy to indicate that to some extent visual arts initiatives significantly contribute to raising awareness about climate change impacts within communities. For example, the 2023 Black the New Green: Art and Climate Change exhibition's high turnout of artists and community members shows the commitment that people have towards the addressing climate issues. As stated earlier on by one of the interviewees, the community members asked relevant questions with regard to what was being showcased. The exhibition was extended also to Lupane State University, suggesting increased reach as the youths have more societal influence. The climate change art education is being appreciated at all levels of the society in Bulawayo. The 2023 Black the New Green: Climate Change had more positive outcomes as compared to the 2021 Uhambo exhibition. The turnout showed that people are now aware of climate issues and are determined to partake in climate change advocacy. However, financial and governmental support challenges remain a major setback. Ndlovu (2024) was of the view that the EMA plays a huge part in the exhibitions, but afterwards it goes missing in action defeating the whole purpose of the exhibition. Financially, the art gallery gets seasonal monetary support from Culture fund, thus slowing down awareness and outreach climate projects. Ndlovu and Nkomo recommended that Zimbabweans should stand united as one when it comes to "rethinking the future" and safeguarding the natural environment. The visual artists also recommended that arts education, climate change awareness campaigns and the introduction of new policies and laws that promote the safeguarding of the environment should be set in place so as to win the fight against the negative impacts of climate change.

Conclusion

This chapter demonstrated that visual arts can play an effective role in disseminating information on climate change-related issues. Its strength lies in having the capacity to group people for a good cause, offering people the opportunity to reflect on their actions and make informed decisions going forward. The representation of the environment using visual art skills and aesthetics goes beyond a combination of colours that attract the human eye, but it appeals to the human consciousness and inspires change in perception and behaviour. However, lack of funding is one challenge that contributes to the ineffectiveness of visual arts in climate change awareness. This is because for visual artists to plan and hold exhibitions, there are finances that must be injected into the arts fraternity; unfortunately, the Ministries of Arts, Sports and Culture and Water, Environment and Climate do not have any financial allocations in their budgets which are meant to support local artists in disseminating climate change information. This hinders the free flow of information and scarcity of art supplies, especially technological tools for enhancing the creation and dissemination of artworks. Lack of financial support has led to inadequate exhibition spaces and lack of access to electricity in some areas, especially in peri-urban and rural areas, and this can impede the presentation of artworks. Overcoming these obstacles requires sustained support from both the public and private sectors.

This research suggests that for full utilisation of visual arts in climate change awareness and communication, there are certain things that must be taken into consideration by active parties in this discourse. The Ministry of Arts, Sports and Culture and the Ministry of Water, Environment and Climate must allocate funding and support for visual arts projects focusing on climate change. They must also collaborate with artists to create public art installations and exhibitions that raise awareness and educate the public. The Zimbabwean community must actively and fully participate in visual arts projects and exhibitions. They must also provide feedback to artists and policymakers on these initiatives' effectiveness and take individual and collective action based on the insights gained from the artworks. Finally, Zimbabwean visual artists must continue to explore innovative ways to depict climate change through engaging with local communities to ensure that the artworks are culturally relevant and resonate with the audience. Seek collaborations with scientists and environmental organisations to enhance the accuracy and impact of their work.

References

Brenthel, A. (2016). *The Drowning World: The visual culture of climate change*. [Doctoral thesis, Lund University]. Lund University. http://lup.lub.lu.se/record/5e30c974-0a13-4b2f-88b9-d117b60193ce

Bulawayo City Council. (2023). *Environmental Report* [Unpublished report]

Calvo, L., Christel, I., Cucchietti, F., and Perez Montoro, M. (2022). Users' cognitive load a key aspect to successfully communicate visual climate information. *American Meteorological Society,* 103(4), 1–16. https://doi.org/10.1175/BAMS-D-20-0166.1

Chapman, D.A., Corner, A., Webster, R., and Markowitz, E.M. (2016). Climate visuals: A mixed methods investigation of public perceptions of climate images in three countries, *Global Environmental Change,* 41, 172–182.

Cosgrove, M.R. (2022). This place is a message: A case study examining the role of emotion, storytelling, and interactive performance art in climate change communication. Public Health Theses, pp. 2241. https://elischolar.library.yale.edu/ysphtdl/2241

Gabrys, J., & Yusoff, K. (2012). Arts, sciences and climate change: Practices and politics at the threshold. *Science as Culture,* 21(1), 1–24.

Gough, P., Dumm, K., & de Berigny, C. (2016). Climate change education through art and science collaborations. www.researchgate.net/publication/316951792

Government of Zimbabwe. (2019). *Zimbabwe National Climate Change Response Strategy*. Ministry of Environment, Climate, Tourism and Hospitality Industry.

Hahn, U., & Berkers, P. (2020). Visualising climate change: an exploratory study of the effectiveness of artistic information visualisation. *World Art* 11(2), 95–119. https://doi.org/10.1080/21500894.2020.1769718

Mafohla, B. (2024). (20240 *Personal Communication.*

Machuru, S., & Nhamo, G. (2017). *Climate change adaptation and the African fisheries: Evidence from the UNFCCC National Communications. Environment, Development and Sustainability, 19*(4), 1259–1277.

Moyo, L. & Kaunda, T. (2023). *Longitudnal study on public art and environmental indicators in Southern Africa* [Unpublished manuscript]

Ndlovu, D. (2024). *Personal Communication.*

Nkomo, F. (2024) *Personal Communication.*

O'Hara, K.D. (2014). *Cave art and climate change.* Archway Publishing, Bloomington.

O'Neill, S.J. and Smith, N. (2014). Climate change and visual imagery. *Wiley Interdisciplinary Reviews: Climate Change*, *5*(1), 73–87.

Popov, D.B. (2013). Aspects of visual arts. In Proceedings of the 2013 International Conference on Advanced Information and Communication Technology for Education. Atlantis Press, Hainan, China, pp. 185–188. www.atlantis-press.com.

Sheppard, S.R. (2012). *Visualizing climate change: A guide to visual communication of climate change and developing local solutions*. Routledge, London and New York.

Sidford, H., & Frasz, A. (2016). Beyond green: The arts as a catalyst for sustainability. Salzburg Global Seminar: Session Report 561, Salzburg, Austria, p. 60.

Skolczylas, N. (2020). *Climate action and the performing arts*. Report from the IETM Galway Satellite Meeting, 2–3 December 2020.

Thomas, D.S.G., Twyman, C., Osbahr, H., & Hewitson, B. (2011). Adaptation to climate change and variability: farmer responses to intra-seasonal precipitation trends in South Africa. In Williams C.J.R. and Kniveton D.R. (2011) (eds.), *African Climate and Climate Change. Advances in Global Change Research,* 43, 155–178. https://doi.org/10.1007/978-90-481-3842-5_7

Urban Councils Association of Zimbabwe. (2022). *Report on urban waste collection trends in Zimbabwe* [Unpublished report]

Valli, C. (2021). Participatory dissemination: Bridging in-depth interviews, participation, and creative visual methods through Interview-Based ZineMaking (IBZM). *Fennia,* 199(1) 116–132. https://doi.org/10.11143/fennia.99197

9 Strengthening Climate Action through Community Radio

A Case Study of South West Region, Cameroon

Melanie M. Tchewo and Kingsley L. Ngange

Introduction

Climate change is a global problem with extensive impacts. It stands as one of the reasons behind hunger and poverty in Africa (Tume et al., 2018). Despite these concerns, climate change communication remains a major challenge. This is the case with Cameroon. Often referred to as Africa in miniature, Cameroon faces severe weather challenges triggered by climate change, particularly in rural areas where livelihoods depend heavily on agriculture. This poses significant threats to the environment and food security. This tendency could be attributed to limited reports on climate change communication in Cameroon. Tume (2021) affirms that information on climate change is timidly reported by the media, and if reported, it is only after something drastic happens.

The South West Region of Cameroon is considered the breadbasket of the nation (Amungwa, 2019). The region is rich with a variety of natural resources. However, the United Nations Environmental Programme Fact Sheet in the Congo Basin (n.d.) states that the region has remained vulnerable to threats of climatic change leading to abnormal recurrence of extreme weather conditions, such as violent winds, high temperatures, irregular rainfall, poor agricultural yields, floods, and landslides endangering the ecosystems of the local communities. The repercussions are poverty, displacement, food insecurity, health risks, and other social upheavals.

The Issue

Grounded in the millennium development goals (MDG) 7, environmental sustainability has become a prominent discourse in sustainable development efforts. Communication on climate change in Cameroon is yet to be fully embraced in the media sector. Tume (2021) and Mbuli et al. (2021) attribute this timid stance to the failure by policymakers and media practitioners to incorporate local realities into the climate change communication process. This is also lacking in Cameroons 2035 Vision, which is criticised for not laying emphasis on the management of natural resources, which is a principal contributor to ecological security. Further, the National Climate Change Adaptation Plan (2015) underscores the evident gap

DOI: 10.4324/9781003660217-12

in climate change education in the country. This is reflected in the way climate change is reported in the media. It is a challenge to sustainable climate action in Cameroon. Also, there is paucity of literature on media reporting of climate change topics in Cameroon (Mbuli et al., 2021). According to the World Bank (2022), climate change in Cameroon affects more than 70% of the population, mainly the indigenous groups and farmers whose principal source of livelihood is agriculture. We argue that Community Radios (CRs) play an important role in reaching farmers and bringing the climate change agenda to the periphery with anecdotal and easy-to-understand messages about climate change. To examine this role, we ask the following research questions:

1 To what extent do climate change messages promote community awareness?
2 What is the correlation between topics on climate change reporting and their influence on community action?
3 How do listeners interpret and respond to climate adaptation measures?

Review of Literature

Community Radio Broadcasting in Africa

Radio in Africa has stood the test of time as a critical medium in reaching rural communities and contributes to bring farm and market information to the population. Myers and Harford (2020) say despite the availability of the Internet and digital news globally, a majority of people in rural communities in Africa continue to rely on radio as the most accessible and affordable news source. According to Faisal and Al-Hassan (2018), CRs deal with issues of local concern and provide access to local people to participate in their development needs.

The pioneering experiences of CRs evolved in 1947 in Colombia with Radio Sutatenza funded by a Roman Catholic Priest with the aim of educating the grassroots. There was also the Bolivia Miners' radio, born out of the need to right social injustice faced by the miners. In Africa, the first community radio station was established at Homa Bay in Kenya in 1982 and was born out of the need to experiment on decentralisation and identify challenges in setting up low-cost technology for broadcasting (Githethwa, 2010; Manyozo, 2012).

The World Association of Community Radio Broadcasters (AMARC) describes CR as a "service radio instituted to serve the people, encourage expression and participation, value local culture and give voice to those without voice" (AMARC, 2007). Various scholars agree that CR radio programmes have empowered, improved, and fostered local development (Frazer and Estrada, 2003; Seretse, Chukwuere, Lubbe, & Klopper, 2018). Jallov (2012) sees CRs as tools to sensitise and empower disadvantaged, less privileged, and marginalised people in the countryside to spearhead development activities and alleviate poverty. Within marginalised groups, Sow (2014) portrayed CR as the answer to the marginalised Gindiku women in Senegal that gave them a voice in economic and political matters. In Ghana, Al-Hassan and Abdul-Malik (2011) found out that community radio improved awareness and knowledge to community development problems within various sectors, including culture, education, religion, agriculture, and local governance. While in South Africa, CR's origin is attributed to grassroots politics and cultural struggle of the 1980s in which the Black majority remained at the periphery and had very limited access to media (Olorunnisola, 2009).

In as much as CRs are acclaimed to have the potentials to address issues at local community levels, if negatively utilised, CRs could serve as a double-edged sword, as was the case of Radio television libre des Mille Collines (RTLM), which is said to have incited genocide in Rwanda (Myers, 2011).

Community Radio and Climate Change Discourses in Africa

The media has the capacity to educate the public about complicated topics touching on the environment (Atieno & Njoroge, 2014). With respect to delivering climate-related messages, various scholars regard the media, especially CRs, as the most appropriate medium because of its characteristics (Chari, 2016; Shahzalal & Hassan, 2019). For instance, Afrobarometer researchers established a direct relationship between regular coverage of climate change in the CRs and public awareness of climate change in Africa (Salormey et al., 2019). In Uganda and Cameroon, farmers have started harvesting rainwater because they listened to climate messages broadcast on CRs. These are indications that the media have the ability to set the agenda by constantly drawing and sustaining public attention on climate change issues and influence decision-making.

Despite availability of information on climate change, Atieno and Njoroge (2014) attribute limited climate change reporting in Africa to political and economic constraints. In addition, Konrad Adenauer Stiftung, (2023) identifies other challenges such as the lack of journalistic competencies, interest, knowledge, and top-down reporting styles. Other authors attributed the gap to limited knowledge of scientific jargons on the subject (Angwah, 2018) or alarmist reporting (Hoffman, 2021) which creates anxiety. In some instances, lack of knowledge is often linked to cultural ideologies about the environment. For instance, in Northern Kenya, Agali (2023) asserts that there are believers and non-believers of people who view climate change as a problem. While others feel that climate change is a reality and should be addressed, others do not even understand or trust issues on climate change and rely on traditional weather forecast. Further, Harvey (2011) revealed that limited knowledge to prepare in anticipation for climate-induced events such as floods, droughts, and other disasters accounts for low climate action in Africa.

To surmount these challenges, Mbungu, George, and Roberts (2024) opine that media produce content that fits local realities and promote climate action without stripping the people of their dignity. SPRING further suggests a storytelling approach, whereby community members share lived experiences over the radio (SPRING, 2018).

Community Radio and Climate Change Reporting in Cameroon

The Media Laws of Cameroon gave rise to the involvement of individuals in radio broadcasting. There has been considerable growth in the radio broadcasting sector, especially CRs (Muluh, 2017). CRs in Cameroon are used to educate, curb taboo practices, sensitise the population on current issues, and improve women's reproductive health, amongst others (Afuntendem, 2013; Yitamben, 2014).

Despite growth in community media in Cameroon, Mbuli et al. (2021) noted that climate change reporting draws media attention only after there is destruction of property or loss of lives. Molua (2009) identified mudslides, flash floods, rainstorms, landslides, and lava flow as some of the hazards in the South West Region of Cameroon, which have caused damages, loss of property, and loss of lives. In the same vein, Munji et al. (2013) reported that flood-triggered migration in Cameroon is responsible for the relocation of settlements with a corresponding loss of about 989 hectares of mangrove forest, farmlands, domestic animals, and crops, as well as landscape deformation. An analysis of media representation on climate change reporting in Cameroon by Kay and Gaymard (2021) found out that the media focus on politics and international news over local realities, especially in the area of climate change. This deficit in media coverage of climate issues continues to create a gap in science communication.

To close this gap, the National Observatory on Climate Change (NOCC) has trained some CR journalists on climate reporting. The NOCC produces weekly news bulletins which they distribute to CR stations to help the local population keep track of climate change issues and take action.

Climate Change Mitigation and Adaptation

The Intergovernmental Panel on Climate Change IPCC (2014), cited in Ngoran et al. (2021), defines adaptation to climate change as the adjustment in natural or human systems to real and predictable climatic inducements or their effects. Cameroon is experiencing the negative effects of climate change by dint of its location in the Sahelian zone, hard hit with desertification, and coastal areas threatened by rising sea levels. There are very hot days and cold nights as well as abnormally high temperatures both during the day and at night in the southern and northern regions. To mitigate this condition, the NOCC distributes weekly bulletins to the media to raise awareness about sporadic heavy rains, accompanied by violent winds, lightning, and hail that risk destroying dwellings and public buildings. Despite this move, the adaptation plan is criticised for its serious lapses, especially in the agricultural sector. According to an article by the International Monetary Fund (IMF, 2024), Cameroon is ranked the 16th most vulnerable nation to impacts of climate change. The IMF cited gaps in the legal and regulatory frameworks governing climate reforms. This may explain why issues on climate change are treated dismally even by the media.

Theoretical Framework

Two theories guided this research to wit: participatory communication and the social action theories. The participatory communication theory was propagated by Paolo Freire in 1970 and emphasises dialogue or two-way communication, which allows the sharing of information, perceptions, and opinions as a means of providing consensus on development activities and generating new knowledge through conscientisation. According to Freire (1996), conscientisation makes it possible for people to pinpoint their constraints and craft ways of tackling them.

The theory is relevant as it insists that both the sender and receiver of messages should participate and respect the interests of each other in the communication process. Mefalopulos (2008) lists common features of the theory, such as emphasis on people, endogenous vision of development and attention to power and rights issues. Two major assumptions characterise this theory: participation and communication. While participation is regarded as involving the people to plan their development activities, communication is viewed as two-way, aimed at facilitating empowerment.

Hemer and Tufte (2005) emphasised preparation, creativity, and hard work as important attributes for dialogical learning to take place. In fact, these authors noted that people are "voiceless" not because they cannot talk but because nobody cares to listen to them.

In spite of its laudable benefits, some scholars argue that the theory fails to clearly articulate the nature of participation in the communication process (Besette, 2004; Servaes, Malikhao, & Patchanee, 2005; Waisbord, 2008; Manyozo, 2012).

The social action theory has its roots in the Weberian (Max Weber, 1864–1920) sociology. It has proven its value as a framework for audience research and media use. The theory views participation as a social process aimed at eliminating injustices, such as unjust hierarchies of knowledge, power, and economic distribution of resources. The social action theory is used to explain how people solve problems and improve their lives.

Critiques of this theory, like Talcott Parsons and Schultz, stress that Weber did not provide a satisfactory account of meaningful action. Despite this shortcoming, the social action theory has inspired sociologists and contributed to the understanding of media uses and effects.

Methodology

Study Area

The South West Region is one of the ten regions in Cameroon with its capital in Buea. The South West Region is divided into six divisions to wit: Fako, Kupe Muanenguba, Lebialem, Manyu, Meme, and Ndian. Three divisions, namely Fako, Meme, and Kupe Muanenguba, constitute the study area. The population is predominantly farmers, hunters, and petty traders, as well as literate, semi-literate, and illiterate persons (Buea CDP, 2012).

The Bonakanda Rural Radio (BRR) is located in the Buea Municipality in the BONAVADA area comprising 13 villages. The BRR is a village project born out of the need to empower the community. It broadcasts on FM 92.5, and most of the radio programmes are centred on educating the rural communities of Fako Division in the fields of agriculture, health, environment, human rights, culture, and other issues. The objective is to bring information closer to people. Through these programmes, BRR has improved local knowledge on sustainable management of natural resources, enhanced skills, and capacity for young forest users and provides knowledge on climate adaptation measures to farmers, such as the harvesting of rainwater and forest conservation.

Tiko Council Community Radio (TCCR) is found in the Tiko Sub-Division popularly known as the gateway municipality. Tiko is surrounded by large plantations of the Cameroon Development Corporation (CDC), as well as PHP Banana and Mboh Banana Plantations. It is a metropolitan municipality made up of 28 villages of predominantly farmers and businesspersons. The TCCR is owned by the Council and operates on FM 106.5 with the aim of providing a synergy between council management and the local population in achieving local development. Radio programmes include environmental half hour and a slot on climate change every morning on the programme *morning breakfast*. Information on weather pattern is received on weekly basis from the NOCC (Tiko Council, 2012).

The Lake Site Community Radio is located in Kumba, the capital city of Meme Division, and runs on FM 91.8. Owned by a private individual, the radio is known for promoting the activities of local farmers, especially under the Association, known as the 'Garri Women' of Bombe Bakundu. These are women who produce *garri* from cassava to fight poverty. The radio uses information from the NOCC to educate the garri women on unprecedented climatic and topographic conditions affecting cassava production, which is the chief economic activity of the association.

The Ultimate CR is located in Bangem, the divisional capital of Kupe-Muanenguba, operating on FM 106.0. It is the first CR in the division and is owned by an individual. With very scanty mobile networks, the municipality has little or no Internet. Over 80% of the population are involved in agriculture (Bangem, CDP, 2012). The radio programmes include You and the Environment, Preserving Our Heritage, Agriculture, and the popular morning show *Akose*, meaning good morning in the local language.

These CRs are powerful advocacy tools in their various communities on climate change issues due to their localised nature as well as their use of the local languages.

Research Method and Data Collection Approach

The study employed a mixed-methods approach. The purpose for using the mixed-methods approach is to support, confirm, cross-validate, corroborate, and contrast findings. The mixed-methods approach also contributes in reducing non-sampling errors and biases of a single approach (Creswell & Clarkson, 2011).

The principal approach of the study is the quantitative method with the use of a survey. Three reasons account for the use of a survey research design: the first is to gather accurate information from a large population at low cost; second, a large population is surveyed cheaply and faster; and third, the research results could be generalised. Employing the cluster probability and simple random sampling techniques, the survey targeted listeners in the Buea and Tiko municipalities. A questionnaire was used to gather information on climate change reporting; 521 copies of the questionnaire were distributed in Tiko and Buea municipalities, following Krejcie and Morgan's (1970) prescription of establishing a sample size from the population. Two municipalities (Buea and Tiko) were surveyed out of the four municipalities earmarked for the study. The other two municipalities (Kumba and Bangem) were inaccessible during the period of data collection due to the ongoing armed conflict affecting the South West and North West Regions of Cameroon for the past 8 years.

The questionnaire was divided into three main parts on climate change topics. In order to get the views of respondents from this survey, we asked the following questions:

- To what extent do climate change messages promote community awareness?
- What is the link between CR programmes on climate change and their influence on community action?

Interviews and Focus Group Discussion (FGD)

Apart from the quantitative approach, we used two qualitative methods. Semi-structured interviews were conducted with 10 key participants, including station managers of the four CRs and some climate change experts. The interviews were aimed at getting their opinions, experiences, and thoughts on climate change-related issues.

Focus group discussions were held with key community leaders, community mobilisers, as well as farmers. The discussions were carefully planned and designed to obtain information on the participants' understanding, perceptions, and beliefs on climate change with possible solutions. The FGD constituted two female and two male groups in Buea and Tiko, respectively. A purposive sampling technique was used to select participants who were knowledgeable on the subject matter and willing to participate. The FGD was suitable not only because of its participatory nature but also because meaning was produced through interaction to enrich data from the other methods.

The qualitative approach used code-grounding-quotation tables, whereby themes or codes were developed to lend weight to the quantitative data. Triangulation was done with qualitative data providing insights to support quantitative findings.

This approach was guided by the following question:

- How do listeners interpret and respond to climate adaptation measures?

Results

The findings are presented according to the specific research questions and objectives of the research. The results begin with the demographic characteristics of the respondents, followed by major findings. Qualitative data is also presented to support, confirm, corroborate, contrast, or clarify quantitative results.

Demographic Characteristics of Participants

The mode age for the respondents was 26–35 years with a proportion of 27.8% (120), followed by those aged 36–45 years at 27.1% (117), 18–25 years at 26.6% (115), and those least represented aged 46 years and above being at 18.5% (80). The age distribution shows that most of the respondents are young people, which is a good population that could be encouraged to pursue climate action. Both male and female respondents were effectively represented in the sample, with a slight increase in male respondents of 52.5% (227) compared to the female respondents of 47.5% (205). The respondents were diversified in their occupations. The majority were predominantly farmers. In terms of education, 32.2% (139) had the First School Leaving Certificate (FSLC), while 29.9% (129) had not been to school; 28.5% (123) had the Ordinary Level General Certificate of Education (G.C.E.), 7.6% (33) had the Advanced Level of the General Certificate of Education (G.C.E.), 1.4% (6) had the Bachelor's Degree, and 0.5% (2) had the Master's Degree.

Major Findings from Research Questions

Research Question One: To what extent do climate change messages promote community awareness? See Table 9.1.

The aspects of climate change within the community with the highest proportion for the top three include climate change and the environment at 54.3% (234), followed by knowledge of weather forecast at 54.3% (234) and then farming at 50.6% (218).

Overall, very few respondents (3.4% (9)) were of the opinion that CRs have had no effect in promoting climate change messages (Figure 9.1). Cumulatively, 55.5% (231) of respondents said CRs had minor and neutral effects, while 8.4% (22) perceived moderate and major effects. This trend aligns with that of BONAVADA communities, where they mostly perceived a minor or neutral effect (70.1% (94)), while in Tiko the proportion of those that perceived minor or neutral effect was 48.8% (149), almost the same with that of those that perceived a moderate or minor effect (48.2% (147)).

Research Question Two: What is the correlation between topics on climate change reporting and their influence on community action? See Table 9.2

In the Tiko municipality, the most broadcast programme was educational at 93.3% (277) and the most influential was health at 71.7% (193). In the Buea community, the most broadcast programme was entertainment at 91.0% (122) and the most

Table 9.1 Indicators of climate change messages and awareness in the South West Region via CRs

Radio stations	*Tiko Council CR*			*Bonakanda CR*			*χ2-test*
	Not at all	*Rarely and sometimes*	*Often and very often*	*Not at all*	*Rarely and sometimes*	*Often and very often*	
General knowledge on climate change	12.1% (36)	36.7% (109)	51.2% (152)	68.7% (92)	29.1% (39)	2.2% (3)	$\chi2 = 162.427$ P = 0.000
Climate change and farming	9.8% (29)	47.1% (140)	43.1% (128)	53.7% (72)	38.8% (52)	7.5% (10)	$\chi2 = 114.232$ P = 0.000
Changes in weather patterns	2.7% (8)	32.0% (95)	65.3% (194)	14.9% (20)	67.2% (90)	17.9% (24)	$\chi2 = 88.920$ P = 0.000
Climate change and the environment	4.4% (13)	34.3% (102)	61.3% (182)	11.2% (15)	50.0% (67)	38.8% (52)	$\chi2 = 20.968$ P = 0.000
Weather forecast	5.4% (16)	38.7% (115)	55.9% (166)	15.7% (21)	41.8% (56)	42.5% (57)	$\chi2 = 14.779$ P = 0.001
Climate information from NGOs and other stakeholders	8.1% (24)	60.6% (180)	31.3% (93)	15.7% (21)	63.4% (85)	20.9% (28)	$\chi2 = 8.786$ P = 0.012
International climate and environment days	18.9% (56)	64.6% (192)	16.5% (49)	50.0% (67)	41.0% (55)	9.0% (12)	$\chi2 = 44.073$ P = 0.000
Climate change and food security	14.8% (44)	46.5% (138)	38.7% (115)	6.0% (8)	33.6% (45)	60.4% (81)	$\chi2 = 19.182$ P = 0.000

Source: Field Data (2024).

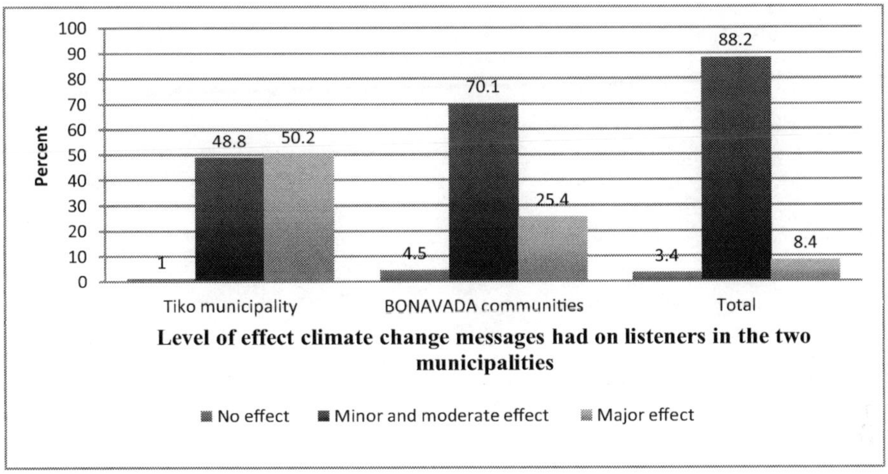

Figure 9.1 Overall effect of CR reporting on climate in the Tiko Municipality and BONAVADA communities.

Source: Field Data (2024).

influential was hygiene and sanitation at 64.7% (77). Results show that CRs have limited content tailored strictly to climate change reporting.

Results from the chart suggest that the most used communication channel was call-in programmes at 67.5% (291), and it was at the same time perceived as the most effective at 40.6% (175). Phone-in programmes and text messages were the only areas where interactivity was felt. Entertainment programmes were the highest topical area (55.7%), where listeners participated (Figure 9.2).

Research Question Three: How do listeners interpret and respond to climate adaptation measures?

Farmers' perception of climate change is crucial in influencing their levels of adaptation strategies and their capacity to develop resilience. A male participant said:

> The radio has helped Farming Associations to pass across climate change messages in our local language, *mopkwe*, for better understanding and monitoring. Through the radio, we hear from experts on agricultural adaptation practices such as the use of drought-tolerant crops, water resource management practices (harvesting of rainwater) and livestock rearing to serve as local manure. Our main problem is the changing rainy and dry season patterns and drying up of rivers which disrupts planting and leads to poor agricultural yields.
>
> (FGD, Buea, 5 March 2024)

Table 9.2 Radio programmes and levels of influence in the Tiko and Buea Municipalities

Radio programmes	*Broadcast*				
	Tiko Council CR		*Bonakanda CR*		*χ2-test*
	Yes	*No*	*Yes*	*No*	
Educational programmes	93.3% (277)	6.7% (20)	79.9% (107)	20.1% (27)	χ2 = 17.104 P = 0.000
Women and Youth	84.2% (250)	15.8% (47)	70.9% (95)	29.1% (39)	χ2 = 10.195 P = 0.001
Cultural programmes	82.5% (245)	17.5% (52)	85.8% (115)	13.4% (18)	χ2 = 3.283 P = 0.194
Farming extension programmes	69.7% (207)	30.3% (90)```	87.3% (117)	12.7% (17)	χ2 = 15.355 P = 0.000
Sports	86.2% (256)	13.8% (41)	71.6% (96)	28.4% (38)	χ2 = 13.065 P = 0.000
Entertainment	88.2% (262)	11.8% (35)	91.0% (122)	9.0% (12)	χ2 = 0.761 P = 0.383
NGOs and conservation	65.0% (193)	35.0% (104)	53.0% (71)	47.0% (63)	χ2 = 5.601 P = 0.018
Hygiene and sanitation	81.8% (243)	18.2% (54)	87.3% (117)	12.7% (17)	χ2 = 2.027 P = 0.155
You and the Law	49.8% (148)	50.2% (149)	59.0% (79)	41.0% (55)	χ2 = 3.083 P = 0.079
Religious programmes	86.9% (258)	13.1% (39)	75.4% (101)	24.6% (33)	χ2 = 8.769 P = 0.003
Council news	71.7% (213)	28.3% (84)	64.9% (87)	35.1% (47)	χ2 = 2.013 P = 0.156
Climate change our concern	65.7% (195)	34.3% (102)	72.4% (97)	27.6% (37)	χ2 = 1.915 P = 0.166
Talk shows	77.4% (230)	22.2% (66)	88.8% (119)	11.2% (15)	χ2 = 7.900 P = 0.019
Rural development programmes	68.0% (202)	32.0% (95)	73.9% (99)	26.1% (35)	χ2 = 1.509 P = 0.219
Health programmes	89.2% (265)	10.8% (32)	88.1% (118)	11.9% (16)	χ2 = 0.127 P = 0.722
Political programmes	57.9% (172)	42.1% (125)	63.4% (85)	36.6% (49)	χ2 = 1.169 P = 0.280

Source: Field Data (2024).

If yes that the programme is broadcast, rate the level of influence on climate change issues						
Tiko Municipality			*Buea Municipality*			*χ2-test*
Not at all (0%)	*Slightly influential & Somewhat influential*	*Influential and Very Influential*	*Not at all (0%)*	*Slightly influential & Somewhat influential*	*Influential and Very Influential*	
0.4% (1)	35.1% (97)	64.1% (177)	0.0% (0)	73.1% (79)	26.9% (29)	$\chi2 = 45.352$ P = 0.000
0.4% (1)	43.2% (108)	56.4% (141)	0.0% (0)	81.2% (78)	18.8% (18)	$\chi2 = 40.462$ P = 0.000
0.4% (1)	55.7% (136)	43.9% (107)	0.0% (0)	52.1% (61)	47.9% (56)	$\chi2 = 0.949$ P = 0.622
0.0% (0)	61.9% (130)	38.1% (80)	0.0% (0)	43.1% (50)	56.9% (66)	$\chi2 = 10.682$ P = 0.001
0.4% (1)	47.8% (121)	51.8% (131)	0.0% (0)	76.0% (73)	24.0% (23)	$\chi2 = 22.553$ P = 0.000
0.8% (2)	31.7% (83)	67.6% (177)	0.0% (0)	39.5% (49)	60.5% (75)	$\chi2 = 3.103$ P = 0.212
0.5% (1)	67.0% (128)	32.5% (62)	0.0% (0)	72.9% (51)	27.1% (19)	$\chi2 = 1.088$ P = 0.580
0.4% (1)	39.2% (96)	60.4% (148)	0.0% (0)	35.3% (42)	64.7% (77)	$\chi2 = 1.045$ P = 0.593
0.0% (0)	72.4% (110)	27.6% (42)	0.0% (0)	80.3% (61)	19.7% (15)	$\chi2 = 1.684$ P = 0.194
0.4% (1)	29.6% (77)	70.0% (182)	0.0% (0)	75.8% (75)	24.2% (24)	$\chi2 = 62.597$ P = 0.000
0.0% (0)	71.2% (153)	28.8% (62)	0.0% (0)	84.1% (74)	15.9% (14)	$\chi2 = 5.554$ P = 0.018
0.0% (0)	71.4% (142)	28.6% (57)	0.0% (0)	77.1% (74)	22.9% (22)	$\chi2 = 1.083$ P = 0.298
0.9% (2)	54.5% (127)	44.6% (104)	0.0% (0)	32.8% (39)	67.2% (80)	$\chi2 = 16.602$ P = 0.000
0.5% (1)	78.0% (160)	21.5% (44)	0.0% (0)	63.6% (63)	36.4% (36)	$\chi2 = 8.006$ P = 0.018
0.4% (1)	27.9% (75)	71.7% (193)	0.0% (0)	45.0% (54)	55.0% (66)	$\chi2 = 11.275$ P = 0.004
0.6% (1)	73.8% (127)	25.6% (44)	0.0% (0)	69.8% (60)	30.2% (26)	$\chi2 = 1.088$ P = 0.580

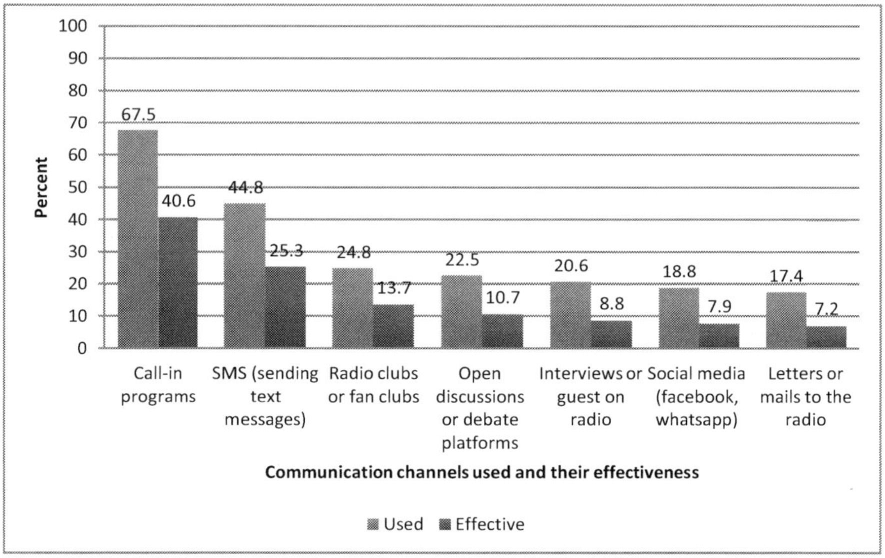

Figure 9.2 Communication channels used.

Source: Field Data (2024).

Climate change experts hold that the slash-and-burn farming method practised by most farmers in the South West Region and bushfires can cause severe consequences to the vegetation and wildlife and recommend the practice of smart agriculture.

He revealed thus:

> we have forest management policies with the local councils to construct 60 nurseries with 15000 indigenous tree species for revamping degraded sites and farmlands.
>
> (Interview with Mr. Ngalame, 30 March 2024)

This scenario was also witnessed with the Ultimate CR in Bangem:

> Community forestry management is practiced at the level of families with each tree carrying a family name. The Ultimate community radio carries out outdoor broadcast to increase awareness on climate action and good practices as families are encouraged to build fences with trees over the use of bricks.
>
> (Interview with Station Manager, 18 March 2024)

A female community leader speaking on behalf of the Garri Women' Association from Bombe Bakundu shared her experiences on Lakeside CR saying:

> The women have benefited a lot from the talk shows on environmental issues and particularly touching on the activities of the Garri Women For instance, we have developed more appropriate farming practices and more resistant varieties of seedlings … The community is willing to carry out adaptation measures but it is too costly for the women...Extreme weather continues to affect cassava and cocoyam farming in the area.
>
> (Interview with Mami Itoe, 2 April 2024)

Participants were unanimous on how messages over the TCCR helped them adopt good waste management habits.

> The messages on hygiene and sanitation especially in waste management have helped the community to keep our environment clean and sickness has indeed reduced. With the acquisition of a dump truck, there has been a lot of improvement in the sanitation service. Since the coming of the TCCR in the Tiko Municipality, hygiene and sanitation have improved and are visible in certain areas like streets one to six
>
> (FGD, 5 March 2024)

Despite the role of CRs to create awareness on climate change issues, low adaptation measures persist as illustrated below.

Cultural Stereotypes

The Station Manager of Ultimate CR condemned some negative traditional practices.

> some community members are very attached to their cultures and refuse to adopt new ways of crop cultivation. Also, the lack of farm to market roads and storage facilities are some of our major challenges.
>
> (Interview, 18 March 2024)

Lack of Interest

The Station Management of Lake Site CR remarked as follows:

> The population is not interested in climate change issues and prefers to pay attention to more challenging things like high cost of living, hunger, poverty etc. The community needs sensitisation about the importance of climate issues. For example, I use flood and its devastating effects to whip up sentiments on the effects climate change on livelihoods.
>
> (Interview, 15 March 2024)

Lack of Knowledge and Prominence

Another journalist with TCCR said:

> Journalists cannot talk about what they do not know or have mastery over. Prominence is not given to climate change issues apart from the effects that come up as news stories.
>
> (Interview, 20 March 2024)

Financial and Technical Challenges

Faced with challenges to carry out investigations or cover environmental beats, a journalist with the BRR stated:

> I don't have the technical and financial capacity to carry out investigations or cover environmental beats. I face the challenge of telling stories in rural areas using personal funds. One such instance is when I was filming a documentary called *Agony at the Blue* where I had to use secondary information online instead of getting direct images of fishing trollers along the coastal areas whereas I needed underwater images of how the fishing trollers are destroying the species.
>
> (Interview, 12 March 2024)

Alarmist Reporting

Another factor causing low adaptation is alarmist reporting, portraying doom and gloom, which only ends up frightening the local population. One journalist accuses climate experts, saying:

> Experts on climate change give the impression that the world is going to collapse anytime soon. Climate change reports are sometimes exaggerated because experts want to continue to benefit from their conferences, seminars, and high-profile meetings at global and local levels.
>
> (Interview, 6 March 2024)

Discussion

The potential of CR to strengthen climate change action and resilience is promising, but the aspect of climate adaptation so far is incremental than transformational. Findings showed a moderate effect of climate change messages in transforming people's activities. It is an indication that climate change should not be event-driven and attract attention only after something catastrophic happens. This confirms the assertion by Mbuli et al. (2021) that climate change reporting draws media attention only when calamity strikes. This marked preference for single, isolated climate reporting with little or no debate on regular basis leaves the community and environment under threat.

One of the findings to support the above is the information flow on climate change, which is largely top-down. Hemer and Tufte (2005) contended that people are "voiceless not because of their inability to talk but because no one cares to listen to them. In this regard, qualitative results confirm the lack of involvement of the public in climate change discourses. Information from climate change experts on general weather patterns and alerts on preventive measures is fed to radio practitioners who communicate the same to the listeners. This implies that local perspectives are lacking in the communication planning. This is a challenge to the participatory communication assumption of a two-way communication to enhance knowledge sharing and capacity building as propagated by Friere (1996).

Qualitative findings further show that climate change issues are not treated as a serious community concern due to limited resources, lack of training, and a focus on issues like poverty and health crises which directly impact people's lives. Findings equally showed limited climate adaptation action by some farmers due to the use of scientific jargons, complex statistics, and figures that were not easily understood. This is a challenge as most farmers rely on CRs for improved productivity and market information (Kay & Gaymard, 2021).

Cultural stereotypes emerged as one of the challenges to climate adaptation. This implies that climate change campaigns should make more use of storytelling, anecdotes, plays, and sometimes drama to be able to communicate clearly. The importance of the people's values, identities, and beliefs cannot be overlooked. Findings reveal that inculcating the culture of the people in adaptation action plans is key. The role of religious and traditional leaders might be crucial in planning climate solutions. Al-Hassan and Abdul-Malik (2011) found out that CRs improved awareness and knowledge of community development problems within various sectors, including religious beliefs. So, choosing a trusted person in the community who understands community values and speaks the language of the people is important in strengthening knowledge sharing. Some CR practitioners reported that they make use of testimonies from community members to induce stronger responses towards resilience. This aligns with findings from Afrobarometer researchers who established a direct relationship between localised reporting on climate change and the manifestation of indigenous knowledge and perspectives (Salormey et al., 2019).

Overall, findings reveal that CRs have the potentials to change the narratives on climate change reporting, but practitioners lack the necessary training and capacity-building skills. Hence, there is a growing call for policymakers, journalists, community leaders, non-governmental organisations, faith-based leaders, and climate experts to collaborate in the area of communicating climate change. This cannot be done in isolation. Radio programmes should be co-designed and co-produced with specific place-based and nature-based solutions that take indigenous knowledge into consideration. In this way, the media would be creating content that humanises climate change news to fit local contexts.

Conclusion

This study set out to assess the role of CRs in strengthening climate action in the South West Region of Cameroon. It concludes by emphasising the critical role of community radio in addressing climate change challenges in Cameroon's South West Region. The findings illustrate that CRs have great potentials to bring climate change discussion closer to the people and mobilise localised solutions. Though climate reporting is seemingly complex with scientific terminologies, CRs can make use of local languages to communicate best practices. A participatory approach is critical for quality climate reporting. Also, the local councils can use CRs to provide a direct channel for representation and contribute to community resilience, thereby strengthening the role of CRs in official communications.

We recommend concerted efforts from the Ministry of Communication, the Cameroon Community Media Network, and other stakeholders to develop a community radio broadcast model that integrates climate action and adaptation into local governance to build capacities for resilient communities. The challenges faced by CRs are within reach, if the medium is well organised with solid representation from the community it serves.

References

Afutendem L. N. (2013). *Community Radio and Health Development in Rural Cameroon: An Impact Evaluation of COL-Cameroon Link Program in Lebialem South West Region.* Cameroon. Commonwealth of Learning. Retrieved October 15, 2024, from http://oasis.col.org/handle/11599/1909

Agali, R. A. (2023). Salience of community radio on climate change risks in rural Kenya: A case of Migori County. *African Social Science and Humanities Journal*, 4, 100–111.

Al-Hassan, S. & Abdul-Malik, A. (2011). The role of community radio in livelihood improvement: The case of Simli Radio. *The Journal of Field Actions Science Reports,* 5, 1–6.

Amungwa, F. A. (2019). Strategizing for Rural Development Stakeholders' Networking and Linkage Possibilities: Impacts and Recommendations, *Conference Paper Presented at the South West Regional Development Forum,* Limbe. November 14–15.

Angwah J. (2018). Stances in West African and Asian Climate Change Discourses: A Corpus-based Appraisal. Unpublished Ph.D., University of Yaounde I.

Atieno, L. & Njoroge, J. M. (2014). Climate change impact representation in Kenya's news media. *European Journal of Business and Social Sciences*, 3(8), 7–20.

Bangem Council (2012). Bangem Communal Development Plan (CDP). Bangem Council.

Besette, G. (2004). Involving the Community. A Guide to Participatory Development. International Development Centre, Canada/Malaysia: Southbound, Canada.

Buea Council (2012). *Buea Communal Development Plan (CDP): Elaborated with the Technical Support of GREMPCO.* Available at https://docslib.org/doc/7168703/buea-communal-development-plan-cdp

Chari, T. (2016). Rethinking climate change communication strategies in Africa: The case of indigenous media. *Indilinga African Journal of Indigenous Knowledge Systems*, 15(2), 217–232.

Creswell, J. & Clark, V. L. P. (2011). *Designing and Conducting Mixed Methods Research.* Sage, New Delhi.

Faisal, A. M. & Alhassan, A. (2018). Community access and participation in community radio broadcast: Case of Radio Gaakii, Ghana. *Journal of Development and Communication Studies*, *5*(2), 85–102.

Freire, P. (1996). *Pedagogy of the Oppressed.* Penguin Books, London.

Githethwa, N. (2010). *Milestones, Challenges and Proposals in the Development of Community Radio in Kenya.* A discussion paper presented at the AUF –ACDM meeting on February 21, 2010 as framework of discussions and a guide to further actions.

Harvey, B. (2011). Climate airwaves. Community radio, action research and advocacy for climate justice in Ghana. *International Journal of Communication,* 5(1), 2035–2058.

Hemer, O. & Tufte, T. (eds.) (2005). *Media and Global Change: Rethinking Communication for Development.* CLACSO and NORDICOM, Buenos Aires and Gothenburg.

Hoffman, G. (2021). New Report Reveals Media's Climate Alarmism Irks Readers. Independent Women's Forum. Retrieved from http://www.iwf.org

International Monetary Fund (IMF), African Department (2024). *Climate Change in Cameroon: Key Challenges and Reform Priorities.* International Monetary Fund. Available at www.elibrary.imf.org/view/journals/002/2024/052/article-A003-en.xml

Jallov, B. (2012). *Radio Empowerment: Voices of Building a Community.* Empower house, Denmark.

Kay N. & S. Gaymard (2021). Climate change in the Cameroonian press: An analysis of its representations. *Sage Journals,* 30(4). 417–433. https://doi.org./10.117096366252 0976013

Krejcie, R. V. & Morgan, D.W. (1970). Determining sample size for research activities. *Educational and Psychological Measurement*, 30(3), 607–610.

Manyozo, L. (2012). *People's Radio: Communicating Change across Africa.* Southbound, Penang.

Mbuli, C. S., Fonjong, L. N. & Fletcher, A. J. (2021). Climate change and small farmers' vulnerability to food insecurity in Cameroon. *Sustainability*, 13(3), 1523.

Mbungu, G., George, S., & Roberts, I. (2024). *Telling climate change stories: Perspectives from African Journalists*. Africa Policy Research Institute, Berlin, Germany. DOI: https// doi.org/10.59184/ccr024.02.

Mefalopulos, P. (2008). Development Communication Source Book: Broadening the boundaries of Communication. International Bank for Reconstruction and Development: The World Bank, Washington.

Ministry of Environment and Nature Protection and Sustainable Development of Cameroon (2015). National Climate Change Adaptation Plan. https://www4.unfccc.int/sites/NAPC/Documents/Parties/PNACC_Cameroun_VF_Valid%C3%A9e_24062015%20-%20FINAL.pdf

Molua, E. L. (2009). Accommodation of climate change in coastal areas of Cameroon: selection of household-level protection options. *Mitigation and Adaptation Strategies for Global Change*, 14, 721–735 https://doi.org/10.1007/s11027-009-9194-5

Muluh, H. Z. (2017). Trends in private radio practice in Cameroon. *African Journal of Social Sciences,* 8(7), 1–18.

Munji, C. A., Bele, M. Y., Nkwatoh, A. F., Idinoba, M. E., Somorin, O. A. & Sonwa, D. J. (2013). Vulnerability to coastal flooding and response strategies: the case of settlements in Cameroon mangrove forests. *Environmental Development*, 5, 54–72.

Myers, M. (2011). Voices from villages: Community radio in the developing world. *Center for International Media Assistance*, 3, 271–292.

Myers, M. & Harford N. (2020). Local Radio Stations in Africa: Sustainability or pragmatic Viability? Accessed on April 18, 2024, on www.cima.ned.org

Ngoran, H. A., Lambi, C. M. & Gur A. S. (2021). Agro pastoral adaptations to climate change in Bui Division, Northwest Region of Cameroon. *African Journal of Social Sciences,* 12(7), 101–114

Olurunnisola, A. A (2009). Community radio: Participatory communication in post-apartheid South Africa. *Journal of Radio Studies*, 5046(9:1), 126–145.

Retrespo-Estrada, S., & Fraser, C. (2003). Community radio for change and development. *Sage,* 6370(200212), 69–73.

Salormey, E. E., Dome, M. Z., Osse, L. & Logan, C. (2019). Change Ahead: Experience and Awareness of Climate Change in Africa. Afrobarometer Policy Paper, No. 60, Accra, Ghana.

Seretse, M., Chukwuere, J., Lubbe, S. & Klopper R. (2018). Problems around accessing information in rural communities in South Africa. *Alternation,* 25(1), 214–244.

Servaes, J. & Malikhao, P. (2005). Participatory Communication: The New Paradigm? *Media and Global change: Rethinking Communication for Development.* Edited by Oscar Hemer and Thomas Tufte. Sweden/Argentina. NORDICOM/CLASCO.91-103.

Servaes Shahzalal, M., & Hassan, A. (2019). Communicating Sustainability: Using Community media to influence rural people's intention to adopt sustainable behaviour. *Sustainability*, 11(3), 812. http://doi.org/103390/su11030812.

Sow, F. (2014). *Women's Community Radio in Africa: The Case of Gindiku FM in Senegal.* CODESRIA, Dakar.

SPRING (2018). *Community Media for Social and Behaviour Change: Using the Power of Participatory Storytelling to Improve Nutrition.* www.spring-nutrition.org/sites/default/files/publications/reports/spring_community_media_report.pdf

Stiftung, K. A. (2023). *Media Perceptions of Climate change in Sub-Saharan Africa.* Edited by Ruth Michaelson and Anja Berretta. Konrad Adenauer Foundation, Nairobi, Kenya. ISBN 978-3-98574-144-1.

Tiko Council (2012). *Tiko Communal Development Plan (CDP).* Tiko Council, Tiko, Cameroon.

Tume, S. J. P. (2021). Impact of climate change on domestic water accessibility in Bamenda III Sub-Division, North West region, Cameroon. *Journal of the Cameroon Academy of Sciences*, 17(2), 131–145.

Tume, S. J. P., Jumbam, M. S., Nsoseka, N. A., Nyarka, N. D., Yenla, L. J. & Njodzeka, N. G. (2018). Role of the media in climate change communication in the northwest region of Cameroon. *Handbook of Climate Change*

Communication: Vol. 2: Practice of Climate Change Communication, Springer, Cham, Switzerland, 47–60.

United Nations Environment Programme (n.d). *Convention on the conservation of migratory species of wild animals* (CMS). The Congo Basin and Climate Change. Available at www.cms.int/en/publications/factsheets

Waisbord, S. (2008). The Institutional Challenges of participatory communication in international aid. *Social Identities*, 14(4), 505–522.

Weber, Max (1864–1920*). The Protestant Ethic and the Spirit of Capitalism* translated by Talcott Parsons. Charles Scribner' Sons, New York, 1958.

World Association of Community Radio Broadcasters, AMARC (2007). Community radio social impact assessment: Removing barriers, increasing effectiveness. *The Democracy and Governance Network.* www.comminit.com/democracy-governance/content/community-radio-social-impact-assessment-removing-barriers-increasing-effectiveness (Accessed 10 January 2024).

World Bank Group (2022): Toward a people-centered green and resilient Cameroon. Available at www.worldbank.org

Yitamben, G. M. (2014). *Reviving the Power to Community: How Radio Rurale Femme de Mbalmayo in Cameroon Became a Catalyst for Equality and Democracy*. Zed Books Ltd, London.

10 Leveraging Indigenous Media Strategies in Climate Change Communication in Ghana

Akwasi Bosompem Boateng

Introduction

Communication and media play a key role in shaping people's perceptions of the environment and climate change (Cox 2013). Climate change has long-term effects on temperature and weather patterns (UNFCCC, 2007). Climate change includes the rise in temperature of the planet and its consequential effects on the climate of the Earth system, which includes the melting of glaziers, rising sea levels, dying cloud forests and scrambling wildlife and habitats (Schuurmans 2021; Cox 2013). Over the years, efforts such as the United Nations Framework Convention on Climate Change (UNFCCC) and the Conference of the Parties (COP-21) in Paris, among others, have been organized for collective global response and climate fight to adapt to the impact of climate change. These efforts have resulted in increased funding to developing countries to assist them in improving their response to climate change threats (Abbass et al. 2022). With increasing concerns about environmental and climate change, innovative communication and media strategies are needed to streamline and improve collective climate action. Indigenous communication and media strategies could reveal new opportunities and possibilities for environmental stakeholders, such as government agencies, to engage the public in addressing issues related to climate change. The quest and search for indigenous knowledge, communication and media strategies that are relatable to people have become a global trend, as their application in climate change communication has increased in popularity in recent years. Indigenous knowledge, communication and media strategies are becoming important complementary options to traditional media. Environmental stakeholders, including civil society organizations and particularly environmental agencies, are seeking indigenous media and communication strategies as important ways to reach larger populations, especially rural and local communities with messages about climate change.

As Cox (2013) noted, issues related to the environment and climate change have been communicated through various channels and messages to address risks and challenges. Other scholars (e.g. Odonkor, Dei & Sallar 2020; Jima 2022; Chari 2016) who also investigated the impact of climate change and action posit that indigenous media strategies can contribute to the promotion of climate change

DOI: 10.4324/9781003660217-13

awareness and information on adaptation and mitigation, among others. This is especially true in the sense that the public and audiences are now looking for more engaging and culturally sensitive communication activities that promote participation rather than traditional mass media channels that passively disseminate climate change information to the public (Moser & Dilling 2012). For this reason, environmental agencies are compelled to explore new approaches and ways to communicate about climate change to address expectations. Indigenous communication and media strategies and practices can assist in the promotion of awareness and sustaining relationships with the public and stakeholders for climate action.

Issues related to climate change have contributed to debates and research on the global environmental crisis. However, there are many studies on climate change in the Western context (e.g. Schuurmans 2021; Davidson 2016), and others largely focus on adaptation, mitigation and variability (Odoom & Fosu 2023; Fosu, Quashigah & Kuranchie 2019). Other studies have focused on communication about climate change through traditional media, including radio, television and newspapers (Hussain, Bajwa & Bukhari 2022; Odoom 2024). Research related to the use of indigenous knowledge and media strategies in communicating about climate change, especially in Ghana, is limited. The few studies on climate change in Ghana have not given adequate attention to culturally relevant dimensions and indigenous media strategies in climate change policy and communication activities of the Environmental Protection Agency (EPA) for climate action. This study attempts to address this gap by investigating how the EPA in Ghana utilizes indigenous media strategies to communicate about climate change and facilitate adaptation, mitigation and collective climate action. In doing so, this study argues that incorporating indigenous and culturally relevant media and communication strategies, for example, folksongs, drama, theatre, poetry and storytelling, can enhance relationships and understanding of climate change information and messages to people, especially individuals in rural and local communities.

This study makes a key contribution to the climate change literature and extends indigenous media strategies by examining the communication aspect of the subject, which is largely understudied in the context of the Global South, particularly Africa and Ghana. With the growing quest for the use of indigenous knowledge and media in participatory engagements, as well as scholarly interest in climate change communication, this study makes empirical contributions by collecting primary data from the EPA, a lead agency of the state responsible for addressing climate change and action, in the underresearched and unique situation of Ghana. Following the justification for this research, the study organizes the article as follows: literature review, including theoretical underpinnings; research questions; methodology; findings and data analysis; discussion and implications; and conclusion. Limitations and suggestions for further studies are also provided to encourage more research in this area of media and communication. To address the issues and gaps identified in the background, this study attempts to answer the following research questions.

Research Questions

- How does the EPA communicate about climate change in Ghana?
- What indigenous media strategies does the EPA use to communicate about climate change in Ghana?
- What are the policy implications of indigenous media strategies in climate change communication?

Literature Review

Climate Change Communication in Sub-Saharan Africa

The United Nations Framework Communication on Climate Change (2007) posits climate change as a climatic crisis attributed directly or indirectly to the activities of humans, which alters the composition and condition of the global atmosphere, in addition to the natural variability of the climate over comparable periods of time. Climate change communication is the process of educating, informing and warning, as well as persuading, mobilizing and solving critical problems about the environment (Cox 2013; Chari 2016). These explanations have implications for debates, policy decisions and responses, especially communication for awareness promotion and participation in climate action, mitigation and adaptation (Odoom & Fosu 2023; Fosu et al. 2019). Despite the significance of the issues related to climate change, there is a lacuna in climate change communication and research in sub-Saharan Africa, which has generated concerns about public awareness and sensitization of climate change for inclusive, engaging and participatory action to address the crisis (Ifeanyi-Obi et al., 2017; McGahey & Lumosi 2018). These discussions have intensified arguments and search for diverse ways of promoting the understanding and dissemination of technical climate change information via innovative and inclusive indigenous strategies that consider increased public knowledge and promote broader dialogue beyond social, cultural and political differences to change attitudes and behaviours towards climate action (Kasuli 2022; McGahey & Lumosi 2018).

The situation is not different from that in Ghana, where there have been increasing calls for new, inclusive and participatory approaches to climate change communication because local and rural communities have been affected by activities related to the environment, including mining, agriculture and oil exploration, among others, with concomitant effects on the lives of the people and the climate (Odoom & Fosu 2023). The low level of knowledge and understanding of climate change by many people make communicating its urgency for action challenging. Considering the participatory features of indigenous media and communication, they could serve as inclusive and useful communication strategies for leveraging broader engagement and participation in climate change action to address environmental issues and promote public awareness of the crisis. According to scholars (Jima 2022; Chari 2016; Asante, Appau & Boadu 2017), indigenous media strategies in climate change communication not only preserve traditional knowledge and practices, but also enhance the participation and engagement of stakeholders and communities. Indigenous communication and media strategies facilitate adaptation and resilience to climate change by relating information and messages in culturally relevant and sensitive ways to influence action, behaviours and attitudes (Nyadzi, Ajayi & Ludwig 2021; Sereenonchai & Arunrat 2020). This enables the issues of climate change to be told and retold in engaging and participatory ways for meaningful actions to eliminate the barriers to climate action (Awuah-Nyamekye 2013; Stoknes 2015).

There are various African indigenous communication and media strategies, which include the use of indigenous language; oral traditions and storytelling; traditional music and dance; community gatherings, including town hall meetings; folk media; plays; proverbs; poetry; puppet shows; riddles; proverbs; theatre; community radio and other mobile phone-based communication, among others (Mda 1993; Mushengyezi 2003; Asimeng-Boahene 2009; Abdi 2010; Nigussie 2016). The usefulness, inclusive and participatory features of indigenous communication support the argument of this chapter for the incorporation of indigenous media strategies in climate change communication initiatives and for broader acceptance and participation to address the global climate crisis. Despite the opportunities for the use of indigenous media and communication in addressing climate change issues, there are concomitant challenges. Indigenous media and communication strategies have significant influence because they integrate traditional knowledge into policies regarding climate, in addition to their ability to use it in specific climate situations and contexts. Indigenous climate change communication and media strategies empower communities and people to take action through local cultures and languages, which enhance climate resilience, mitigation and adaptation. However, there could be issues related to language barriers, such as limited access to climate information, limited financial resources, social and cultural vulnerabilities, and competing priorities (Jima 2022; Odoom & Fosu 2023; Moser 2010).

Theoretical Framework

The participatory communication theory underpins this study. Participatory communication theory refers to a communication approach that is based on dialogue, hence allowing the sharing of information, opinions and perceptions among stakeholders on particular issues (Pringle & Conway 2012; Tufte & Mefalopulos 2009). It involves the exchange of experiences and information, as well as the exploration of knowledge and result-oriented ways to address situations, such as climate change, that need to be addressed by the world. This begs more inclusive and participatory engagements in communication about climate change. This makes the participatory communication framework relevant to this study of indigenous strategies in addressing climate change issues. The participatory communication theory is applicable to local, regional, national and international policies; initiatives; campaigns; and information related to issues that affect diverse groups of people. As Pringle and Conway (2012) suggest, organizations that deliver their activities and programmes via participatory communication and engagements are able to engage stakeholders and seek wider consensus on issues, which offers the public and stakeholders a sense of ownership and belonging to contribute to environmental and climate sustainability.

The appreciation of climate change issues revolves around understanding and participating in collective action (McGahey & Lumosi 2018). Collective climate action can largely occur when the public and stakeholders relate and understand messages from environmental agencies through participatory approaches. The attitudes and behaviours of individuals towards the climate can be influenced if information is clearly presented through local language and indigenous media strategies to many people to increase awareness and participation in climate action. The promotion of participatory communication by the EPA through indigenous media strategies strengthens the capacity of people and communities to act responsibly towards the environment. This is especially true when the messages in the indigenous media effectively communicate the implications of changes in climate conditions through local cultural values and norms. By using the participatory communication framework to support this study, the chapter is able to argue for the leverage of indigenous media and communication strategies to complement the traditional popular exogenous media in extending and improving communication about climate change. The theory helps the study explain how the EPA is taking advantage of the inclusive and participatory nature of indigenous communication to promote climate action and environmental sustainability.

Methodology

This research adopted a qualitative approach and collected data from interviews and document analysis to investigate how the EPA leverages indigenous media strategies to communicate about climate change in Ghana. The in-depth case study interviews allowed this study to critically examine the experiences and perspectives of personnel at the EPA with respect to their use of indigenous media strategies to communicate about climate change to gather qualitative data for analysis (Cleland 2017; Denzin & Lincoln 2011). The EPA is sampled for the case study because it is the lead agency under the Ministry of Environment, Science, Technology and Innovation that is responsible for the coordination of Ghana's efforts and actions to address climate change. The participants from the agency were involved in climate change initiatives, including communication programmes and activities. The study interviewed eight personnel of the environmental agency who had worked with some experience of communication and media activities to improve the knowledge, awareness, attitudes and behaviours of people towards climate change and action, especially its mitigation and adaptation, among others. The participants included two personnel at the national level, two personnel at the regional level and four at the district level, who were interviewed extensively in their offices. The study discussed with participants the subject of climate change, indigenous communication and media strategies, and policy implications, among others, in the interviews. The recorded interviews were conducted between June 2024 and August 2024, each lasting between 45–60 minutes, and were transcribed by the study for discussion and analysis under recurring themes to address the research questions.

The study received the consent of the participants for the interviews to be conducted, recorded and transcribed. The semi-structured interviews ensured consistency and flexibility, which gave participants adequate time to provide their detailed perspectives on how the EPA communicates about climate change at various levels through different media strategies to the public. This study also analysed documents as a secondary data collection method to complement the in-depth interviews. Document analysis entails the systematic review and analysis of written texts, for example, minutes of meetings, reports, policies and guidelines (Dalglish, Khalid & McMahon 2020; O'Leary 2014; Bowen 2009). Through document analysis, this study conducted an in-depth and critical examination of Ghana's National Climate Change Policy (NCCP) in 2013, which is an important foundational policy document that guides Ghana's climate change efforts and actions of the EPA. This provides valuable insights into the communication policies and strategies that the agency employs to promote the inclusion, capacity building and participation of stakeholders in climate-resilient action.

The study used the thematic framework of Caulfield (2023) and Braun and Clarke (2006) to analyse the qualitative data from in-depth interviews and national climate change policy documents to identify the patterns and recurring themes for the discussion of the findings with respect to the research questions, literature and theory. The six-step approach of Braun and Clarke (2006) was followed to analyse the data collected from interviewees with personnel and climate change policy documents that guide communication about climate change. This involved becoming familiar with the data, generating initial codes, searching for themes, reviewing themes, defining themes and writing the report (Caulfield 2023; Braun & Clarke 2006). The study read through interview transcripts and national climate policy documents on several occasions for detailed insight into data from different data collection sources for coding. These codes were clustered into key themes for the discussion of findings and issues.

The following section provides details of the study's findings with respect to the research aim and questions that it sought to address. The findings and discussion are conducted under themes and subthemes to make the responses from the participants and critical examination of Ghana's climate change policy documents more meaningful and understandable. For example, the findings on the first research question are thematized as follows: communicating about climate change, exogenous media and national communication report, while data regarding the second question are analysed under themes such as folk music and songs, drama and theatre performances, folklore, proverbs and poetry, and considerations in indigenous media integration;

and finally, policy implications and responses are presented as the themes under which the discussion with respect to the third research question.

Findings and Discussion

This section presents the findings and analysis of the study. The data collected are discussed with respect to the research questions, which are analysed under key themes to provide an understanding of communication about climate change by the EPA to the public for climate action.

How Does the Environmental Protection Agency (EPA) Communicate About Climate Change in Ghana?

Communicating About Climate Change

The EPA is a government institution in Ghana that has made various interventions to address climate change for climate action, adaptation and mitigation. The national climate change policy suggests the need for innovative communication and media strategies to educate and promote extensive awareness among the Ghanaian population. According to the participants, the agency's climate change communication initiatives and activities are guided by international and local frameworks. This is because of Ghana's commitment to and contribution to global climate action, which is a sign of several international conventions and campaigns. This signifies Ghana's efforts in the global campaign and action to address climate change. Therefore, in its climate change communication and awareness initiatives, the agency has relied on policies such as the Ghana NCCP of 2013, which was developed in line with the UNFCCC, and important protocols of the Intergovernmental Panel in Climate Change (IPCC). In other words, communicating about climate change by an EPA is underpinned by global frameworks. With respect to how this information informs communication about climate change, participant 3 noted the following:

> Ghana's commitments to international protocols compel our agency to ensure that climate change communication initiatives are consistent with the country's national climate change policy and the United Nations Framework Convention on Climate Change, among other goals, for the country to follow global trends in addressing climate change.

The use of international guidelines by the EPA to communicate about climate change influences the type of media and communication strategies and messages to diffuse to the public. Given that these guidelines were developed in Western contexts that prioritized exogenous media strategies, their application in climate change communication in the Ghanaian context relegates culturally relevant and indigenous media strategies to the background. This means that the potential of indigenous knowledge and communication has not been harnessed to transform climate change communication.

Exogenous Media

The responses of the study participants revealed that the EPA has largely concentrated on exogenous media and communication strategies, which have been challenged by inadequate infrastructure and a lack of access. This prevents many people, especially those in local and rural communities, from accessing exogenous media through mediated technologies, including television, newspapers, radio and internet-supported platforms such as social media. The high level of illiteracy among people discourages them from relying on exogenous communication channels and media avenues for discussing the socio-economic, cultural, political, religious and environmental impacts and consequences of a global crisis such as climate change. Hence, the focus on traditional exogenous initiatives and mass media could limit climate change communication efforts, particularly in local and rural communities. Respondent 6 explained how they used exogenous avenues to communicate about climate change:

> The agency organizes several climate change initiatives as part of Ghana's Action for Climate Empowerment (ACE). These include the promotion of awareness, adaptation and mitigation of the effects of climate change via conventional mass media, such as television, radio, newspaper and social media platforms, to communicate largely about climate change to citizens. In addition to the conventional media, we use national communication reports to inform the public and stakeholders about progress and challenges to our efforts in climate action.

From the response, it is evident that since the agency aims at reaching larger populations, it largely relies on traditional media to passively disseminate climate change information and messages to the public. The findings indicate the prioritization of exogenous and conventional mass media over indigenous communication and media strategies, which were briefly mentioned by some participants during the in-depth case study interviews.

National Communication Report

The EPA communicates about change through National Communication Reports, which also relegate indigenous media strategies to the backdrop of climate change communication. This was evident in Ghana's Fourth National Communication Report published by the EPA, which was cited by participants to elaborate how the agency promotes access to information to climate change information and messages, public participation, and awareness and communication. According to participants, the agency has made efforts to promote access to public information on climate change and crisis. For example, participant 1 stated:

> We have communicated and raised awareness about climate using our various Climate Change Sharing Platforms; Climate Change Data Hub; National

> Drawing Contest on Climate Change; Installing Automatic Weather Stations in Senior High Schools; National and African Climate Change Week; REDD Eye Campaigns; Climate Change Summit; Renewable Energy Fair; Climate Change and Population Conference in Africa; Climate Change Dialogue with Political Parties and Parliamentarians; and National Climate Change and Green Economy Learning Strategy, among others. Activities in these initiatives are highlighted in our agency's fourth national communication report.

The statements above reiterate the over reliance on exogenous mass media strategies by the agency, which are not situated in the cultures of Ghana, relegating indigenous communication and media to the backdrop of communication efforts and strategies in addressing climate change. The participants mentioned the use of the agency's social media platforms, especially websites, to report climate news, where the agency shares the latest information about climate change and its impact on human lives. This climate news is linked to the agency's *Climate Change Data Hub*, where important climate change policy documents and reports are uploaded for public access. The exclusion of indigenous media and communication strategies in the national communication report confirms the assertions of Moser and Dilling (2012) and Chari (2016) that governments and environmental agencies have given more attention to mass media and stipulated exogenous guidelines in climate change communication, which has not adequately considered indigenous strategies and local participation in climate action. This calls for rethinking and reassessment of conventional climate change communication approaches to embrace indigenous media strategies for inclusive participation, especially local and rural populations, in climate action.

What Indigenous Media Strategies Does the Environmental Protection Agency (EPA) Employ to Communicate About Climate Change in Ghana?

Folk Music and Songs

The participants mentioned folk music and dance as some of the indigenous ways that the EPA communicates the risks associated with climate change to the public. The information about climate change, including repercussions and obligations towards addressing the crisis, is expressed in songs and performed with dance to make the issues culturally relatable and engaging in shaping the attitudes and behaviours of the public towards climate action. The agency's use of folk music and songs gives credence to Chari (2016) and Awuah-Nyamekye's (2013) postulation that singing and dancing are important social and cultural events that could be performed at gatherings of people to inform and educate them about critical issues, including climate change. These can be performed at festivals, funerals and marriages, among others, to engage and share information on climate change issues. For example, participant 8 elaborated as follows:

> In some situations, we have used cultural music and songs to provide entertainment, education and information to individuals who are not aware of issues such as climate change. We use indigenous Ghanaian music, including cultural,

> dance drums and bell ensembles, to perform and tell climate change stories at events to educate and sensitize people about climate change risks. This draws many people to climate change information and messages.

The potential of folk music and songs as indigenous communication lies in their ability to create and recreate verses through narration and dialogue. The EPA engages musicians to compose songs to narrate the dangers of the climate change crisis to change the behaviours and attitudes of people in two ways: educating the public about climate change issues and reforming individuals who act irresponsibly towards the environment.

Drama and Theatre Performance

The EPA communicates about climate change through dramatic performance, which has enormous potential for changing behaviours and attitudes towards the environment and climate change. As Mda (1993) and Chari (2016) contend, it is easier to communicate social issues, the environment and climate change in an interesting manner through theatre and dramatic performances. The participants indicated that the agency uses theatrical performances to dramatize issues, causes and effects of climate change to the public. The participatory nature of theatre and drama enables communities and audiences to be engaged in communication performances (Mda 1993; Chari 2016). This allows issues to be addressed because it gives individuals the opportunity to express their views about climate change rather than relying solely on technical experts to talk about the risks to the public. The following is how participant 7 described their use:

> The agency promotes climate change information, especially regarding mitigation and adaptation through play and other performances, where actors, audiences and agencies hold discussions on climate change to brainstorm and promote participation in climate action and efforts to address the impact of the changing climate. In some cases, we use travelling theatre in rural areas to educate people about the effects of climate change.

The response gives credence to Chari's (2016) study that highlights how the use of indigenous dramatic and theatrical performances can help environmental agencies connect with rural communities and local people because they relate with them, which encourages them to participate. This is because members of local communities become performers rather than observers of performances, which makes such indigenous media and communication strategies effective due to their participatory features (Mda 1993; Mushengyezi 2003).

Folklore, Proverbs and Poetry

The EPA promotes climate change communication through popular beliefs, customs, myths, proverbs and storytelling, which relate to particular activities and places, groups of individuals and activities related to the environment and climate

(Asimeng-Boahene 2009; Asante et al. 2017). The participants indicated proverbs as indigenous folk media that the agency used to communicate about climate change. This was done in the form of short folk sentences, which entail wisdom, truth, morals and traditional views in metaphorical, fixed and memorable sayings that are widely known and passed on from generation to generation, which scholars, such as Awa (1995), Mushengyezi (2003) and Awuah-Nyamekye (2013), posited as useful in communication with local communities and indigenous people. The usefulness of proverbs is described by Asimeng-Boahene (2009:6) as a "treasured lubricant of expressions". In highlighting how proverbs as part of Ghanaian cultures serve as credible sources of climate change communication, this is what respondent 6 reported: "In communicating about climate change, we sometimes use indigenous proverbs in local languages to conscientize people about the environment. Such proverbs address factors that contribute to climate change and its consequences using wise sayings". This assertion highlights Asante et al.'s (2017) notion of power and influences culturally centred expressions in the creation of awareness of climate change issues.

In addition to proverbs, participants highlighted poetry as another form of indigenous media strategy that the agency used to address climate change issues. Like proverbs, poetry is used in praising, criticizing or emphasizing something or an issue (Nigussie 2016). According to the participants, the EPA uses oral poetry to communicate information about different climate change issues and topics. For example, participant 5 stated the following:

> When the agency communicates and engages with communities and local people, we use Ghanaian poetry related to the environment and climate change in some cases, which consists of vocal compositions recalled by traditional poets related to what people know about the past to the present. These are usually composed and performed on the basis of the gatherings, programmes and events we organize. The situation of poets in face-to-face communication with audiences makes climate change communication traditional and local for easy understanding, appreciation, access and consumption of messages.

This finding validates the notion of Abdi (2010) that information sent through poems, riddles or verbal puzzles is more powerful than other forms of indigenous oral communication. This study supports the argument that the value of poetry and poems to social change lies in their capacity to provide artistic embodiments and representations to generate new feelings, thoughts, ideas and ways of being and seeing, especially in changing the belief structures and feelings of individuals and groups (Nigussie 2016; Asimeng-Boahene 2009). This is especially true because the language, including idiomatic expressions used in poetry, is congruent with the worldviews and knowledge of rural communities and local people which makes communication easier through them. Although the respondents were committed to the use of indigenous media strategies in communicating about climate change, there was no specific mention or statement in Ghana's NCCP in 2013 and Fourth National Communication Report (2020). These documents emphasize the use of exogenous mass media such as radio and TV interviews, jingles, documentaries, community durbars, information vans and the dissemination of education materials to communicate about climate change to the public. The EPA has not adequately utilized these useful tools for the indigenization and localization of climate change communication and action (Asante et al. 2017). These factors limit inclusion and local participation in climate change campaign actions.

Considerations in Indigenous Media Integration

Scholars, including Chari (2016) and Moser and Dilling (2012), suggest the importance of shifting the African perspectives of climate change communication and concomitant issues associated with information dissemination via exogenous sources. This can encourage and empower people by exploring the inherent powers of the local and indigenous features of communities to ensure inclusion and participation in climate change discussions and action. Through indigenous elements, for example, cultural orientation, language translation and interpretation of climate change communication into local and relatable settings are necessary to draw stakeholders', communities' and people's attention to public participation in responsible acts towards the environment. One participant in the interviews highlighted that the agency continues to make efforts to integrate indigenous elements and content in climate change communication initiatives. This prevailed in the response of participant 5, which was as follows:

> As part of integrating indigenous media strategies in the agency's climate change initiatives, programmes and activities to make them more accessible to many communities and people as possible, we sometimes interpret and translate some important information and documents into widely spoken Ghanaian languages so that many indigenous and local people across the country can understand and appreciate the communication messages to act responsibly as environmental citizens and protect the climate, mitigate and adapt to the changing climatic conditions and weather patterns.

Although the response indicates some efforts towards climate change communication, there was no specific mention of the national climate change policy or reports on how indigenous media and communication strategies should be incorporated into programmes and activities. This calls for further consideration of the use of culturally relevant and indigenous approaches in climate change communication in Ghana.

What Are the Policy Implication(s) of Indigenous Media Strategies in Communicating About Climate Change in Ghana?

Policy Implications

The climate change communication efforts are situated in national policy documents, particularly Ghana's NCCP from 2013, which serves as the framework for the EPA in communication about climate change. Ghana's NCCP highlights the importance of climate change communication and raising awareness. The climate change policy guides the EPA in exploring broader and encompassing transmission and exchange of information and discussion, dialogue and debates about climate change. The EPA is responsible for Ghana's national communication reports in line with the United Nations Convention on Climate Change, which highlights climate change policies, action and progress. The agency has promoted capacity-building initiatives, including climate change learning, that aim to improve stakeholders' capacity to address critical issues-related climate change.

As the foundational policy document, Ghana's 2013 NCCP guides communication efforts and activities related to climate change. The policy document highlights the importance of indigenous knowledge and communication in addressing climate change issues. From Ghana's NCCP, it is evident that while the document does not explicitly state the particular role of indigenous media and communication, it highlights the need for the integration of traditional knowledge with technical and current knowledge in addressing climate change. The policy document recommends the establishment of a climate change centre that can focus on indigenous knowledge and localized communication about changing climatic conditions.

Policy Response

Ghana's NCCP in 2013 highlights priority areas, such as disaster preparedness and response, agriculture and food security, and natural resource management, which may entail indigenous media strategies to communicate about climate change as a critical issue. This implies that not only policymakers but also organizations and agencies that are involved in climate change communication must explore nuanced, innovative and engaging approaches to sharing knowledge about risks, adaptation and mitigation. The incorporation of indigenous and culturally relevant media strategies, such as folksongs and storytelling, among others, can enhance understanding of climate change information and messages from environmental agencies in relatable and participatory communication, especially among rural and local populations (Odonkor et al., 2020).

The lack of explicit mention and framework relating to indigenous media in climate change action and strategies in Ghana's NCCP implies the need for environmental authorities and climate change agencies to explore result-oriented dialogue and innovative indigenous media strategies in policymaking for broader engagements in climate change action. This should involve indigenous communities in the development, production and communication of messages about climate change by responsible authorities and agencies. In addition, it is important for policymakers and the EPA to recognize the importance of indigenous knowledge, cultural protocols and culturally relevant messages and information for the communication, inclusion and participation of people in environmental and climate action.

Conclusion

This study highlights the potential of indigenous media and communication strategies in addressing climate change in the context of Ghana. Considering the minimal use of indigenous media strategies in climate change communication by environmental agencies, this chapter argues for their incorporation to complement exogenous media for broader and effective communication and participation in climate action. By leveraging indigenous media strategies and focusing on the particular information needs and conditions of climate change in Ghana, the EPA can enhance climate change communication outcomes via culturally relevant and contextually sensitive approaches and information to improve awareness, mitigation, adaptation and climate action in Ghana. Although evidence suggests that the agency uses indigenous media strategies, for example, folk music and dance, proverbs and poetry, and theatre and dramatic performances to communicate about climate change, its deployment is still inadequate compared with how it uses traditional exogenous channels and platforms. This calls for collaborative effects between policymakers, environmental protection agencies, stakeholders and the public to ensure equitable and effective deployment of indigenous media strategies in climate change communication. Indigenous media and communication strategies are result-oriented because they resonate with cultures and communities, especially local and rural areas. These factors facilitate indigenous participatory approaches and effectiveness in changing people's attitudes and behaviours towards climate change to sustain the environment.

Despite the potential and opportunities that indigenous media and communication strategies present, there are issues that need to be addressed. These include logistical issues and the need for a robust policy framework to integrate indigenous media strategies in climate change communication and initiatives in Ghana. By acknowledging and addressing these issues, Ghana, especially the EPA can make significant strides towards achieving its climate change communication goals, ensuring that indigenous communication and media strategies serve as important avenues, tools and channels for enhancing and addressing climate change and its concomitant issues. The popular exogenous media and communication strategies widely used have peculiar issues related to ethical, illiteracy and language barriers, which can impede communication about climate change in Ghana. The EPA must integrate traditional and indigenous knowledge into climate change communication and consider the development of context-specific climate issues. This includes the use of more local and indigenous Ghanaian languages and media to promote understanding of the risks and dangers associated with climate change. The EPA should support community-led climate action and initiatives in climate change communication, which can make messages more sensitive and relatable to individual experiences. This has to be coupled with the provision of support to indigenous communication and media infrastructure, such as indigenous language programmes, especially on community radio stations, among others, to communicate the risks of climate change to local and rural areas.

References

Abbass, K., Qasim, M., Z., Song, H., Murshed, M. & Mahmood, H. 2022, A review of the global climate change impact, adaptation, and sustainable mitigation measures. *Environmental Science and Pollution Research 29*, 42539–42559.

Abdi, A., A, 2010, Clash of oralities and textualities: The colonization of the communicative space in Sub-Saharan Africa. In: Kapoor, D. & Shizha, E. (Eds.), *Indigenous knowledge and learning in Asia/Pacific and Africa* (pp. 147–164). New York: Palgrave Macmillan.

Asante, E., Appau, A., S. & Boadu, K., B. 2017, The use of indigenous cultural practices by the Ashantis for the conservation of forests in Ghana. *Sage Open 7*(1), 1–7.

Asimeng-Boahene, L., 2009, Educational wisdom of African oral literature: African proverbs as vehicles for enhancing critical thinking skills in social studies education. *International Journal of Pedagogies and Learning 5*(3), 59–69.

Awa, N., 1995, The role of indigenous media in African social development. In: Nwosu, P., Onwumechili, C. & M'Bayo, R. (Eds.), *Communication and the transformation of society: A developing region's perspective* (237–251). Maryland, USA: University Press of America.

Awuah-Nyamekye, S., 2013, *Managing the environmental crisis in Ghana: The role of African traditional religion and culture—A case study of Berekum traditional area.* Retrieved from http://etheses.whiterose.ac.uk/id/eprint/5780

Bowen, G., A., 2009, Document analysis as qualitative research method. *Qualitative Research Journal 9*(2), 27–40.

Braun, V. & Clarke, V. 2006, Using thematic analysis in psychology. *Qualitative Research in Psychology 3*(2), 77–101.

Caulfield, J., 2023, *How to do thematic analysis: Step-by-step guide and examples*. Retrieved on 2 February 2024 from http://scribbr.com/methodology/thematicanalysis/

Chari, T., 2016, Rethinking climate change communication strategies in Africa: The case for indigenous media. *Indilinga-African Journal of Indigenous Knowledge Systems 15*(2), 217–232.

Cleland, J., A., 2017, The qualitative orientation in medical education research. *Korean Journal of Medical Education 29*(2), 61–71.

Cox, R., 2013, *Environmental communication and the public sphere*. Thousand Oaks, CA: Sage.

Dalglish, S., L., Khalid, H. & McMahon, S., A., 2020, Document analysis in health policy research: The READ approach. *Health Policy and Planning 35*(10), 1424–1431.

Davidson, D. 2016, Gaps in agricultural climate adaptation research. *National Climate Change 6*(5), 433–435.

Denzin, N., K. & Lincoln, Y., S., 2011, *Handbook of qualitative research*. Thousand Oaks, CA: Sage.

Environmental Protection Agency, 2020, *Ghana's Fourth National Communication to the United Nation's Framework Convention on Climate Change*. Retrieved on 25 June 2024 from www.epa.gov.gh

Fosu, M. Quashigah, T. & Kuranchie, P. 2019, Broaching agenda for climate change in Africa: A perspective on media engagement with climate issues in Ghana. In: Pinto, J., Gutsche, R., E. & Prado, J., P. (Eds.), *Climate change, media and culture: Critical issues in global environmental communication* (93–111). Leeds, England: Emerald Publishing Limited.

Hussain, M., Bajwa, A., M. & Bukhari, S., F. 2022, Climate change and university students: Traditional media, new media, and interpersonal communication sought for obtaining informational gratifications. *Journal of Journalism Media Science & Creative Arts 3*(1), 82–100.

Ifeanyi-Obi, C., C., Togun, A., O., Lambell, R., Adesope, O., M. & Arokoyu, S., B. 2017, Challenges faced by cocoyam farmers in adapting to climate change in Southeast Nigeria. *Climate Risk Management 17*, 155–164.

Intergovernmental Panel on Climate Change, 2019, Emission factor database. Retrieved on 11 June 2024 from www.ipcc-nggip.iges.or.jp/EFDB/main.php

Jima, A., O. 2022, Significance and restraint of indigenous of indigenous knowledge inclusion in Ethiopian higher education curriculum: In focus Gadaa system. *Cogent Education 9*(1), 1–16.

Kasuli, M., F., 2022, *The role of climate change communication on the adoption of climate-smart agriculture by smallholder farmers in the Northern Region of Ghana.* MPhil thesis: University of Development Studies, Tamale.

McGahey, D., G. & Lumosi, C., K. 2018, Climate change communication for adaptation: Mapping communication pathways in semi-arid regions to identify research priorities. *Journal of Sustainable Development in Africa 20*(1), 87–107.

Mda, Z. 1993, *When people play people: Development communication through theatre.* London: Zed Books.

Moser, S., C., 2010, Communicating climate change: History, challenges, process and future directions. *Wiley Interdisciplinary Reviews: Climate Change 1*(1), 31–53.

Moser, S., C. & Dilling, L., 2012, Communicating climate change: Closing the science-action gap. In R. B. Norgaard & Scholberg, D. (Eds.), *Oxford handbook of climate change and society* (pp. 161–174), Oxford: Oxford University Press.

Mushengyezi, A., 2003, Rethinking indigenous media: Rituals, 'talking' drums and orality as forms of public communication in Uganda. *Journal of African Cultural Studies 16*(1), 107–117.

National Climate Change Policy (NCCP), 2020, *Ghana's fourth national communication to the United Nation's Framework Convention on Climate Change.* Accra, Ghana: Environmental Protection Agency. https://unfccc.int/sites/default/files/resource/Gh_NC4.pdf.

Nigussie, H. 2016, Indigenous communication forms and their potential to convey food security messages in rural Ethiopia. *Indian Journal of Human Development 10*(3), 414–427.

Nyadzi, E., Ajayi, O., C. & Ludwig, F. 2021, Indigenous knowledge and climate change adaptation in Africa: a systematic review. *CAB Reviews: Perspectives in Agriculture, Veterinary Science, Nutrition and Natural Resources.* https://doi.org/10.1079/PAVSNNR202116029.

Odonkor, S., T., Dei, E., N. & Sallar, A., M., 2020, Knowledge, attitude and adaptation to climate change in Ghana. *The Scientific World Journal 1*, 1–9.

Odoom, D. 2024, Public perception on the level of effectiveness of climate change communication channels in Ghana: A cross-sectional survey. *African Journal of Empirical Research 5*(1), 24–35.

Odoom, D. & Fosu, M. 2023, Towards adaptation in Africa: A Ghanaian perspective. *Journal of Policy and Development Studies 14*(1), 156–174.

O'Leary, Z., 2014, *The essential guide to doing your research* (2nd Ed.). Thousand Oaks, CA: Sage.

Pringle, P. & Conway, D., 2012, Voices from the frontline: The role of community-generated information in delivering climate adaptation and development objectives at project level. *Climate and Development 4*, 104–113.

Schuurmans, C., 2021, The world heat budget: Expected changes. In: Eisma, D. (Ed.), *Climate Change* (1–15), Florida, USA: CRC Press.

Sereenonchai, S. & Arunrat, N., 2019, Practical agricultural communication: Incorporating scientific and indigenous knowledge for climate mitigation. *Kasetsart Journal of Social Sciences 41*(1), 60–67.

Stoknes, P., E., 2015, *What we think about when we try not to think about global warming: Toward a new psychology of climate action*. Vermont, USA: Chelsea Green Publishing.

Tufte, T. & Mefalopulos, P., 2009, *Participatory Communication: A Practical Guide*. World Bank Publications - Books, Number 5940, April. Washington DC: The World Bank Group.

UNFCCC, 2007, *Climate change: Impacts, vulnerabilities and adaptation in developing countries*. Bonn, Germany: United Nations Framework Convention on Climate Change.

11 Effective Communication on Climate Change in Africa

Global Noise, Local Expression

Jakub Siwak

Introduction

"Effective and appropriate communication has been identified as an essential step" in relation to affecting acceptance, understanding and behavioural change amongst populations in relation to climate change (McGahey and Lumosi 2018). This work agrees with the assertion that effective and appropriate communication is essential in the above regard but argues that the term (and discourse surrounding) 'climate change' is applied in a global, decontextualized sense, and in being applied this way has little efficacy in terms of said acceptance, understanding and behavioural change. Indeed, as within any other localized geographical context, in Africa, the reception of a global message (climate change) that comes across as ethereal and vague, and that further, is politicized, often polarizing, and furthermore (and most importantly) often cast strongly as Western-centric, will struggle to find purchase.[1]

As a counter to this, this study wishes to – by way of two historical examples – demonstrate how communication of environmentally oriented concern that is rooted in local context and that speaks to markers of local identity is a far more effective and appropriate mode of communication. In service of this, the study turns first to the work of Yves-Valentin Mudimbe, who, in his Foucauldian overview of Christianity in Africa, demonstrates how communication is a dynamic process in which a system of thinking associated with a particular set of ideas (in this case religious doctrine wedded to an exploitative politico-economic and socio-cultural system) can be reappropriated in service of the needs of local communities. This sets the scene for a discussion on a communication that was explicitly environmental in orientation. That is, and second, the study wishes to demonstrate how Rachel Carson's *Silent Spring*, published in 1962 – a smash hit and considered the seminal text in the contemporary environmental movement – was so strongly effective precisely because of its highly localized focus and because of its evocative tone that touched on the deep religious sensibilities (latent or expressed) of a large segment of the American population at the time. Similarly, any communication on climate change across the multi-varied contexts that constitute Africa requires both an openness to localized subjectivities and localized communication on matters of immediate environmental concern.

DOI: 10.4324/9781003660217-14

This study makes the case for localization of climate change discourse by working through the two aforementioned works. And this is in service of searching for a more effective manner in which to communicate issues of pressing environmental concern – in this instance ideas that coalesce around the now ubiquitously employed term 'climate change.'

This is a qualitative[2] paper that makes use of discourse analysis, drawing on Foucault's genealogical method. For Foucault, "it followed that discourse cannot be analysed only in the present, because the power components and the historical components create such a tangled knot of shifting meanings, definitions and interested parties over periods of time" (Powers 2022: 26). As such, the work is particularly interested in adding a historical contextualization (or dimension) to the discussion on climate change communication in Africa.

Climate Change Communication

Literature on climate change communication suggests that communicating effectively on the topic of climate change to diverse audiences is a serious challenge. For instance, while Nerlich et al. speak of "the importance and difficulties inherent in talking about climate change to different types of publics using various types of communication tools and strategies" (Nerlich et al. 2010: 97), Markowitz and Guckian discuss "key challenges communicators face, including features of the problem itself, cultural and political conflict, and psychological barriers that shape how people engage with the issue" (Markowitz and Guckian 2018: 35).

To further complicate the already difficult matter of communicating a concept such as climate change to heterogeneous audiences, Ballantyne argues that "the field of climate change communication is characterized by diverging and incompatible understandings of communication as a theoretical construct." Furthermore, "In some instances, communication theory appears reduced to an 'ad hoc' toolbox, from which theories are randomly picked to provide studies with a fitting framework" (Ballantyne 2016: 329).

What are we to make of this multi-faceted communication challenge from the perspective of engaging with climate change communication from the specific (and indeed heterogeneous) context that is Africa?

Firstly, we should heed Ballantyne's argument and state that this work does not wish to reproduce the errors identified by her, in doing so, cheaply taking on some or other communication theory and then applying it to climate change discourse within Africa. Rather, the work seeks to understand *why* climate change communication may (and does) find misunderstanding, apathy or opposition expressed to it. Further, while there may exist divergent and incompatible understandings of communication as a theoretical construct amongst writers in the field of climate change communication, an argument on why a message on a critically important topic may not be landing speaks to a well-established concept in communication theory, that of *noise*. Noise, quite indisputably, is anything that interferes with an attempt at communication. And indeed, it follows that "the more interference in the signal, the less information is received" (Lister et al. 2009: 383) Lastly, as R. Haven

Wiley notes in *Noise Matters: The Evolution of Communication*, "[Noise] is not just a conditional aspect of communication, but an essential element" (2015: 1). It follows then that this work does not speak to the eradication of the noise associated with climate change communication (or any communication for that matter), but rather attempts to understand how its minimization could be achieved.

Secondly then, in the African context, we could see the "difficulties inherent in talking about climate change to different types of publics" spoken of by Nerlich and colleagues and the "cultural and political conflict" regarding climate change spoken of by Markowitz and Guckian as a specific noise. And that specific noise is a global mantra attempting to imbed itself into a local context. That is, it is the argument of this work that the term 'climate change' is generally applied in a global, decontextualized sense, and in being applied this way has little efficacy in terms of acceptance, understanding and behavioural change. Indeed, as within any other localized geographical context, in Africa, the reception of a global message (climate change) that comes across as ethereal and vague, and that further, is politicized, often polarizing, and furthermore cast often enough as Western-centric, will struggle to find purchase.

Noise: Western-Led Communication on Climate Change

A communicated matter (in this instance climate change) that is associated as stemming from a specific context can encounter some justifiable suspicion and thus resistance to its message:

Consider as an opening point that the person most associated with the topic of climate change over recent years has been the Swedish environmental activist, Greta Thunberg. While this may seem an unfair point, and we will soon discuss deeper structural and political ones, the optics of this association in a new media-saturated meme-rich hyper-speed communication environment should not be underestimated. For ordinary consumers of the news and social media, an environmental activist from one of the most prosperous nations on Earth is the most immediate association with the topic of climate change. For ordinary consumers of the same in Africa (and indeed, in other less affluent parts of the globe), such optics may come across as acutely discordant. Catherine Happer, writing for *The Conservation*, rightly points out that "The environmental movement has always had a middle-class aura to it and, in spite of attempts to use the language of inclusivity, it has never quite lost this tag" (Happer 2019). And we could add to this that, with Greta Thunberg as its symbol (for a time at the very least), for climate change, this tag feels stickier and more expansive – the concern of prosperous wealthy Western nations and their citizens.

Beyond such arguably superficial optics, we could look to deeper structural (and often thus politicized) issues that bedevil the efficacy of communication on climate change to recipients located outside of the wealthy nations.

In "Development," Graham Huggan and Helen Tiffin perform an overview of the work of the German sociologist Wolfgang Sachs in combination with the complimentary work of the Colombian anthropologist Arturo Escobar. Both critics of

Western-led notions of development, it is pointed out by Escobar that fashionable terms of Western genesis or orientation such as sustainability should be (and are) regarded with some suspicion. With specific reference to sustainability (although we could update this to newer projects like Net Zero),[3] Escobar writes that this is "a pet project of the global-capitalist development apparatus in which, under the aegis of 'planetary management,' the 'Western scientist continues to speak for the [entire] Earth'" (in Huggan and Tiffin 2015: 35). Further, "planetary management" (under which responses to climate change – such as Net Zero – could be classified) is described by the cultural theorist Andrew Ross as "continuous with the scientific perspective of quantitatively dominating the physical world," in which "questions of ecology [are not] answered by social theory or social action [but] resolved at the level of 'resource management' by the level of the multinational corporate state" (Ross 1991: 208–210).

To further complicate the matter, this resource management for the benefit of the multinational corporate state by that very same actor may well have terrible downstream effects for geographical locales outside of its centres of power. As an example, in "The decarbonisation divide: Contextualizing landscapes of low-carbon exploitation and toxicity in Africa," Benjamin Sovacool and his colleagues, working off "extensive fieldwork" (2020: 1) in the Democratic Republic of Congo and Ghana respectively, notice that

> Much academic research on low-carbon transitions focuses on the diffusion or use of innovations such as electric vehicles or solar panels, but overlooks or obscures downstream and upstream processes, such as mining or waste flows. Yet it is at these two extremes where emerging low-carbon transitions in mobility and electricity are effectively implicated in toxic pollution, biodiversity loss, exacerbation of gender inequality, exploitation of child labor, and the subjugation of ethnic minorities.
>
> (Sovacool et al. 2020: 1)

These kinds of unfortunate processes they characterize as part of an emerging "decarbonisation divide" (Sovacool et al. 2020: 1). When viewing such duplicitous and hidden costs (costs which in this case directly and disproportionally affect African countries) it is little wonder then that although climate change is an evidently global problem, climate change communication driven by actors associated with the West may be met with sceptical and resistant audiences in Africa.

This level of scepticism is indeed confirmed in research completed by Julia Lück and her colleagues, who, in "Counterbalancing global media frames with nationally colored narratives: A comparative study of news narratives and news framing in the climate change coverage of five countries," found that "in five democratic countries (Brazil, Germany, India, South Africa, and [the] United States)" (Lück et al. 2018: 1635) characterizations of and interpretations of climate change varied sharply from Western characterizations and interpretations of the same. Explaining the above, Petrescu-Mag et al. argue that

> On the one hand, in German and US newspaper coverage, both the victims and the villains of climate change are less individualized, encompassing developing countries and small island states, and humankind and past generations, respectively … On the other side, in South Africa, Brazil, and India, their home countries are portrayed as victims and heroes. At the same time, the image of the villains is also unambiguous—rich and developing countries are responsible for climate change problems.
>
> (2022: 9)

Beyond such oppositional viewpoints, there is one further dimension to consider as a possible driver for climate change communication being undermined in non-Western contexts by the noise of its Western association. Diego Andreucci and Christos Zografos explore this dimension and levy the charge in "Between improvement and sacrifice: Othering and the (bio)political ecology of climate change." Using a Foucauldian framework, they make the case that "othering is central to the government of climate change" (Andreucci and Zografos 2022: 1). In their argument, measures to mitigate climate change are less about the well-being of various populations, and more about the continued maintenance of self-serving economic configurations and mechanisms. Of this continent, they write:

> This is particularly visible in constructions of Africa as a climatic "heart of darkness" of unruly environments and backwards, conflict-prone populations (Verhoeven, 2014). Such discourses mirror Eurocentric self-images of the West as rational, scientific and altruistic. They represent a contemporary reactivation of the colonial "white man's burden" logic, whereby the inability of people in much of the "developing world" to cope with climate disasters (their lack of "adaptive capacity") infuses adaptation interventions with a sense of necessity and urgency.
>
> (Andreucci and Zografos 2022: 5)

While this may be a harsh charge, again we are exposed to the notion that climate change communication cannot be read in a decontextualized sense. Certainly, for the African context, climate change communication must consider the baggage present in communication originating in Western settings.

As such, this section of the work has attempted to display why climate change communication is often not efficacious in its global reach. Its optics, structural concerns, political challenges, real-world downstream effects and the potential waft of implicit othering all serve as noise that interferes with the intended message (if we were to read the senders of the message as sincere in their intentions). And while the display has been built up of academic sources, one would not have to venture far into public popular spaces featuring African voices (understood here as social media and video sharing sites) to witness a confirmation of messages of climate change mitigation being met with some considerable challenge.

Minimizing the Noise: The Re-articulations of Valentin-Yves Mudimbe and Rachel Carson

While we can concede to Wiley's earlier point that noise is an essential element of communication, we should try to consider how we could minimize the noise associated with climate change communication in Africa. If global – or more accurately Western-led messages – on the urgency of a response to climate change are obstructed by the fact that local audiences may often-enough respond with alienation and suspicion to them, how do we find a mode of communicating climate change in a more resonant way? As opposed to constructing some kind of model in an 'ad hoc toolbox' (Ballantyne) manner to try to answer this question, this work wishes to point to two historical works that demonstrate instances when disputed communication has overcome noise to cut through and find resonance. In this regard, the work turns to the writings of the Congolese philosopher Valentin-Yves Mudimbe and to the writings (and reception thereof) of the American writer Rachel Carson. In the first instance, with Mudimbe, we find an example of how a concept evolves significantly over time within an African context. In the second instance, with Carson, we find a text on environmental communication that was so strongly effective precisely because of the highly localized focus it adopted.

Mudimbe and the Long View on Communication

In "African Gnosis Philosophy and the Order of Knowledge: An Introduction," Mudimbe sets about contending with the entry of Christianity into Africa, as well as its continued existence and popularity within the continent. For Mudimbe, it is deeply interesting that a religion inextricably associated with – and variously implicitly or explicitly supportive of – an extractive (and often violent) colonialism can still find great purchase in contemporary postcolonial African societies. And in his investigation of this phenomenon, we can see that communication (in this case, communication of the tenets of a religion) only becomes internalized, affecting and effective when it is expressed with agency within a local framework. Said differently, imposition of a particular concern (in this case the Christian belief system, or in our case Climate Change) may be (at best) begrudgingly accepted but local expression of concern along the same lines is the truer form of communication on that subject.

To encapsulate Mudimbe's vast article in summative form, it could be said that what is tracked is a deep and long production of an alternative discourse to an initially imposed one. That is, the article tracks the arrival of Christianity in Africa as part of, and as a justifier of, the colonial project. But we should ask, as the author does, as to what happens to the discourse of Christianity *over time* in Africa. It is, as all things are, subject to change. And this change – seen as a *dynamic* encounter over a long period of time between Africans and Christianity – led to an indigenization of the Church (in itself a process that generated varied iterations and developments). Indeed, in Mudimbe's argument, African Christians adopted Christianity and *changed* it. It became something *other* than it initially was. While initially a top-down imposition in service of colonial power, Christianity as a concept was reconfigured and repurposed through local agency and sensibility.

To track the article in a slightly less summative manner, and thus to demonstrate the mechanics of this argument (as well as its resonance for climate change communication in Africa), it is worth working through the four stages of encounter and change identified by Mudimbe more carefully:

He begins his investigation at the point of genesis, tracking the original (and obviously prejudicial) idea that the continent was a blank slate awaiting partition and dominion. Indeed,

> The missionaries, preceding or following a European flag, not only helped their mother country to acquire title to new lands, but also accomplished a "divine" mission ordered by the Holy Father, *Dominator Dominus*. It is in God's name that the Pope considers the planet as his franchise and sets the basic principles of *terra nullius* which denies all the "non-Christian natives" the right to an autonomous political existence, and the right to own or to transfer ownership (De Witte, 1958).
>
> (Mudimbe 1985: 152)

From this, we can deduce, as the author himself does, that "The missionary was an essential part of this universe." Indeed, the religious fervour and message were "allied, rather than opposed to commercial motive" (Mudimbe 1985: 153). Further, to demonstrate that the communication of Christian principles housed imperatives beyond the strictly devotional, Mudimbe points out that

> Obviously, the missionary's objectives could but be co-extensive with political and cultural perspectives on colonization, as well as with the Christian view of his mission. With equal enthusiasm, he served as an agent of a political empire, a representative of a civilization, and an envoy of God. There is no essential contradiction between these titles. All of them implied the same purpose: the conversion of African minds and space.
>
> (Mudimbe 1985: 153)

While this genetic kernel of a blank slate to be taken hold of remains evident in the communications of the subsequent missionaries (Romano, Crowther and Tempels) studied by Mudimbe, what is picked up on is a gradual softening of the rhetoric towards the African populations encountered by these said missionaries. Despite the adoption of a gentler rhetoric over time, what remained was a "common ideology" (Mudimbe 1985: 156) in which "commentaries on African people are arranged with regard to the prospect of their possible conversion; on the other hand, specific socio-cultural symbols design the meaning of the passage from primitiveness to civilization" (Mudimbe 1985: 158).

While this status quo held for a long time, and while it faced some minor challenge throughout, it was with the momentum of African independence movements following World War II that Christianity would have to face existential questions regarding its presence on the continent. Already deeply rooted in many regions, and internalized by many within those regions, it was implausible that Christianity would *leave*. Rather, a newly constituted local adaptation was necessary, and so here, Mudimbe tracks two phases in the Africanization (or localization) of Christianity. He writes that "From the 1950s onward, new orientations appear for the indigenization of the Church". And he sees this as occurring in two phases – the first phase being relatively superficial and the second phase containing a far more rigorous ontological and epistemological challenge. Regarding relatively superficial changes, Mudimbe speaks of the "first step of adaptation" as "insisting on the Africanization of some external aspects (music, hymns, etc.)." Far more powerful than this move (which occurred at the level of optics) was the second phase which heralded "a reflection on the content of Christianity in an African setting" (Mudimbe 1985: 160). As an example of this phase, Mudimbe points to the work of two Jesuit priests from Cameroon, Mveng and Hebga, who were active and influential in reconciling Christianity and Africanity. Of their work, he writes:

> A theology of incarnation [was] being promoted with a particular emphasis on new premises … African history with its own symbols as a preparation to Christianity … and finally the experience of slavery, exploitation and colonization as signs of the suffering of God's chosen ones.
>
> The most striking feature of these intellectual positions resides in the clear between the program of political liberation which should permit a transformation of the traditional civilization and that of rethinking Christianity as an integral part of the local culture.
>
> (Mudimbe 1985: 162)

Such re-articulation and localization was not met without consternation and challenge from the traditional centre of the religion. Contrary positions were in conflict throughout the 1960s, 1970s and 1980s, and yet, this new discursive construction constituted by African theologians remained strongly in play.[4] And it is plausible to suggest that part of the reason that they did so was because a locally communicated matter does not contain the noise that accompanies a communication coming from the outside (an outside that furthermore is associated with a self-serving *modus operandi*).

Carson and the Right Communication at the Right Time

While Mudimbe's work tracks a contest of discourse over centuries, the example of Rachel Carson's strongly influential book *Silent Spring* (1962) demonstrates a specific point in time in which the local context allows for the greatest possible global receptivity to a message. Indeed, with Carson's work, we find a text on environmental communication that was so strongly effective precisely because of the highly localized focus it adopted in response to a troubled local context with global parallel.

By topic, Carson's book does not strike one as particularly evocative – "this book focused international attention on the deadly effects of DDT and other chemical pesticides." But the topic in itself is not the reason for the book's success in having had "a profound influence on the public's attitude toward chemical pollution and environmental protection" (Des Jardins 2006: 3). We can identify two salient reasons for the breakthrough. Firstly, the language employed by the author spoke to a sensibility embedded within the American reading publics that encountered the book. Secondly, those same publics matched these evocative words with a series of localized environmental disasters, as such seeing the text as both a serious contention with the local and the global consequence of man-made alteration to the environment.

To the first point: Much like with Mudimbe's overview of how Christianity was re-articulated by those challenging its implicit colonial logic, so Carson's book re-articulated a serious but potentially dry and scientific matter through her employment of a style of rhetoric that leant on evocative and historical modes of writing. As Frederick Buell argues in *From Apocalypse to Way of Life: Environmental Crisis in the American Century*, even though her work may "not [have been] the first to raise the spectre of imminent human-made environmental crisis … [it] had a decisive effect on account of its evocative articulation of the problem" (Buell 2003: viii). To speak specifically to the language employed, Greg Garrard, in *Ecocriticism*, explains that Carson made heavy use of both "poetic parable" and of "the literary genres of pastoral and apocalypse, pre-existing ways of imaging the place of humans in nature that may be traced back to sources such as Genesis and Revelations" (Garrard 2004: 2). One can imagine the powerful effect that this had on American readers of the time (specifically considering American society was on the whole was far more religious in the 1960s than it is today). Indeed, "in its recourse to pre-industrial and Christian tropes, Carson's book struck an intuitive and emotive chord among many of its American readers" (Siwak 2017: 14). Much like with the tussle between African theologians and the traditional power centre of Christianity over what constituted acceptable practice for the Church in Africa, so Carson's poetic and pastoral text which had generated a significant interest also faced challenge. As Clapp and Dauvergne explain, "the scientific community lost no time in attacking Carson in the media, accusing her … of overstating the case against DDT (dichlorodiphenyl-trichloroethane) and other chemicals" (Clapp and Dauvergne 2005: 49). Again, much like with those theologians posing a counter-vision, so Carson's work maintained its standing with the public. Indeed, "after Carson's work, the long-term consequences to both humans and the natural world, as well as the political and ethical implications of pesticide use, came to the forefront" (Des Jardins 2006: 4). Arguably, they then never subsequently left the public discourse, although, to be sure, concern over environmental matters may have waxed and waned at different periods.

To the second point: Evocative language that challenged scientific discourse around the pressing topic of DDT aside, Carson's challenge may well have fallen flat if a number of environmental disasters (covered by the media at the time) hadn't come to public attention – to further localize Carson's poetic plea for a reorientation to nature. "A future spring without the songs of birds" (Clapp and Dauvergne 2005: 49) certainly hit harder when the American public subsequently witnessed environmental destruction within its own space. As Gottlieb notes,

> Environmental crisis seemed to be written for all to see in such disparate events of the late 1960s as the burning of the Cuyahoga River in the center of Cleveland, the eutrophication of Lake Erie, and the dying birds washed up on the oil-slicked shores of Santa Barbara.
>
> (in Buell 2003: viii)

To offer a brief explanation for each of these visually affecting disasters (all amplified by national media):

The Cuyahoga River in Cleveland had for a long time been a highly industrialized waterway which had caught fire on several occasions as a result of the toxic waste clogging its channels (Dutka 2014: 65–68). "On June 22, 1969, a patch of the Cuyahoga River below a railroad bridge burst into flames," with this fire "reaching a height of five stories" and lasting "twenty-six minutes" (Dutka 2014: 64). What brought the flame to broad American public attention was a *Time* magazine article which beside writing up the event published stunning images of the blaze (Dutka 2014: 64).

The "rapid eutrophication" of Lake Erie was primarily as a result of industry and mechanized agriculture releasing pollutants such as phosphorous into a lake that was relatively shallow. This led to a rapid proliferation of algae which had the effect of killing existing life in the lake. This then had a further effect in that the water had a wretched smell, affecting residents and tourists alike (*Sea Wind* 1991: 8). As with the Cuyahoga River incident, the situation of Lake Erie received significant media attention and thus public exposure to matters of environmental degradation. The phrase "Lake Erie is dead" was repeated across "national publications" (Rotman 2016: 1).

And as if lakes, flaming rivers, and rotting lakes where not striking enough to a public increasingly alarmed by growing exposure to such disaster, an oil spill of the coast of Santa Barbara in 1969, which destroyed a significant amount of bird life in the area (all captured by television stations and broadcast to shocked views) (Farmer 2006: 108), functioned as a sordid addition. In response to this, and as Erik Loomis points out in *Out of Sight: The Long and Disturbing Story of Corporations Outsourcing Catastrophe*, this specific incident "united citizens of all political persuasions in a truly non-partisan cause" (2015: 30). Further, "coverage of the 1969 oil slick fire on the Cuyahoga River and the Santa Barbara Channel oil spill were key moments in the growing media coverage of the human impact on the environment" (Haq and Paul 2012: 77).

As such, Carson's work, if one can forgive the unfortunate phrasing, was a local plea that came about at the *right time* followed by the *right* sequence of disastrous environmental events.[5] It was a communication of the problem that relied on an unfashionable usage of language that was then combined by the power of the press, photography and television covering harrowing events. It was, in effect, an

American pastoral image conjured in her writing operating within a subsequent context of the image of industrial nightmare. Said differently, it was an impassioned and irrefutable communication coming from a local source warning against matters that were localized via disaster caught by and on camera (and yet also generalizable to a global context).

Conclusion

While "Effective and appropriate communication has been identified as an essential step" in relation to affecting acceptance, understanding and behavioural change amongst populations in relation to climate change (McGahey and Lumosi 2018), quite often climate change communication coming from a global context struggles to translate through to local sensibilities. In terms of optics, political-structural relations and history, such communication is often-enough regarded as disingenuous, foreign and alienating. The challenge then is for the localization of a global matter.

While historical examples of such localization have been discussed in this work, whether a contest over centuries (as per Mudimbe) or whether a well-timed message of immediate and local concern that ultimately found global reach (as per Carson), it would be up to us to amplify the voices of local initiatives that attempt to articulate climate change communication in their own terms. And those terms may, furthermore, not always comport with the sensibilities and communications on offer from global players offering us their version of how climate change should be contended with. In this regard, we could turn to Okoliko and de Wit who offer more localized communicators on climate change hope when they call for a "problematizing [of] public engagement" in terms of climate change communication. Criticizing outmoded thinking about public engagement, they write:

> Common assumption pitches "the public" on the one side and "the communicator" on the other side. This bifurcated model of "communicating" climate change has import for the forms of subjectivity in climate (in)action, including a weakened citizenship representation in climate discourse and the de-pluralization of ideas.
>
> (Okoliko and de Wit 2020: 36)

While such thinking around public engagement encourages local actors outside of mainstream capitalized communicators, it must also be noted that any such local, (possibly) alternative and pluralized articulation of climate change will still encounter significant obstacle:

Local initiatives attempting to articulate climate change communication in their own terms (via popular social media platforms or via free video sharing sites such as YouTube) will have to compete with the monetary power and algorithmic logic of global (most often Western) actors communicating *their* understanding of climate change and how to go about mitigating its impacts.

Further, climate change communication coming from either local or global actors faces some distinct challenges across various African countries and contexts. For instance, for a South African context, Levi points out that

> In South Africa, one of the world's most carbon-intense economies and a society marked by gross social inequality, climate change is not a popular topic. As of 2018, more than half of the population had never heard of climate change and only one in five South Africans believed that human activities lead to global warming. The communication of climate change in South Africa is influenced by the notorious inequality that the country still suffers decades after the apartheid regime has ended.
>
> (2021)

In addition,

> Environmental NGOs, journalists, and scientists frequently set communication cues on climate change. However, their messages are largely circulated in newspapers catering to an urban and educated readership and resonate less with people living in rural areas or those who rely on employment in the coal and mining sector.
>
> (Levi 2021)

As such then, while significant challenges exist for both global and local communicators on climate change, localization of climate change communication remains a potentially influential communicative force that may inspire a sense of agency in the matter, and that may reach people in a way that global, decontextualized communication does not.

Notes

1 The article acknowledges that care must be taken here. While Africa is referred to here as a "localized geographical context," this is not to suggest that the countries that make up the continent are a uniform mass. Indeed, besides the unique and specific cultural contexts that exist across the many countries that make up the continent, in terms of communication from the Global North, while the continent had the common experience of colonialism, each such experience was variated. As such, different countries (and even different localities within those different countries) would have potentially differing orientations to communication stemming from the Global North.

2 With the purpose of qualitative research understood by this work as being "to analyze how people understand, experience, interpret, and construct the social world." With this, "Instead of analyzing a fixed, pre-established, and pre-determined social reality, qualitative researchers observe the social world, knowledge, meanings, and notions of reality as contingent and dynamic." This is "in order to understand the socio-culturally constructed meaning of the individuals' experiences" (Khan and MacEachen 2021: 1).

3 Net Zero is defined by the United Nations in the following terms: "Put simply, net zero means cutting carbon emissions to a small amount of residual emissions that can

be absorbed and durably stored by nature and other carbon dioxide removal measures, leaving zero in the atmosphere" (United Nations, 2024).

4 And continues to remain in play. There are numerous contemporary examples of Christianity re-articulated through an African lens, with African Christians melding together context-specific traditional belief systems with Christian belief and practice. For instance, Kleinhempel discusses "double faith," "especially in the African Instituted Churches (AIC), which have adopted and transformed elements of mediumist practice and ritual" (2017: 642), while Mofokeng discusses "the advent of neo-prophetism in South Africa" (2024: 1) which features "the claimed interaction of neo-prophetism and sangomaism suggest[ing] a radical epistemic shift with implications for all forms of Christianity in South Africa" (2024: 8).

5 There is a further development that should be at least shortly illuminated. As the media began to place emphasis on environmental reporting (this being paired with affecting visual imagery from relatively new mediums like television), the *Earthrise* photo taken by Apollo 8 astronaut William Anders in 1968 came into public consciousness (Ignatow 2007: 26). For Haq and Paul, "in particular, the view of the Earth for the very first time from space transformed people's understanding of our place in the universe" with "the images of a desolate and lifeless moon contrasted with the blue and white, fertile Earth." For them, this could be seen as "a stark reminder of humankind's fragility" (2012: 77–78). A reminder and new sensibility that ultimately formed part of the impulse toward an "eco-renaissance" (Poole 2008: 3).

References

Andreucci, D. & Zografos, C. 2022. "Between improvement and sacrifice: Othering and the (bio)political ecology of climate change." *Political Geography,* 92, 1–11.

Ballantyne, A. 2016. "Climate change communication: What can we learn from communication theory?" *WIREs Climate Change,* 7(3), 329.

Buell, F. 2003. *From Apocalypse to Way of Life: Environmental Crisis in the American Century.* New York: Routledge.

Clapp, J. & Dauvergne, P. 2005. *Paths to a Green World: The Political Economy of the Global Environment.* Cambridge: MIT Press.

Des Jardins, J. 2006. *Environmental Ethics: An Introduction to Environmental Philosophy.* Stamford: Thomson Wadsworth.

Dutka, A.F. 2014. *Cleveland Calamities: A History of Storm, Fire and Pestilence.* Charlestown: The History Press.

Farmer, B. 2006. *American Political Ideologies: An Introduction to the Major Systems of Thought in the 21st Century.* Jefferson and London: McFarland & Company.

Garrard, G. 2004. *Ecocriticism (First Edition).* London and New York: Routledge.

Happer, C. 2019. "Climate change is too middle class – here's how to fix that" in *The Conversation.* At: https://theconversation.com/climate-change-is-too-middle-class-heres-how-to-fix-that-123231 Accessed on 23 June 2024.

Haq, G. & Paul, A. 2012. *Environmentalism since 1945.* New York Routledge.

Huggan, G and Tiffin, H. 2015. *Postcolonial Ecocriticism: Literature, Animals, Environment* (Second Edition). London and New York: Routledge.

Ignatow, G. 2007. *Transnational Identity Politics and the Environment.* Lanham: Lexington Books.

International Marine life Alliance & Ocean Voice International. 1991. *Sea Wind: Bulletin of the International Marine Life Alliance, Volume 10.* Berkeley: IMA & University of California.

Khan, T. and MacEachen, E. 2021. "Foucauldian discourse analysis: Moving beyond a social constructionist analytic." *International Journal of Qualitative Methods*, 20, 1–9.

Kleinhempel, U. 2017. "Covert syncretism: The reception of South Africa's Sangoma practise and spirituality by "Double Faith" in the contexts of Christianity and of Esotericism." *Open Theology*, 3(1), 642–661.

Levi, S. 2021. "Climate change communication in South Africa." November 29. Published Online). *Oxford Research Encyclopaedia of Climate Science.*

Lister, M., Dovey, J., Giddings, S., Grant, L., & Kelly, K. 2009. *New Media: A Critical Introduction, Second Edition*. London: Routledge

Loomis, E. 2015. *Out of Sight: The Long and Disturbing Story of Corporations Outsourcing Catastrophe.* Atlanta: Bookbright Media.

Lück, J., Wessler, H., Wozniak, A., & Lycarião, D. 2018. "Counterbalancing global media frames with nationally colored narratives: A comparative study of news narratives and news framing in the climate change coverage of five countries." *Journalism*, 19(12), 1635–1656.

Mofokeng, T.R. 2024. "Reconsidering syncretism and contextualization: The Sangoma-Prophet phenomenon in South African Neo-Prophetic Pentecostalism." *Religions*, 15(1), 84.

Markowitz, E. & Guckian, M. 2018. "Climate change communication: Challenges, insights, and opportunities." In *Psychology and Climate Change: Human Perceptions, Impacts, and Responses* (ed. Clayton, S. and Manning, C.). Cambridge, Massachusetts, USA: Elsevier, 35–63.

McGahey, D.J. & Lumosi, C.K. 2018. "Climate change communication for adaptation: Mapping communication pathways in semi-arid regions to identify research priorities." *Journal of Sustainable Development in Africa,* 20(1), 88–107.

Mudimbe, V.Y. 1985. "African gnosis philosophy and the order of knowledge: An introduction." *African Studies Review*, 28(2/3).

Nerlich, B., Koteyko, N., & Brown, B. 2010. "Theory and language of climate change communication." *WIREs Climate Change*, 1(1), 97–110.

Okoliko, D.A. and de Wit, M.P. 2020. "From 'communicating' to 'engagement': Afro-relationality as a conceptual framework for climate change communication in Africa." *Journal of Media Ethics,* 36(1), 36–50.

Petrescu-Mag, R.M., Burny, P., Banatean-Dunea, I., and Petrescu, D.C. 2022."How climate change science is reflected in people's minds. A cross-country study on people's perceptions of climate change." *International Journal of Environmental Research and Public Health*, 19, 4280.

Poole, R. 2008. *Earthrise: How Man First Saw the Earth.* New Haven: Yale University Press.

Powers, P. 2022. "The philosophical foundations of Foucauldian discourse analysis." *Critical Approaches to Discourse Analysis across Disciplines,* 1(2), 18–34.

Ross, A. 1991. *Strange Weather: Culture, Science and Technology in the Age of Limits.* New York: Verso

Rotman, M. 2016. "Lake Erie," in Cleveland Historical, viewed 5 March 2016, from: http://clevelandhistorical.org/items/show/58#.V1AS_fl97IU

Siwak, Jakub. 2017. *Investigating Emerging Deleuzoguattarian Connections to the Environment via Information Technology*. Port Elizabeth/Gqeberha: Nelson Mandela Metropolitan University.

Sovacool, B., Hook, A., Martiskainen, M., Brock, A., & Turnheim, B. 2020. "The decarbonisation divide: Contextualizing landscapes of low-carbon exploitation and toxicity in Africa." *Global Environmental Change*, 60.

United Nations, 2024. "For a liveable climate: Net-zero commitments must be backed by credible action." At: www.un.org/en/climatechange/net-zero-coalition Accessed on 30 June 2024.

Wiley, R.H. 2015. *Noise Matters: The Evolution of Communication.* Cambridge, MA: Harvard University Press

Part III

Digital Media and Climate Change

12 Artificial intelligence in climate change reporting

Emerging journalistic practices in African contexts

Refilwe Whitney Mofokeng and Allen Munoriyarwa

Introduction

Climate change has become a topical issue, dominating contemporary global public discourses (Paterson and Grubb, 1992). This dominance is because, increasingly, scientific research is proving that the planet's climate is changing rapidly (Bulkeley and Betsill, 2013). The evidence of climate change and scientific discourses claim lies in the melting of ice sheets covering the Greenland and Antarctica regions (Keohane, 2015), strange and powerful storms punctuating our seasons, extreme weather evidence (Mirza, 2003), and rising global temperatures heralding serious global warming (National Academies of Sciences, Division on Earth, Life Studies, Board on Atmospheric Sciences, Committee on Extreme Weather Events, and Climate Change Attribution, 2016). Scientists note that climate change will eventually destroy our environment, adjust in our lifestyle patterns, and even cause major disruptions in human life and the natural ecology (Keohane, 2015; Lineman, Kim and Joo, 2015).

The debate about climate change is intricately linked to the future of both the planet and humanity who inhabit it (Goklany, 2012). There is now a consensus that at a global level, humanity needs to mobilise politics, cultures, funds, education technologies, etc. to save the planet (Giddens, 2015; Goklany, 2012; Bulkeley and Betsill, 2013). Every facet and institution should now be mobilised for the battle against climate change (Campbell, Hansen, Rioux, Stirling and Twomlow, 2018; Dunlap and McCright, 2015; Flusberg, Matlock and Thibodeau, 2017; Lombardo, 2019). Therefore, it is agreed that climate change impact has dire consequences on the well-being of humans, the ecosystems, and the economy globally. The most important question for humanity has been: How do we harness our society politically and technologically for sustainable climate practices?

This chapter addresses this question but from a more specific angle. It examines how a specific community of communication practice – climate journalists – are integrating AI into their news-making practices. By climate change journalism, we adopt Schäfer and Painter's (2020:1) definition of the profession thus, "Climate journalism gathers, evaluates, selects, and presents information about climate change, its characteristics, causes, and impacts, as well as ways to mitigate it, and distributes them via technical media to general and specialist audiences." Drawing

DOI: 10.4324/9781003660217-16

on in-depth interviews with climate change reporters sampled from within South African and Zimbabwean journalists, the chapter asks two main questions: How is AI being integrated in climate change news reporting by selected journalists in South Africa and Zimbabwe? In what ways does the integration of AI in climates change news coverage alter news production and journalistic practices? While literature that examines the role of journalists and journalism institutions in the global climate debate is growing, there is still very scant literature that examines how AI, a burgeoning technology that has taken the world by storm, is being integrated into climate change reporting. This chapter contributes to this emerging literature with a stiff focus on climate journalists from two Sub-Saharan African countries, South Africa and Zimbabwe. Existing literature on AI in Southern Africa highlights challenges hindering AI adoption in news organisations in the region. This has been attributed by concerns over skill deficiencies, financial constraints, ethical dilemmas, and the potential impact on democracy(Munoriyarwa, Chiumbu and Motsaathebe, 2021; Munoriyarwa, 2024; Ogola, 2023).

Research in this dimension is crucial for several reasons. First, AI is an emerging technology whose full development cannot be anticipated. This makes exploring its contribution to everyday news reporting important, especially concerning climate change. Second, if we adopt Giddens' (2015), Goklany's (2012), and Bulkeley and Betsill's (2013) view that at a global level, humanity needs to mobilise politics, cultures, funds, education technologies, etc. to save the planet, it means it is productive to understand how a particular range of technologies like AI, can or, is being harnessed by an important group in society – journalists – responsible for the production and distribution of mainstream discourses on the climate. Third, for a continent that largely imports AI technologies, it is important to understand how specific groups within these countries integrate these technologies into their daily lives. The chapter critically contributes to the emerging literature on AI and climate change news reporting for these reasons.

Climate news and journalism: A review of the literature

Earlier on, we pointed out that AI and climate reporting literature is still emerging. However, as the climate change discourses spread worldwide, research on climate reporting and journalism has increased. McKnight (2010) notes that journalism has moved from an earlier era of dissensus, characterised by news coverage that, almost in equal measure, endorsed and simultaneously opposed climate change, to more news that endorses climate science views. McKnight (2010) states that modern journalism has moved away from denialism to directly endorsing scientific consensus on global warming. Brüggemann and Engesser (2017) agree, noting that media coverage across countries and media outlets has moved closer to the basic scientific consensus on climate change. The era of political correctness has ended, and it is being replaced in media houses by a direct acknowledgement of climate science (Brüggemann and Engesser, 2017; McKnight, 2010).

However, other scholars are not entirely happy about how journalism has handled the climate crisis. Rusbridger (2020) has argued that climate reporting is one of the greatest failures in the journalism terrain. He calls for reinvigorated conversations on climate journalism that contribute to public interest climate change journalism, not what interests the public. Painter (2019) also notes the same, arguing that journalism adapts to changing society and discourses when covering the climate. Gess (2012), however, sees these debates differently. Gess (2012) argues that it is not possible for journalism, in its current form, to report the climate crisis effectively. Gess (2012:54), says, "Journalism, in its current form, is perhaps not well suited to reporting climate change, as conventional reporting does not usually run to depth and the story loses its interest value, failing to maintain novelty over an extended period of time". Appelgren and Jönsson (2021) agree but diverge from Gess's argument, stating that in addition to the current structure of journalism, framing climate news is critical to mobilising collective action. But is journalism supposed to mobilise collective action, as Appelgren and Jönsson (2021) argue, or should they just interpret events, as they further argue in their article? Much more recently, Schäfer and Painter (2020) have noted that climate journalism has mutated at a global level, driven by technological changes. They note several changes that have happened to climate journalism. First, they note that climate journalists have moved from gatekeepers to curators. Second, they note that climate news sourcing routines have shifted in the digital age, from focusing on elite sources to a broader range of stakeholders. Third, Schäfer and Painter (2020) note that there is new climate journalism emerging, dominated by the rise of stakeholder public relations on climate journalism. The literature reviewed in this section demonstrates three significant changes that have taken place in climate journalism. These are the age of climate denialism is gradually ending, and climatic hang denialists are increasingly becoming peripheral sources (McKnight, 2010). Second, climate journalism is beginning to be treated as central to news and important to the global public discourses (Brüggemann and Engesser, 2017). Third, and perhaps more important to this chapter, technology is increasingly becoming a defining factor in climate journalism (Schäfer and Painter, 2020). Drawing from the last point, we ask in this chapter, if technology is beginning to influence climate journalists, how are AI technologies shaping and influencing climate news reporting? To examine this, we draw on the conceptual lenses of disruptive news-making ecosystems.

Conceptualising disruptive news-making ecosystems

[New] technologies have led to radical changes in news production (Anderson, 2010) and have led to changing the architecture of local journalism ecologies (Anderson, 2010). Journalism and newsroom technologies like big data interpreting technologies, large language technologies and various versions of AI technologies have disrupted newsmaking in both positive and negative ways (Munoriyarwa, 2024). For example, the algorithmisation of news (see Shin, 2024) changes the way readers/audiences interact with news content and how journalists and institutions pitch their news online (Szews, 2016). Digital technologies, like AI, can also affect the market in terms of what gets to be read and who gets to read it (Szews, 2016). Thus, basically, technologies change news ecologies. Furthermore, mediated technologies affect our social interactions and the diversity of news and news-sourcing routines (Wiard, 2019). Because of technology, many actors get involved in news production (Öberg, 2023). For newsrooms and journalists, technologies can reassure them about news-making opportunities, question their judgements and sometimes unsettle them. Simultaneously, newsroom technologies can make leeway for other unwanted practices. For example, some AI technologies can, through algorithmic manipulation, be weaponised to generate climate disinformation and climate-related deepfakes (Chen, 2024; Rossing, 2023). In this chapter, we argue that AI technologies can offer 'the best of both worlds'. They can be strategic for climate new production. At the same time, they can be hauntingly dangerous in that they may further information malpractices like climate disinformation, misinformation and climate-related deepfakes. Therefore, these technologies offer both positive opportunities for journalists and present challenges that impede in climate news reporting.

A note on methodology

We rely on in-depth interviews with practising climate journalists. Due to geographical limitations in recruiting participants for face-to-face interviews, we draw on telephone interviews that yielded insightful information on emerging AI-driven climate change reporting practices amongst sampled journalists in South Africa and Zimbabwe. This chapter focused on South Africa and Zimbabwe as these countries face pressing issues related to climate change, including extreme weather events like droughts and floods, shifting rainfall patterns and increasing temperatures (Nyahunda, 2024). While telephone interviews obviously have their limitations, like a perceived lack of non-verbal cues and sometimes an inability to ask probing questions, in this chapter, we actively explored the subject by asking follow-up questions. One of our immediate findings is that climate news is a relatively new phenomenon in these two countries. As one journalist said, "There was never a dedicated climate news reporter in our newsrooms until 2019 … I am not even there because I am trained as such … I was asked to report climate change" (Respondent 6, telephone interview from Zimbabwe). We noticed that some newsrooms in these two countries do not have dedicated climate change reporters, although this seems to be changing. As a result of this, our sample was limited to 11 journalists to whom we were snowballed. In the sample, six journalists were from Zimbabwe, and five were from South Africa. One climate reporter snowballed us to another. It was difficult to get more because of three reasons. First, some newsrooms do not have these reporters specialising in climate as a news beat. Second, while some journalists we contacted cover climate news, they flatly refused to identify themselves as climate reporters. Third, we noticed palpable fear amongst journalists at the mention of AI in climate journalism. During the interviews, we realised that climate change was least understood by journalists who practise it. One of them said:

> I am not sure if what I am doing here is what constitutes climate change … I sometimes doubt if it is … but I will talk to you if you say it is.
>
> (Respondent 8, telephone interview from Harare)

Journalists experienced challenges when we moved from asking about climate reporting to AI. Some of the journalists withdrew from the interviews. After initially promising to participate, one journalist said, "I do not think I am the right person for this … I know nothing about AI …". This points to the fact that for climate change journalists, who hardly understand this news beat (because they have spent much of their journalism careers reporting something else), bringing into question AI in climate change took them to unfamiliar territory. Another journalist from South Africa added:

> My knowledge about AI is extremely rudimentary, and I hesitate to offer any opinions/insights on this issue. May I suggest that you make contact instead with my younger and more tech-savvy colleague?
>
> (Email conversation with a South African journalist, 12 July 2024)

Because this is more of a scoping and exploratory research, our dataset presents no problems. These interviews were conducted between April and October 2024. Some of the questions we asked included, amongst others, the AI technologies they (journalists) use in climate change news production, the opportunities these AI technologies bring to climate reporting, the challenges they encounter when using AI, their overall impression of AI in climate news reporting; what they think they possess in terms of skills (their capital) and their vulnerabilities when using AI in climate news production. To analyse the data, we drew out themes using the traditional and often-used thematic analysis. In the next section, we present our findings.

Findings

The first section of our findings answers our first research question: How is AI being integrated in climate change news reporting by selected journalists in South Africa and Zimbabwe? ? In the second and third parts, we answer the second research question: In what ways does the integration of AI in climate change news coverage alter news production and journalistic practices? We conclude the chapter with a discussion.

Our findings show that climate change journalists in both contexts are slowly integrating AI technologies in their everyday climate news reportage practices. We note that this integration of AI technologies is slow for several reasons, chief reason being that the newsrooms themselves have not, at least at the moment, made strategic decisions to equip journalists with these technologies. Interviewed journalists noted that what they use as AI climate change reporting technologies is what is available for free for them. For instance, one journalist noted that

> Simple and rudimentary software that is sometimes embedded on cellphones can help you write a good story on weather … even visualisation AI technologies that produce clear pictures of cyclones and hurricanes and even map it for

> you are helpful … but these are not purchased technologies. Some of them are free, and others embedded in our everyday gadgets like phones and personal computers.
>
> (Respondent 6, telephone interview from South Africa)

Another journalist from South Africa added:

> If it's a simple text-based reporting without data analysis, GPT-3 can be useful for improving drafts of news articles, reports, or summaries related to climate change, and it is free.
>
> (Respondent 10, telephone interview from South Africa)

While this integration continues gradually, it has not been punctuated by strategic integration that would include careful choices of which technologies to include through well-thought-out purchasing decisions on the part of newsrooms.

Integrating AI in climate news reporting

As AI gradually makes its way into newsrooms, climate journalists have begun to use some of these technologies for surveillance and environmental monitoring. Some of the AI technologies, according to journalists, allow them to gather data about the climate in real time and, in the process, allow them to effectively surveil and monitor of the environment. One journalist from Zimbabwe said:

> If you are a climate change reporter, it is important to scan the environment for data that can easily be turned into a story … we used to do it even before AI, but AI has increased the rate at which we can get datafied information that can easily be turned into a story … it has also helped us even with the analysis and interpretation of that data.
>
> (Respondent 6, telephone interview from Zimbabwe)

While a respondent from South Africa also added,

> For heavy Data Analysis and Visualisation types – then tools like Tableau or Google Data Studio with AI add-ons can help journalists analyse and visualize complex climate data.
>
> (Respondent 10, telephone interview from South Africa)

Some climate-related AI tools are predictive tools that help journalists read rainfall, hurricane and storm patterns. However, as some journalists admitted, most of these tools are still experimental, difficult to understand and relatively out of reach for other journalists. One journalist said:

> So far, I think we are largely in the experimental stage of AI tools for climate change communication … as journalists, we understand some AI-driven

> applications. But more are still being rolled out, and some are costly to get, and others we really have not mastered.
>
> (Respondent 7, telephone interview from South Africa)

Another climate change reporter admitted that AI-driven climate communication technologies like Eugenie.ai, Sipremos AI tools and many more are still relatively out of reach for most African climate change communicators (Respondent 4, telephone interview from South Africa). Some respondents agreed, however, that big tech like Google, through their climate AI technologies like GraphCast, a tool produced by Google DeepMind, have been useful and relatively available. GraphCast enables climate change communicators to map, via AI-driven satellite images, the progression of potential disruptive weather events like cyclones and hurricanes about ten days before the event and in less than a minute. This is a machine-learning capability that humanity had not possessed in history. Furthermore, GraphCast models ten-day weather forecasts within a single minute, with unprecedented accuracy, according to DeepMinds.[1] There are similar technologies that are being invented by a Brazilian technology company, Sipremos.[2] One journalist noted:

> Ideally, we would want these technologies; for example, I would like a technology that helps with tracking ocean pollution or one that predicts the level of damage oil exploration would cause to our coasts … I think this would help me in the nature of climate journalism that I am engaged in;
>
> (Respondent 3, telephone interview from South Africa)

There are specific necessities that journalists believe can be achieved through the use of AI. For journalists, AI can be liberating in terms of its newsroom capabilities if it can perform those chores that traditionally could not be performed by journalists but had to be reported in the newsroom. For instance, as noted by the respondent above, these include tracking ocean pollution and oil damage. These are important environmental issues. Yet, they have always been reported from secondary sources with capabilities. For instance, Marine and Ocean organisations can provide information about ocean pollution to journalists. Journalists, however, think that AI can make their jobs easier by providing this information.

Hard news is a very important element of the climate communication ecosystem (Robbins and Wheatley, 2021). It is important because this is the news beat that informs us about rain patterns over time and climate forecasts over time. In the short term and medium term hard news informs us about sudden shifts in weather conditions and provides early warning systems. AI tools have been useful in this regard. Some AI tools, like AI-powered aerial expedition cameras and AI-powered GoPro-cameras, visualise climate issues and generate images and visuals that can be used by journalists. As one journalist explained,

> When you have modern AI technologies, you can visualise data, you can also make use of big climate data, which AI can help interpret because they are fast

> and accurate … and this makes it easy for you to produce quick and accurate hard news climate stories.
>
> (Respondent 9, telephone interview from Zimbabwe)

A very important hidden takeaway from all these is the question of access to these technologies. As noted, some journalists posit that these technologies are not as ubiquitous as prevailing public discourses might make them look. They are still available to a few 'privileged' others. This is an important point, AI can actually exacerbate climate change communication divides if the benefits of these technologies do not filter down. Admittedly, African start-up techs like Kuzi in Kenya are beginning to produce their own technologies that are anchored in African climate challenges and provide unique and 'customised' climate solutions to African countries (see Taylor and Munoriyarwa, 2025). How have these technologies changed climate communication? In the next section, we answer this question.

AI-driven changes to climate news

What is gradually growing is a practice that can be referred to as AI-driven climate reporting. We note from the interviews that AI-driven climate reporting comprises several related practices. The most visible ones are AI-induced data-driven climate change communication and AI-induced climate change news and climate news visualisations. The claim is not that these practices never existed. It is rather that in the age of AI, they have been supercharged and enhanced.

Data journalism has been part of journalism practice for more than a century (Moyo and Munoriyarwa, 2019). It was rather enhanced, and its application by journalists accelerated with new data mining technologies coming into existence (Chattratichat et al., 1997). What AI is doing is accelerating climate communication opportunities through big data. One journalist noted:

> There are several AI-driven software programs that help break down complex data that you can use for climate communication. These technologies have always been there since the advent of data journalism practices. But now they are being driven by AI, and they are making climate communication much easier.
>
> (Respondent 8, telephone interview from Zimbabwe)

For climate communication journalists, the most popular of these technologies include, amongst others, Google Data Studio, which makes it easy for journalists to code large amounts of data. There is also eDiscovery, which allows journalists to organise key documents and provide an AI-driven platform for the analysis of data in these documents. There are also 10+ mapping tools that provide verifiable crowd data. These technologies have been a key staple for data journalists. Now, they have been supercharged by AI, which is making climate reporting much easier. One respondent notes:

> For some time, I have reported other news beats, not necessarily climate issues. I have used some of these tools. Now I realise that the same big data tools now use AI, and it's very important when covering climate news. Climate news produces huge amounts of data. AI tools help analyse these data and interpret climate phenomena that we also need clarity on, like ocean pollution, greenhouse gas emissions.
>
> (Respondent 2, telephone interview from South Africa)

In addition to these, AI provides platforms where climate journalists can access free climate information. For example, already existing platforms like Enigma are now AI-powered and provide a broad collection of public data. One journalist said, specifically of Enigma:

> The platform (Enigma) has been there, and we used to gather public data for data journalism reporting. AI has made it even more useful. On Enigma, you get a huge amount of data. You cannot easily sift through it. But now AI has made it easy to sift through this data and analyse it … so yes, data-driven technologies have become even more useful because they have made it easy for us to write stories;
>
> (Respondent 5, telephone interview from Zimbabwe)

Furthermore, AI technologies now allow for unprecedented practices of data visualisation in climate news reporting. Climate data can now easily be visualised because there is a host of AI tools that perform visualisations. Examples include Qlik, which integrates data and analyses data on its AI-powered platforms. These AI-driven technologies now allow for better visualisation of climate phenomena. One journalist noted:

> I enjoy using AI maps, especially on rainfall patterns, weather systems, climate outlook in the next few years … The visuals are clear … and they are also usable.
>
> (Respondent 7, telephone interview from South Africa)

Another respondent agreed thus,

> There are a lot of exciting AI visualisations images online. You can download and just acknowledge the source. But these visualised images really help explain climate issues. I have used them especially to explain complex processes like the melting of glaciers and its consequences on climate … these images help, and they make it easy to understand such issues, especially to an audience that has not seen glaciers and does not understand the melting and its effects.
>
> (Respondent 5, telephone interview from Zimbabwe)

Data visualisation in communication is very important. Visuals clarify, for better understanding, the complex climate information. This is helpful, for example, to have climate information simplified on an AI-driven visualisation platform like Tableaux for laymen audiences who always see climate communication as a hard science. Thus, the visualisation of climate information can make it more user-friendly to non-experts. Thus AI-driven visualisation technologies may, in the long run, be central to discourses of climate communication by reducing complex information into easy-to-understand formats.

Risks of AI to climate news reporting

The adoption of AI technologies in climate change communication comes with its own risks to journalism practice. In addition to that, AI can alter news production in ways, some of which we have discussed above – that is, the introduction of intensified approaches to data-driven AI news production. Furthermore, AI technologies have significantly altered news production by reducing the overt reliance by climate journalists on experts as news sources to more factual climate change communication. Piatek, Haines and Larson (2024) note that climate change communication is now increasingly vulnerable to misinformation and disinformation. Our respondents agree that disinformation and misinformation are the gravest danger to contemporary climate change communication. One respondent notes:

> It is increasingly becoming difficult to sort facts from fiction, especially on the information you find online.
>
> (Respondent 11, telephone interview from South Africa)

Another Respondent agreed, adding, the most difficult is when you are confronted by images:

> You do not know which one is true, which one was generated online ... and which one was deliberately falsified.
>
> (Respondent 5, telephone interview from Zimbabwe)

The biggest fear amongst journalists is how to separate disinformative images from those that are not. One journalist from South Africa said:

> Images are the most vexatious. You would not know whether, in the age of AI-generated images, the one you would like to use for your story is the right one. People are using AI to generate their own images.
>
> (Respondent 4, telephone interview from South Africa)

A respondent from Zimbabwe recounted his own experience thus,

> I remember when I was covering Cyclone Idai. The images of the destruction brought by the Cyclone varied, yet they were all about the same cyclone. However, I learnt that people were generating and sharing images from the Asian Tsunami and attributing them to the Cyclone. People were editing images,

> and it was hard to establish which ones were from Zimbabwe and which ones were from Asia.
>
> (Respondent 9, telephone interview from Zimbabwe)

This is one of the major challenges of AI-driven climate change journalism: how do you separate climate facts from climate fiction? With the advent of AI-driven deepfakes, it is increasingly hard to separate facts from fiction. The challenge is a big one for journalists. For journalists, this has never been higher.

There is also the growing danger of misinformation about climate change. Journalists note that as disinformation thrives, especially online, gullible audiences, unable to separate facts from fiction, often sharc falsc information unintentionally, and this has a multiplier effect on false news about climate news. One journalist said:

> You always feel sad when you are fact-checking trending information about the climate. You realise that misinformation is actually being shared by unsuspecting audiences. If you are not careful, you end up spending much of your time debunking misinformation and disinformation instead of doing the actual reporting.
>
> (Respondent 4, telephone interview from South Africa)

As more AI-powered digital platforms come to the fore, the dangers of misinformation and disinformation continue to grow. This has a detrimental effect on climate news reporting. Journalists, who would naturally see AI as a natural ally to news reporting, may have to rethink their approaches to integrating these technologies in their everyday practices.

Another significant risk is the issue of representation in AI technologies. These technologies are often built on value systems that do not necessarily reflect the African context. Typically developed and owned by Western companies, AI technologies lack transparency and accountability to users in Africa and other regions (Ogola, 2023). Furthermore, they are frequently trained on data that fails to accurately represent diverse perspectives and the complexities of global climates, making it difficult to achieve fair representation of climate and environmental contexts in various areas. Additionally, Ogola (2023) argues that these AI applications are primarily designed for different contexts, which can hinder their effective adoption in African contexts. One journalist from South Africa concurs:

> Understanding the limitations of AI – the data and algorithms that AI models employed before using AI-driven tools is important. There are biases related to geography, socioeconomic status, political affiliations, etc and it is crucial to address these to ensure fair reporting on climate change issues.
>
> (Respondent 10, telephone interview from South Africa)

It is the existence of these biases that can lead to the production of disinformation and misinformation at an industrial scale (Francisco-a-j-Camacho and Waldman,

2024), with the danger that climate change and environmental reporting are polluted. This danger, however, does not exist in a social vacuum. Climate change itself is a heavily contested issue. In fact, it has generated high levels of polarisation. AI technologies have raised significant concerns around their susceptibility to disinformation and misinformation (Francisco-a-j-Camacho and Waldman, 2024). And, as we noted earlier, becoming a to-go to tools, they can further heighten this polarisation and make climate change reporting much harder for journalists, more polarised and very divisive.

Discussion and conclusion

Our aim in this chapter was to answer two main research questions. How is AI being integrated in climate change news reporting by selected journalists in South Africa and Zimbabwe? In what ways does the integration of AI in climate change news coverage alter news production and journalistic practices? Our paper examined the interactions between AI and news production practices in newsrooms, specifically in the context of climate change reporting. In the first section of our findings, we outlined some of the AI tools that are still emerging and used by journalists in climate communication. The second and third sections note that AI technologies have enhanced and powered data journalism practices. However, we also noted the dangers of AI-driven technologies to climate communication. These dangers include misinformation and disinformation that is also spurred by algorithms.

Furthermore, our findings show that the integration of AI in climate change communication is happening gradually. Emerging AI-driven technologies are being adopted by journalists in communicating climate change. We also demonstrate that most journalists are still taking leaps of faith in the dark. They are slowly learning to apply these technologies without fully understanding them. This creates a conundrum for journalists who have to learn on the job in terms of the efficacy of these technologies and how much value these AI technologies can add to climate reporting.

Our findings reveal a continuous disruption of news production, which was also confirmed by Anderson (2010). AI usage in climate change communication shows that climate news production is no longer a monopoly of journalists. It is now hidden within technological structures. For instance, an AI algorithm can generate an image about glacier melting, which can find itself in the mainstream, as journalists noted, and become part of our mainstream conversations. Therefore, the dangers of disinformation and misinformation in climate communication are real and present. AI, therefore, has a colouring effect on climate change communication through its disruption of climate news ecologies. In this regard, we should pause and think: what are the main sources of our climate news? In an AI-driven media ecology, how far should we continue to trust even mainstream sources of climate change news, considering how journalists are overwhelmed by disinformation and misinformation? One journalist summed it:

> The amount of climate disinformation and misinformation is simply overwhelming. You cannot fact-check everything every day … it just gets to a point where you also get tired … people are so intentional in their bid to spread disinformation and alter the climate change agenda.
>
> (Respondent 1, telephone interview from Zimbabwe)

Currently, there is an added danger of deepfakes in climate communication. This often comes in images – for example, altered images of ocean pollution, forests burning down in the Brazilian Amazon and waves of destruction by hurricanes. These new disruptions in the digital ecosystem present danger to climate communication. Shin (2024) noted that deepfakes, for instance, that arc algorithmically manipulated and weaponised (Chen, 2024: Rossing, 2023) present a great threat to our news media ecosystem. All these developments influence what we consume as climate communication news. However, there is a silver lining despite the challenges. As Öberg (2023) notes, digital platforms have led to variations in news sources. We can argue the same for AI in climate communication. Climate statistics and other forms of data are no longer a monopoly of experts. They can be sourced through crowdsourcing and on other AI-driven platforms.

It is hard to predict how the future of climate communication in the AI age will likely morph. This is because, at the moment, we do not fully understand the range of capabilities the current AI technologies have, even though we explored current capabilities in this chapter. As the interviewee noted, while we celebrate certain advantages, which are already noted in this chapter, there are obvious dangers that we should be wary of. The more climate-related disinformation and misinformation we interact with, the more we undermine public trust in climate information. This has a long-term negative effect on climate change communication (Szews, 2016). More so, disinformation, misinformation and deepfakes will negatively expose mainstream actors, like journalists, which has a detrimental effect on our news ecosystem.

The strength of this paper lies in its detailed analysis of the gradual integration of AI technologies by journalists in two Southern African countries: South Africa and Zimbabwe. These technologies have empowered journalists in both nations to generate more data-driven insights on climate news than ever before. The predictive capabilities of various AI tools have enhanced storytelling techniques in newsrooms. Most respondents highlighted the convenience of these tools, which align well with the fast-paced nature of newsrooms. The deployment of AI tools helps save time, as they can execute certain tasks more quickly than a human. However, the integration of AI also brings certain challenges, including issues related to misinformation, disinformation and representation. These challenges can negatively impact public discourse on climate change reporting and may undermine the credibility of climate journalists striving to communicate climate issues effectively. The paper's reliance on a limited set of interviews is a limitation. As noted in the methodology, some participants were reluctant to engage when AI was mentioned, and others felt unqualified to participate despite their experience in climate reporting. Nevertheless, despite these challenges, the sample remained effective, as the research prioritised depth of understanding over broad representation.

We recognise that AI technologies can enhance climate action; however, it is essential to regulate them properly and make systematic, context-informed usage of these technologies to mitigate the risks associated with their use in news-making practices. The opacity layers of AI (see Carabantes, 2023, p. 424) – that is, AI's black-boxed algorithms, intentional secrecy and the still widespread technological illiteracy of most journalists – make us caution against a very deterministic way, without taking cognisance of their technical limitations and technical affordances. Many of the tools mentioned by respondents are easily accessible and do not require significant financial investment, which raises the potential for misuse by journalists. Therefore, organisations should develop policies to guide journalists on the appropriate use of these tools. Such policies should be part of a broader ethical framework for AI tool usage. Future research could expand on this study by specifically examining the regulation of AI tools in climate change communication, potentially including a larger sample of journalists from various Southern African countries to gain a more comprehensive understanding of AI technology regulation.

Notes

1 This can be followed here: https://t.co/ygughpkdeP.
2 The inventions by Supremos can be followed here: https://www.sipremo.com/

References

Anderson, C. W. (2010). Journalistic networks and the diffusion of local news: The brief, happy news life of the "Francisville Four." *Political Communication*, *27*(3), 289–309. https://doi.org/10.1080/10584609.2010.496710

Appelgren, E., and Jönsson, A. M. (2021). Engaging citizens for climate change—Challenges for journalism. *Digital Journalism*, *9*(6), 755–772.

Brüggemann, M., and Engesser, S. (2017). Beyond false balance: How interpretive journalism shapes media coverage of climate change. *Global Environmental Change*, *42*, 58–67.

Bulkeley, H., and Betsill, M. M. (2013). Revisiting the urban politics of climate change. *Environmental Politics*, *22*(1), 136–154.

Campbell, B. M., Hansen, J., Rioux, J., Stirling, C. M., and Twomlow, S. (2018). Urgent action to combat climate change and its impacts (SDG 13): Transforming agriculture and food systems. *Current Opinion in Environmental Sustainability*, *34*, 13–20.

Carabantes, M., 2023. Why artificial intelligence is not transparent: A critical analysis of its three opacity layers. In Lindgren, S. (ed.), *Handbook of Critical Studies of Artificial Intelligence* (pp. 424–434). London: Edward Elgar Publishing.

Chattratichat, J., Darlington, J., Ghanem, M., Guo, Y., Huning, H., Kohler, M., ... and Yang, D. (1997). Large scale data mining: The challenges and the solutions. In *Proceedings of third international conference on knowledge discovery and data mining (KDD-97)*.

Chen, L. (2024). Combatting climate change misinformation: Current strategies and future directions. *Environmental Communication*, *18*(1–2), 184–190.

Dunlap, R. E., and McCright, A. M. (2015). Challenging climate change. In Dunlap, R. E., and Brulle, R. J. (ed.), *Climate Change and Society: Sociological Perspectives* (pp. 300–332). Oxford, UK: Oxford University Press.

Flusberg, S. J., Matlock, T., and Thibodeau, P. H. (2017). Metaphors for the war (or race) against climate change. *Environmental Communication*, *11*(6), 769–783.

Francisco-a-j-Camacho, F. A. J., and Waldman, S. (2024). Climate misinformation is rampant. AI may be able to stop it. Accessible at: www.scientificamerican.com/article/can-ai-stop-climate-misinformation/. Accessed on 10 November 2024.

Gess, H. (2012). Climate change and the possibility of 'slow journalism'. *Ecquid Novi: African Journalism Studies*, *33*(1), 54–65.

Giddens, A. (2015). The politics of climate change. *Policy & Politics*, *43*(2), 155–162.

Goklany, I. M. (2012). Is climate change the number one threat to humanity? *Wiley Interdisciplinary Reviews: Climate Change*, *3*(6), 489–508.

Keohane, R. O.(2015). The global politics of climate change: Challenge for political science. *PS: Political Science & Politics*, *48*(1), 19–26. doi: 10.1017/S1049096514001541

Lineman, M., Do, Y., Kim, J. Y., and Joo, G. J. (2015). Talking about climate change and global warming. *PloS One*, *10*(9), e0138996.

Lombardo, G. (2019). The fight against climate change. *SEER: Journal for Labour and Social Affairs in Eastern Europe*, *22*(2), 181–188.

McKnight, D. (2010). A change in the climate? The journalism of opinion at News Corporation. *Journalism*, *11*(6), 693–706. https://doi.org/10.1177/1464884910379704

Mirza, M. M. Q. (2003). Climate change and extreme weather events: Can developing countries adapt? *Climate Policy*, *3*(3), 233–248.

Moyo, D., and Munoriyarwa, A. (2019). Data journalism in the age of big data: An exploration into the uptake of data journalism in leading south African Newspapers. In Mutsvairo, B., Bebawi, S. and Borges-Rey, E. (eds.), *Data Journalism in the Global South* (pp. 85–105). Cham: Palgrave Macmillan.

Munoriyarwa, A. (2024). Unravelling socio-technological barriers to AI integration: A qualitative study of southern African newsrooms. *Emerging Media, 2*(3), 474–498. 10.1177/27523543241288814.

Munoriyarwa, A., Chiumbu, S., and Motsaathebe, G. (2021). Artificial intelligence practices in everyday news production: The case of South Africa's mainstream newsrooms. *Journalism Practice*, *17*(7), 1374–1392. https://doi.org/10.1080/17512786.2021.1984976

National Academies of Sciences, Division on Earth, Life Studies, Board on Atmospheric Sciences, Committee on Extreme Weather Events, & Climate Change Attribution. (2016). *Attribution of extreme weather events in the context of climate change*. Washington D.C.: National Academies Press.

Nyahunda, L. (2024). Integration of indigenous knowledge systems (IKS) into climate change mitigation and adaptation endeavours: Milestones and gaps in South Africa and Zimbabwe's climate policy frameworks. *Climatic Change, 177*, 162.

Öberg, C. (2023). Disruption and the ecosystem: The changing roles of ecosystem stakeholders in the course of disruption. *Technological Forecasting and Social Change*, 194, 1–12, https://doi.org/10.1016/j.techfore.2023.122679

Ogola, G. (2023). AI, Journalism, and Public Interest Media in Africa. International Media Support, 1–25. www.mediasupport.org/publication/ai-journalism-and-public-interest-media-in-africa/

Painter, J. (2019). Climate change journalism: Time to adapt. *Environmental Communication, 13*(3), 424–429.
Paterson, M., and Grubb, M. (1992). The international politics of climate change, *International Affairs,* 68(2), 293–310. https://doi.org/10.2307/2623216
Piatek, S. J., Haines, A., and Larson, H. J. (2024). We need to tackle the growing threat of mis-and disinformation about climate change and health. *BMJ, 387*, 1–2.
Robbins, D., and Wheatley, D. (2021). Complexity, objectivity, and shifting roles: Environmental correspondents march to a changing beat. *Journalism Practice, 15*(9), 1289–1306.
Rossing, J. (2023). The disinformation landscape of the future: A qualitative perspective (Master's thesis, University of Twente).
Rusbridger, A. (2020). *News and How To Use It: What to Believe in a Fake News World.* Edinburgh, Scotland: Canongate Books.
Schäfer, M. S., and Painter, J. (2020). Climate journalism in a changing media ecosystem: Assessing the production of climate change-related news around the world. *Wiley Interdisciplinary Reviews: Climate Change, 12*, 1–20. 10.1002/wcc.675.
Shin, D. (2024). Algorithms and journalism: A case study of an AI-driven news ranking and recommendation system. In Pavlik, J.V. (ed.), *Milestones in digital journalism* (pp. 178–196). Routledge.
Szews, P. (2016). Algorithms of the Web. *Acta Universitatis Lodziensis. Folia Litteraria Polonica, 35*(5), 199–216. earch/sociology/projects/watchingthewatchersstrengtheningpublicoversightofintelligencedrivensurveillance/
Taylor, R., and Munoriyarwa, A. (2025). The Climate Imperative: How AI Can Transform Africa's Future. Available at SSRN: https://ssrn.com/abstract=5233767 or http://dx.doi.org/10.2139/ssrn.5233767
Wiard, V. (25 February 2019). News ecology and news ecosystems. *Oxford Research Encyclopedia of Communication.* Retrieved 6 October 2024, from https://oxfordre.com/communication/view/10.1093/acrefore/9780190228613.001.0001/acrefore-9780190228613-e-847

13 Datafication of climate change news in Zimbabwean Daily Newspapers

Albert Chibuwe and Takunda Maodza

Introduction

The study of human-induced climate change is not new, but the huge interest in climate change study amongst communication scholars is recent (see Olteanu, Castillo, Diakopoulos, and Aberer, 2015). As Olteanue et al. (2015) note, climate change has generated great interest from scholars globally. Matsilele and Tshuma (2023, p. 2) also note that environmental journalism, though still under-researched, "is now attracting interest from different players within the broader field of mass communication and journalism studies". In contrast, the idea of data journalism and its practice are still new (Appelgren, Linden, and Van Dalen, 2019). The practice is enabled by digitization (see Lomborg, Dencik, and Moe, 2020; Chibuwe, Munoriyarwa, and Maodza, 2025) and the advent of big data, algorithms and Artificial Intelligence. Digital media facilitate datafication (Mayer-Schönberger and Cukier, 2013, as cited by Lomborg, Dencik, and Moe, 2020, n.p.). Unlike traditional journalism, data is the primary source of news in data journalism (Chibuwe et al., 2025), whilst human beings are secondary sources (see Kalantzis, Bratsas, and Veglis, 2018).

Academic interest in data journalism is equally recent; the area is still generally under-researched in Africa (see Sackey, Asiamah, and Osei-Mensah, 2022; Mutsvairo, 2019a,b). Following on from the above, it is arguable that the study of data-driven climate change journalism is even more recent than the study of big data and climate change. A Google Scholar search of climate change news coverage and big data in Africa and in Zimbabwe yielded nothing. Instead, the search yielded research articles that focused either on media coverage of climate change or on data journalism. There were no articles that straddled the divide between data-driven journalism and climate change journalism. Specifically, the few available studies, on one hand, focus solely on media's coverage of climate change, and on the other hand, on data journalism. But there were fewer studies on data journalism than on climate change journalism. The low academic interest in climate change journalism could be because climate change is not given the same wide coverage by the media as other news beats such as sport and politics (see Matsilele and Tshuma, 2023).

DOI: 10.4324/9781003660217-17

Even though data journalism has been mainly associated with digital media, this chapter focuses on mainstream newspapers. This is because mainstream media have also been datafied as they use Artificial Intelligence to track and profile people and use algorithms to create personalized content (Van den Bulck and Moe, 2018, as cited by Lomborg, Dencik, and Moe, 2020, n.p.). It is crucial to note that this chapter is located at the intersection of three key journalism concepts, namely, investigative journalism, data journalism, and climate journalism. First, even though there is no agreed definition of data journalism (Chiumbu and Munoriyarwa, 2023), for purposes of this chapter, data journalism is defined as the processing of seemingly meaningless information, largely statistical, into an explicit news item to hold power to account (Kalatz et al., 2018; Jamil, 2019). Second, climate journalism gathers, processes, and mass mediates climate change-related information, its causes, impacts, mitigation, and adaptation measures (Schäfer and Painter, 2021). Third, investigative journalism – often wrongfully confused with leak journalism – is a journalism form "that goes beyond the daily news cycle" through exposure of hidden truths (Galvao, 2023). Its main agenda is to expose wrongdoing and enable corrective action to be instituted in the public interest (Chambers, 2000; Pilger, 2005; Carson, 2019; Konow-Lund, 2023).

In light of the foregoing, this chapter seeks to fill this notable void in research on datafication of investigative climate change news. This is significant considering that, "as a phenomenon and area of study, 'datafication' is typically qualified in relation to a specific domain: datafication of health, work, the public sector, the city, or everyday life" (Lomborg et al., 2020, n.p.) or climate change news. We thus seek to achieve this goal by interrogating the use of big data in climate change reporting in selected Zimbabwean newspapers. The chapter seeks to answer the following questions: How has big data been deployed in climate change investigative news reporting? In what ways has the datafication impacted climate change news reporting? And, what new, if any, traits in climate change news reporting has the datafication of climate change news ushered in?

Data journalism, investigative journalism, and climate change

There is a corpus of literature on data journalism (Milan and Trere, 2019; Moyo, 2019; Mutsvairo, 2019a; Chiumbu and Munoriyarwa, 2023; Chibuwe et al., 2025; Kalender, 2024), but there is a dearth of scholarship on how journalists derive investigative climate change-related news articles from big data. This chapter addresses this lacuna by interrogating how journalists from selected newspapers in Zimbabwe deploy data in investigative climate change reporting. Whereas extant studies explore the state of data journalism in Africa by embracing a newsroom-centric approach (Moyo, 2019; Mutsvairo, 2019a; Mutsvairo et al., 2020; Munoriyarwa, 2022; Chibuwe et al., 2025) or problematize the concept of datafication, which they equate to data colonialism (Milan and Trere, 2019), this contribution seeks to understand the extent to which journalists at the selected newspapers derive climate change-related news articles from big data in a way that exposes wrongdoing for corrective action to be instituted.

The chapter goes beyond deriving empirical evidence from journalists on the extent they appreciate data journalism by investigating how reporters actually originate investigative climate change articles from data. Extant literature on data journalism emphasizes how it enhances investigative reporting (Moyo, 2019) or how data journalism augments the media's watchdog role on the governing elites (Kalatzi et al., 2018; Cheruiyot et al., 2019; Jamil, 2019). While scholars (Mutsvairo, 2019a; Chiumbu and Munoriyarwa, 2023; Chibuwe et al., 2025) also suggest that data journalism empowers citizens to demand good governance, transparency and accountability, there is a poverty of scholarship interrogating the potential value of big data in investigative climate change journalism in Africa. Conversely, this chapter opens new avenues into understanding how data – treatable as evidence in news production – aids investigations in climate change reporting. We appreciate that data journalism is a complex and multifaceted field demanding expertise from various fields including data science, social sciences and computer sciences (Heravi and Lorenzi, 2020). The chapter investigates the extent to which journalists at the selected newspapers are conversant with data journalism and how they apply it as an investigative journalism tool when reporting climate change.

The chapter is cognizant of the fact that in the era of heightened digitalization, data is an essential news source. Consequently, it seeks to understand how journalists derive investigative climate change stories from open-source data. Extant scholarship on the state of climate change journalism in Africa notes that it is event driven; journalists lack expertise in climate reporting; editors claim climate stories are a "hard sell" (Cramer, 2008; Nassanga, 2013); and it is less popular and less profitable (Matsilele and Tshuma, 2023). In most instances, climate change-related stories are eclipsed by news articles on politics, entertainment, and sports, a situation Boykoff and Boykoff (2007) blame on journalistic norms (news values), which give saliency to drama and novelty at the expense of serious issues such as climate change. Climate change, as Kovarik (2020) notes, is given less prominence by news media, and its coverage is episodic and event driven (see Kovarik, 2020).

This chapter argues that the concepts – data journalism, investigative journalism, and climate change – have not received a combined scholarly gaze, while there is also a scarcity of scholarship specifically examining how climate change journalists extract investigative stories from data. The chapter fills this void by examining how journalists at *The Herald* and *Daily News* deploy big data in investigative climate change reporting.

Theoretical framework

Theoretically, the study deploys insights gleaned from the sociology of news production and the concept of datafication. It utilizes the conceptualization of "journalism as a set of social interactions and ritualistic behaviors" (Belair-Gagnon and Revers, 2018, n.p.). These interactions and behaviours are largely influenced by journalists, the media's structure, political-economic-socio-cultural and technological factors, news sources, and political culture (Benson, 2004; Belair-Gagnon and Revers, 2018). The categories above, however, "hide almost as much as they reveal" (Benson, 2004, p. 279). Nevertheless, this chapter is interested in the interactions (between journalists and editors, news sources, environmental activists, government etc.) and the day-to-day news-making rituals or news production processes and how these impact the deployment of data journalism in reporting climate change. In other words, what relations and considerations (e.g. knowledge, ethics, political, economic, socio-cultural and technological factors) influence how climate change news is constructed? Here the chapter considers both the news sourcing routines and production processes involved as they relate to investigative climate change reporting. News items are viewed as a social construction (see Schudson, 1989). The chapter is thus interested in the interactions and ritualistic behaviours that shape this construction. It interrogates both the macro-level (political, economic, technological, and socio-cultural) and micro-level (news routines, bureaucratic pressures) influences (see Benson, 2004).

But to make up for what the categorization of influences above hides, the chapter also examines the space between the macro- and micro-levels. Benson (2004) calls it "the mezzo-level 'journalistic field' [which] represents an important [news] shaping factor … largely ignored" by researchers. It is in this mezzo-level space where knowledge and context are located, factors that the chapter considers crucial in investigative data-driven climate journalism. The categorizations listed above hide knowledge and context. As Schudson (1989, p. 266) notes, political economy, social organization, and culturological categorizations of influences "have thus far fallen short of a comparative and historical social science of news production". Benson (2004) concurs and states that these three ignore context and are ahistorical. Thus the insertion of knowledge and context into the mezzo-level is significant considering the inadequacies of the categorizations largely used in studying the sociology of news production. It is also significant given observations by Chibuwe, Munoriyarwa, and Maodza (2025) that Zimbabwean journalists betray a lack of knowledge of what data journalism is. Beyond that, the chapter is interested in establishing the state of investigative climate change reporting in Zimbabwe.

The chapter is also informed by the concept of datafication. Datafication, a process that involves quantification and huge data sources (see Mathieu and Jorge, 2020; Baack, 2015), is enabled and accelerated by digital media but is not new. Datafication helps in decision-making on what content to produce and to whom to deliver the content to etc. Following on from the above, the chapter argues that it is possible to unravel the existence or absence of the process of datafication or quantification in climate journalism by subjecting climate news articles to qualitative content analysis. It is crucial to note that even though data-driven journalism generates stories from largely statistical information (see Kalatz et al., 2018; Jamil, 2019), numbers are not the beginning and end of data-driven journalism. Data-driven journalism is not exclusively quantitative or statistical. Qualitative data sources can also be utilized in data journalism. The term "data" does not refer to numbers only but to both quantitative and qualitative information. The data or publicly available seemingly meaningless information (see Chibuwe et al., 2025) may as well be qualitative. The stories from such data sources are equally data-driven. Data activists acknowledge that raw data might be meaningless to citizens;

consequently, there is a need for "data intermediaries" (Baack, 2015). Journalists can be considered as "data intermediaries".

The study of datafication in the media can, arguably, look at the three main areas that traditional media research has always been interested in, namely: audiences (Mathieu and Jorge, 2020); production and distribution; and text. At both the production and audience levels, datafication has been used in decision-making (see Cinnamon, 2019) on audience consumption patterns; on surveilling audiences; and on what to produce and deliver to them. This current work, located at the textual level, examines the climate-related stories in Zimbabwean newspapers to unravel the datafication processes that went into the writing of the stories.

Research methodology

This chapter adopts the qualitative research paradigm. It is a descriptive and interpretive multiple case study seeking a nuanced understanding of how journalists at *The Herald* and *Daily News* use data to derive investigative climate change stories. Qualitative studies capture people's lived experiences (Streubert and Carpenter, 1999; Polit and Beck, 2004), enabling extraction of meaning. *The Herald* and *Daily News* are the selected case studies, as they are Zimbabwe's most vibrant daily newspapers pursuing distinct editorial policies. Published by the Zimpapers Media Group, *The Herald* is pro-ZANU–PF government and practices patriotic journalism, whilst the *Daily News* traditionally engages in adversarial or oppositional journalism that holds power to account (Ranger, 2005). The chapter is a comparative analysis of similarities and dissimilarities in *The Herald* and *Daily News*' use of data-driven investigative climate change reporting during the period January 2024 to December 2024. Twenty-five (25) climate change news articles were purposively sampled from *The Herald*, whilst seven (7) were purposively sampled from the *Daily News*. The selected stories are presented in Tables 13.1 and 13.2.

Table 13.1 Daily News stories

Daily News	*Date*
Women leading battle against climate change. By *Staff Writer*	29 December 2024
Faith leaders unite against climate change. By Shamaine Chirimujiri	6 October 2024
Climate change is turbo-charging Somalia's problems – but there's still hope. By Justin Rowlatt, BBC/*Staff Writer*	18 September 2024
Disaster at the intersection of cholera, climate change. By *Staff Writer*	3 March 2024
Mobilise resources to fight climate change. By Shamaine Chirimujiri	31 October 2024
Climate change needs gender-sensitive policies. By Hazel Marimbiza	23 August 2024
Coltart Challenges youth on climate change. By Hazel Marimbiza	11 August 2024

Table 13.2 The Herald stories

The Herald	*Date*
Funding key to climate change fight: President	15 November 2024
No room for half measures in climate fight	13 November 2024
Zimbabwe unwavering in commitment to fight climate change: SADC chair	12 November 2024
President urges diaspora to share innovative ideas	12 November 2024
Climate smart agriculture pays off	12 November 2024
Respect developing nations' views: President	11 November 2024
Global climate indaba: What's at stake	11 November 2024
Farmers start collecting Pfumvudza inputs	22 October 2024
Beneficiaries applaud timely distribution of Pfumvudza inputs	22 October 2024
First Lady launches Pfumvudza4She programme in Mt Darwin	24 October 2024
Tavengwa launches 2024/25 Pfumvudza/Intwasa scheme in Epworth	29 October 2024
200K Harare residents benefit from Pfumvudza scheme	30 October 2024
Over 8 million Pfumvudza plots ready for planting	31 October 2024
Embrace Pfumvudza/Intwasa, farmers urged	01 November 2024

Source: Authors' compilation.

Interviews were also conducted with purposively selected *Daily News* and *The Herald* journalists specializing in climate change journalism. Purposive sampling enables selection of information-rich sources (Marshall, 1996; Yin, 2011). Whereas initially the intention was to interview all journalists who generated climate change-related stories during the period under study, the researchers eventually settled for those specializing in climate journalism. This enabled the extraction of rich insights. Ultimately, four journalists – two from each newspaper – were interviewed based on their vast experience in climate change journalism. Journalists interviewed at *The Herald* are identified as Respondents A and B, while those at the *Daily News* are identified as Respondents C and D. Respondent A has been reporting on agriculture and climate for a decade and a half now (as of 2024). Respondent B has over two decades' experience in climate journalism. Respondents C and D are also well versed in climate journalism. The intention was to interview editors also, but they were unwilling. The data from interviews was subjected to thematic analysis.

Finally, the chapter also subjected climate change stories published by *The Herald* and *Daily News* to qualitative content analysis to gauge the extent to which journalists make use of data-driven investigative climate data journalism when gathering and processing climate news. Qualitative content analysis enables collection and analysis of "content of a text" (Nueman, 1997, p. 272). The analysis focused on identifying: the use of numbers/statistics; the data sources of the stories; and the use of infographics, amongst other manifestations of data-driven journalism.

The following sections thematically present the findings of this chapter.

The Herald *and* Daily News' *event-based coverage of climate change*

The findings show that the *Daily News* and *The Herald*'s coverage of climate change news was mainly event-based. *The Herald* had more (25) climate change stories than the *Daily News* (7) during the study period. But *The Herald's* stories were neither investigative nor data-driven. *The Herald's* climate stories were, just like the *Daily News'*, derived from events such as climate conferences, press conferences, and press releases. The *Daily News* relied on a multiplicity of sources including government sources, ordinary people, non-governmental organizations (NGOs), and data sources, amongst others. In contrast, *The Herald* largely relied on government sources and government-owned companies. It also praises the president's efforts at climate change proofing Zimbabwe. For example, *The Herald* published seven stories in five days based on President Mnangagwa's attendance at the United Nations' 2024 Climate Change Conference COP29 hosted by Azerbaijan. The stories were based on his speeches and interviews. The following are examples of some of the stories: "Funding Key to Climate Change Fight: President" (15 November 2024); "President Urges Diaspora to Share Innovative Ideas" (12 November 2024); "Respect Developing Nations' Views: President" (11 November 2024) etc.

The president was the story and is, in some stories such as "Climate Smart Agriculture Pays Off" (12 November 2024), the only source, and his claim that the country broke all records in wheat production is taken as truth. It is neither substantiated nor disputed with statistics or infographics. There was no attempt by *The Herald* to deploy data journalism to illustrate Zimbabwe's wheat production patterns and confirm the president's remarks. Single sourcing in journalism results in one-sided narratives (Wilcox, 2011), whilst multiple sourcing establishes facts and achieves balance and fairness. Accuracy and accountability are arguably unachievable under a single-sourcing newsgathering regime.

In contrast, out of the *Daily News'* seven stories, one was sourced from the BBC, whilst three did not use any data journalism in their reportage. The stories were event based, specifically deriving from media briefings and addresses made by officials at seminars or conferences. For example, the story titled "Mobilise Resources to Fight Climate Change" was based on President Mnangagwa's address at a Climate and Health Africa conference in Harare. The *Daily News*, just like *The Herald*, did not go beyond the speeches. The episodic and event-based nature of climate change news reporting in both newspapers under study confirms arguments by scholars that climate journalism in Africa is event driven and/or episodic (Cramer, 2008; Nwabueze et al., 2015; Okoliko, 2021; Pointer and Matsiko, 2023). The stories in both newspapers also betray a lack of in-depth knowledge of data-driven investigative climate journalism, something that data gathered through interviews also confirms, as shall be discussed below. For example, the assumption that mere use of statistics (this shall be discussed in detail later), even unreliable ones provided by officials, is data-driven journalism is faulty.

The foregoing notwithstanding, the *Daily News'* stories demonstrate a better deployment of data journalism than *The Herald*. This is despite the fact that the *Daily*

News does not have, as Respondents C and D stated, a dedicated climate news desk. Those that write climate news do it out of passion. Three *Daily News* stories made use of data journalism. The stories made use of statistics and documents from government departments and organizations such as the World Health Organisation (WHO). They also relied on research published in academic journals. For example, the story titled "Disaster at the Intersection of Cholera, Climate Change" (Staff reporter, 3 March 2024) relied on statistics, documents, articles, and research from the following: the Ministry of Environment, Water and Climate; the United Nations; WHO; the United Nations Educational and Scientific Organisation (Unesco); the National Library of Medicine; the Journal of Law, Policy and Globalization; the Multidisciplinary Digital Publishing Institute; the International Institute of Science, Technology and Education; and the Climate Centre. Of all the *Daily News* stories, this is the story that made extensive use of data journalism in reporting climate change. The *Daily News* thus made use of both statistical data and qualitative data in generating data-driven stories.

Use of statistics to generate stories

Both newspapers used, to an extent, statistics to generate or substantiate their stories. However, *The Herald's* sole reliance on single government sources to generate stories is problematic. This is because the figures or numbers they relied on were apparently manipulated statistics. Interviewed journalists averred that the government has a tendency of manipulating statistics such that their data-driven stories end up being based on doctored statistics. The story titled "Over Eight Million Pfumvudza Plots Ready for Planting", is an example of the use of unreliable statistics. The story, whose only source was a government official, claimed that "2 807 263 households have prepared at least one Pfumvudza plot and 1 824 720 have prepared three or more plots". It further states that "in total 8 175 023 plots have been prepared and ready for planting". The official was from a ministry one interviewee specifically mentioned as providing inflated statistics. Respondent B stated that

> … there are challenges when it comes to statistical information given by government. For example, the ministry of agriculture when they generate statistics, say on wheat production, *they inflate the figures* [emphasis ours]. They have a lot of inconsistencies, and it is meant to achieve a certain agenda. As for us (journalists at *The Herald*), we must comply with the information given even if you doubt it.
>
> (Interview with Respondent B, 18 December 2024, Harare)

Respondent B further asserted that there is always conflict between data on crop assessments availed to them by government and NGOs. This was corroborated by Respondent A, who asserted:

> It's difficult to verify data/information that you are given by the government. We usually get crop assessment reports … but then the problem is with verifying the

> information because … you can get different information from them than that given by non-governmental organisations (NGOs). NGOs will have done their own assessments whose *results do not match* [emphasis ours] those generated by the government.
>
> (Interview with Respondent B, 18 December 2024, Harare)

The disparity in statistics given to *The Herald* by government and NGOs must have prompted the journalists to go beyond the normal news cycle to expose the hidden truth (Galvao, 2023). Instead, the journalists simply regurgitate the doctored statistics, thereby missing an opportunity to make data the source of the story. As Respondent A observed, "Under such scenario, we write what government would have given us". This is a consequence of government control over *The Herald* and is a typical example of what Ranger (2005) calls patriotic journalism. The stories become mere propaganda as they turn a blind eye to the "inconsistencies", which, if further investigated, would expose government wrongdoing.

The Herald also merely summarized the scientific jargon-laden speeches of experts and officials and made no effort to break down the jargon to enable ordinary readers to understand the climate issues. The article titled "Research Institute Tackles Climate Change" (11 November 2024) is an example. It was based on an address by an expert from the Ministry of Lands, Agriculture, Fisheries, Water and Rural Development. The stories largely portrayed the government in a positive light, with nine (9) of the 25 stories praising the president and government's alleged achievements in ensuring food security in the country.

In contrast, the *Daily News'* journalists believed that data journalism involves use of statistics (only). This belief is reflected in the newspaper's climate change news' use of statistics. The statistics are however not the source of the news. But they are merely used to support the point the article is advancing. Their deployment of statistics, though linked to data journalism, is contrary to scholars' views on how statistics are utilized in data journalism. The statistics, just like the climate jargon, were not processed into investigative climate news items. As scholars argue, data journalism is the processing of seemingly meaningless information, largely statistical, into an explicit news item holding power to account (Kalatz et al., 2018; Jamil, 2019). Furthermore, the statistics were not used as springboards for investigative journalism. They were not used to unravel hidden truths; instead, they were used, by *The Herald*, to legitimize government policies and, by the *Daily News*, to show the gravity of climate change effects. For example, the *Daily News* provided statistics of cholera outbreaks' deaths and infections to demonstrate the effects of climate change-induced cholera outbreaks. This is contrary to what investigative journalism does; it goes beyond ordinary everyday newsgathering and writing through exposure of hidden truths (Galvao, 2023). Its main agenda is to enable corrective action to be instituted in the public interest (Carson, 2019; Chambers, 2000). In this regard, the statistics utilized by *The Herald* were meant to serve elite interests through legitimizing their rule, whilst in one of its stories, the *Daily News* utilized statistics to legitimize the initiatives by NGOs. The story was titled "Women Leading Battle Against Climate Change" (29 December 2024). In another titled "Disaster at the Intersection of Cholera, Climate Change" (3 March 2024), the *Daily News* tried to link climate change and cholera, and the corrective action that it wants taken by Harare municipality is the provision of adequate safe portable water. But the link is not forcefully established nor are the other underlying causes of the shortage of water and cholera outbreak investigated.

Journalists' misunderstanding of investigative climate data journalism

The findings show that *Daily News* and *The Herald* journalists have a fair understanding of climate change, climate journalism, and data journalism but lack understanding of investigative journalism in general and data-driven investigative climate journalism in particular. For example, Respondent D defined climate journalism as "writing and reporting impact of climate change", whilst Respondent B stated that it is "reporting the causes, impact and coping mechanisms employed by communities" as they fight global warming. The respondents also argued that climate change, which at times is conflated with global warming, was vast and entails reporting national response strategies taken by governments towards adaptation to climate change. Respondent C defined climate change as "significant and lasting changes in earth's climate patterns caused by humans", while Respondent D claimed that it was the "rise in the earth's temperature causing problems like rising sea levels and hot weather".

Journalists from both newspapers covering the climate news beat understand what data journalism is. For example, Respondent A stated,

> Data journalism…involves I think use of statistics. We do statistics... We now make use of graphs and pie charts to illustrate a story. You can decide not to write the full story but just write a paragraph and everything is illustrated in graphs and charts. Here we call them "particles". I can use statistics, let us say you compare this farming season with the previous one in terms of yields. We are shifting to digital (media) and that is why we have intensified the use of data when writing stories.
>
> (Interview, 18 December 2024, Harare)

It is apparent that this journalist knows what data journalism is. Whereas Respondents A and B from *The Herald* indicated that they have written data-driven climate change news, Respondents C and D from the *Daily News* seemed well versed in climate journalism but indicated that they have never written data-driven climate change stories. They also lack skills in generating data-driven investigative climate change stories.

Ironically, the *Daily News*, which had fewer climate news stories (7) than *The Herald* (25) and journalists who said they have never written a data-driven climate change story, had more stories (3) that utilized data journalism than *The Herald*, which had none. This is despite the fact that *The Herald's* two selected journalists indicated that they have written data-driven climate change stories and their management encourages deployment of data journalism. Respondent B stated that "there is an instruction from management that we take data journalism seriously". The findings from the analysis of stories from *The Herald* do not substantiate the claims by Respondents A and B that they deploy data-driven climate journalism. There were no data-driven stories. *The Herald's* stories that deployed statistics cannot be considered as data-driven climate news stories because data was not the primary source.

In contrast, even though the *Daily News* did not use numbers as the primary sources, they made extensive use of publicly available data. Interestingly, both newspapers had zero investigative data-driven climate news stories despite interviewees' claims that they deploy investigative climate data journalism. The stories did not go "beyond the daily news cycle" through exposure of hidden truths (Galvao, 2023) as investigative journalism should. For example, Respondent B admitted to lacking depth in generating investigative climate stories from data. They blamed the dearth of climate experts in Zimbabwe for the situation. Respondent B argued that "there are not so many [experts] in that area" and "we lack critical scientists".

The response below by Respondent A to whether they have at any point in their career indulged in investigative data-driven climate change journalism demonstrates a worrying lack of understanding of data-driven investigative climate change journalism:

> I report on agriculture. The Tobacco Industry and Marketing Board (TIMB) at times sends us statistics on tobacco exports, tobacco registrations, farmer registrations, sector by sector from A1 farmers, commercial farmers to communal farmers. They give us such data according to provinces to say so many farmers registered this farming season in this province. *So, I used to write stories from such information which would have been provided in form of just graphs. I used to write detailed stories using the information without even interviewing anyone* [emphasis ours].
>
> (Interview, 18 December 2024)

The apparent lack of in-depth understanding of investigative climate data journalism amongst the selected climate news journalists confirms Chibuwe et al.'s (2025) observation that Zimbabwean journalists in general require schooling in data journalism to enable them to instrumentalize it in investigative journalism.

All the journalists that were interviewed had limited understanding of investigative journalism. Respondent A partly claimed it is the extraction of a news story from graphs and pie charts given to them by sources such as the meteorological department. Respondent B suggested investigative climate reporting partly meant telling a story in visual form, as the contemporary young audience lacks interest in reading the newspaper, hence the need to save time. Presentation of stories pictorially, further claimed Respondent B, helps attract audience interest – especially young ones. Asked how often they derive investigative stories from seemingly meaningless online data, Respondent B revealed that they do that "especially when I come across interesting titbits from critics and other observers like NGOs", adding, "they throw titbits and then you generate a story" (Interview, 18 December 2024, Harare). Again, Respondent B's belief that investigative journalism derives from "titbits" and "NGOs" suggested limited appreciation of the concept. They suggestively confuse investigative journalism with leak journalism, and they are prone to manipulation by those throwing "titbits" at them.

The Zimbabwean political environment is highly politically polarized. On the one hand is the governing party (ZANU–PF) and on the other hand, the opposition (Movement for Democratic Change/Citizens' Coalition for Change). "Critics" and "NGOs" align with opposition politics. In such a political context, reliance on critics and NGOs risks biased narratives. The situation is worsened by their reliance on government sources, who routinely feed the journalists falsified data for propaganda purposes. Respondents A and B confirmed the habit of some government portfolios to falsify data, which they feed journalists at *The Herald*, fully cognizant that such information gets published as given. This speaks of the extent to which *The Herald's* political economy structures and shapes narratives on climate change.

Respondent A noted that the fact that the government has a stake in *The Herald* meant journalists cannot question information released to them by its arms. This speaks not only to the political economy influencing narratives but also to the sociology of news production as enunciated by theorists including Breed (1955). Journalists are cultured into knowing the publishable and unpublishable stories without anyone pointing a gun at them. They automatically internalize the acceptable and unacceptable, and consequently, they practise self-censorship. Self-censorship kills investigative journalism. Investigative journalism must expose wrongdoing for corrective measures to be instituted by responsible authorities (see Carson, 2019; Chambers, 2000). The polarization mentioned earlier is evident in Respondent B's assertion that there is always conflict between data on crop assessments availed to them by government and by NGOs.

The conflicting data must have prompted the journalists to investigate the disparity. Instead of single sourcing, as is the case in *The Herald's* stories that were analysed, investigative journalism encourages multiple sourcing of information. This is because single sourcing is the source of news evil (Wilcox, 2011), as it makes it difficult to achieve accuracy and accountability (Colon, 2022). It also stifles diversity of opinion (see Mellado and Scherman, 2020, p. 17). Investigative journalism is also in-depth, original and systematic inquiry, which cannot be done in a short period of time. On the basis of the foregoing, *The Herald* stories do not qualify as investigative data-driven articles seeking justice and accountability. The same is true of the *Daily News* stories. However, *The Herald's* stories are a mere regurgitation of information given by the state and its arms. The situation is worsened by the reluctance of climate change experts to process data for the media, as noted by all the respondents.

Discussion

In *The Herald* stories, government officials were the main sources of climate change news and not the data. The stories did not utilize data journalism, and where statistics were used, these were provided by the human sources such as the president and experts from government departments. The statistics were unreliable as well. In light of scholars' observation that, unlike traditional journalism, the source of news in data journalism is data (Chibuwe et al., 2025), it is plausible to argue that climate change news reporting in *The Herald* does not utilize data journalism. *The Herald's* failure to use publicly available data to verify statements and statistics by government officials is a result of its long-observed pro-government stance (see Ranger, 2005). The climate change news stories, as observed, were neither investigative nor data-driven. The lack of knowledge about data-driven investigative climate journalism explains the absence of data-driven investigative climate news stories in *The Herald* and the *Daily News*. The journalists' location in their respective newspapers and Zimbabwe's current political-economic context also explain the apparent absence of data-driven climate journalism. The foregoing confirms the significance of studying the space between the macro- and micro-levels; the space Benson (2004, n.p.) calls "the mezzo-level 'journalistic field' [which] represents an important [news] shaping factor ... largely ignored" by researchers. It is a space that categories of factors that influence social relations and rituals in journalism identified by Benson (2004) and Belair-Gagnon and Revers (2018) do not explain. Knowledge is located in this space: the mezzo-level. Both factors are crucial in investigative data-driven climate journalism.

But the sourcing routines of *The Herald* demonstrate an over-reliance on human sources or single government sources as primary sources. The absence of expert sources on climate change was attributed to the experts' reluctance to interact with journalists. In contrast, the *Daily News'* sourcing routines in the data-driven stories reflected a balance between human sources and data sources. For example, the story titled "Disaster at the Intersection of Cholera and Climate Change" (3 March 2024) is testimony that the *Daily News* used a multiplicity of data and human sources. Specifically, the story cited seven human sources and nine data sources that included academic journal articles.

The journalistic interactions and rituals, insofar as climate change is concerned, are influenced by both macro- and micro-factors. For example, they are influenced by journalists, news organizations' structure, political-economic-socio-cultural and technological factors, news sources, and political culture (Benson, 2004; Belair-Gagnon and Revers, 2018). *The Herald's* pro-government stance and its unquestioning use of inflated government-provided statistics are a consequence of control; it is state-controlled. This state-control also influenced *The Herald's* news sourcing routines, as it relied heavily on government officials as sources. Furthermore, it also influenced the newspaper's tendency to rely on single sources since critical journalism is not high on its agenda; it serves as a mere government mouthpiece.

In contrast, the *Daily News'* private ownership freed it from this obligation to positively cover the government. Consequently, it relied on a multiplicity of sources and also utilized data journalism despite not having a dedicated desk. Those that cover the beat do it out of passion, and this perhaps explains why the newspaper only had seven stories as compared to *The Herald's* 25 during the study period. The newspaper's structure and individual journalists' (Benson, 2004; Belair-Gagnon and Revers, 2018) interest were the factors that influenced the production of data-driven climate news in the *Daily News*.

Conclusion

In light of the foregoing, the chapter concludes that selected Zimbabwean journalists do not understand the related concepts of investigative journalism and data-driven investigative climate journalism. This has had a negative bearing on their reportage of climate change news reporting. Climate change news reporting remains event-driven and episodic despite the seismic shifts in journalism wrought by big data and Artificial Intelligence. The event-based or episodic nature of climate change journalism in Zimbabwe confirms Kovarik's (2020) observation that climate change news reporting is episodic and peaks when disaster strikes or when there is a major global event. In addition, the journalists' lack of knowledge explains the absence of data-driven investigative journalism in *The Herald* and the *Daily News*. It also demonstrates the significance of the oft-ignored space between the macro- and micro-levels of the journalistic field (see Benson, 2004). The "mezzo-level", as Benson (2004) calls it, also influences journalistic interactions. State control of *The Herald* arguably hampers the deployment of data-driven investigative climate journalism. Finally, it is arguable that beyond the lack of knowledge, Zimbabwe's post-2000 context in which journalists have been harassed, arrested, and threatened by the state has had a chilling effect on investigative journalism in general and data-driven journalism in particular. This is plausible given the COVID-19 era arrest of a journalist, Hopewell Chin'ono, for allegedly exposing corruption involving the procurement of personal protective equipment.

References

Appelgren, E., Lindén, C., and Van Dalen, A. (2019). Data journalism research: Studying a maturing field across journalistic cultures, media markets and political environments. *Digital Journalism*, 7(9), 1191–1199. https://doi.org/10.1080/21670811.2019.1685899

Baack, S. (2015). Datafication and empowerment: How the open data movement re-articulates notions of democracy, participation, and journalism. *Big Data and Society*, *2*(2), 1–11. https://doi.org/10.1177/2053951715594634

Belair-Gagnon, V. and Revers, M. (2018). The sociology of journalism. In: Vos, T. P. (ed.) *Journalism. Handbooks of communication science* (Vol. 19, pp. 257–280). Berlin: De Gruyter.

Benson, R. (2004). Bringing the sociology of media back. *Political Communication*, 21(3), 275–292, https://doi.org/10.1080/10584600490481299

Boykoff, M.T. and Boykoff, J.M. (2007). Climate change and journalistic norms: A case study of US mass-media coverage. *Geoforum*, 38(6), 1190–1204. https://doi.org/10./1016/j.geofurm.2007.01.008

Breed, W. (1955). Social control in the newsroom: A functional analysis. *Social Forces*, 33(4), 326–335.

Carson, A. (2019). *Investigative journalism, democracy and the digital age* (1st edn). New York: Routledge. https://doi.org/10.4324/9781315514291

Chambers, D. (2000). *Critical approaches to the media: The changing context for investigative journalism*. Routledge.

Cheruiyot, D., Baack, S., and Ferrer-Conill, R. (2019). Data journalism beyond legacy media: The case of African and European civic technology organizations. *Digital Journalism*, 7(9), 1215–1229.

Chibuwe, A., Munoriyarwa, A., and Maodza, T. (2025). Data journalism, accountability and transparency in Zimbabwe's "New Dispensation": Some empirical reflections. In: Mutsvairo, B. and Orgeret, K.S. (eds) *The Palgrave handbook of global digital journalism*. Cham: Palgrave Macmillan. https://doi.org/10.1007/978-3-031-59379-6_22

Chiumbu, S. and Munoriyarwa, A. (2023). Exploring data journalism practices in Africa: Data politics, media ecosystems and newsroom infrastructures. *Media, Culture & Society*, 45(4) 841–858.

Cinammon, J. (2019). Attack the data: Agency, power, and technopolitics in South African data activism. *Annals of the American Association of Geographers*, 110(1), 1-17, https://doi.org/10.1080/24694452.2019

Colon, A. (2022). A question of ethics: How long do you give a source to respond?, Journalism Institute, National Press Club. www.pressclubinstitute.org (accessed 17 January 2025).

Cramer, C.M. (2008). *The framing of climate change in three daily newspapers in Western Cape Province of South Africa* (Degree of Master of Science). Stellenbosch University, South Africa.

Galvao, G. (2023). The power and importance of investigative journalism: Uncovering truths for a better world. *Journal of Mass Communication and Journalism*, 13(4), 533.

Heravi, B.R. and Lorenz, M. (2020). Data journalism practices globally: Skills, education, opportunities, and values. *Journalism and Media*, 1(1), 26–40.

Jamil, S. (2019). Increasing accountability using data journalism: Challenges for the Pakistani journalists. *Journalism Practice*, 15(1), 19–40.

Kalatzi, O., Bratsas, C., and Veglis, A. (2018). The principles, features and techniques of data journalism. *Studies in Media and Communication*, 6(2), 36–44.

Kalender, A. (2024). The datafication of newsrooms: A study on data journalism practices in a British newspaper. *Journalism and Media*, 5, 48–64. https://doi.org/10.3390/journalmedia5010004

Konow-Lund, M., Park, M., and Bebawi, S. (2023). *Hybrid investigative journalism*. Cham: Springer Nature.

Kovarik, B. (2020). The rise of environmental journalism in Asia, Africa, and Latin America. In Sachsman, D.B. and Valenti, J.M. (eds) *Routledge Handbook of Environmental Journalism* (pp. 52–69). London: Routledge.

Lomborg, S., Dencik, L., and Moe, H. (2020). Methods for datafication, datafication of methods: introduction to the special issue. *European Journal of Communication*, 35(3), 203–212. https://doi.org/10.1177/0267323120922045

Marshall, M.N. (1996). Sampling for qualitative research. *Family Practice*, 13(6), 552–526. https://doi.org/10.1093/fampra/13.6.522

Mathieu, D. and Jorge, A. (2020). The datafication of media (and) audiences: An introduction. *Journal of Media and Communication Research*, 36(69), 1–10. https://doi.org/10.7146/mediekultur.v36i69.122585

Matsilele, T. and Tshuma, L. (2023). Through the media looking glass: Journalists' perceptions on South Africa's funded environmental journalism. *African Journalism Studies, 44(1), 58-72*. https://doi.org/10.1080/23743670.2023.2280269

Mellado, C. and Sherman, A. (2020). Mapping source diversity across Chilean news platforms and mediums. *Journalism Practice,* 15(7), 974–993, https://doi.org/10.1080/17512786.2020.1759125

Milan, S. and Trere, E. (2019). Big data from the South(s): Beyond data universalism. *Television and New Media*, 20(4), 319–335. https://doi.org/10.1177/1527476419837739

Moyo, L. (2019). Data journalism and the Panama papers: New horizons for investigative journalism in Africa. In: Mutsvairo, B., Bebawi, S., and Borges-Rey, E. (eds) *Data journalism in the Global South. Palgrave studies in journalism and the Global South.* Cham: Palgrave Macmillan. https://doi.org/10.1007/978-3-030-25177-2_2

Munoriyarwa, A. (2022). Data journalism uptake in South Africa's mainstream quotidian business news reporting practices. *Journalism*, 23(5), 1097–1113.

Mutsvairo, B. (2019a). A new dawn for the "developing" world? Probing data journalism in non-western societies. In: Mutsvairo, B., Bebawi, S., and Borges-Rey, E. (eds). *Data journalism in the Global South.* 2020 (pp. 1–20). Cham: Springer Nature.

Mutsvairo, B. (2019b). Challenges facing development of data journalism in non-western societies. *Digital Journalism*, 7(9), 1289–1294.

Mutsvairo, B., Bebawi, S., and Borges-Rey, E. (eds) (2020). *Data journalism in the Global South.* Cham: Springer Nature.

Nassanga, G.L. (2013). African media and global media discourse. Implications for sustainable development. *Open Society Repository Communication Journalism*, Art. e23050451, 1–17.

Neuman, W. (1997). *Social research methods: Qualitative and quantitative approaches.* Needham Heights, MA: Allyn & Bacon.

Nwabueze, C., Nnaemeka, F. and Umeora, D. (2015). Nigerian newspapers coverage of climate change issues. *European Scientific Journal, 11*(17), 171–184.

Okoliko, D.A. (2021). Media(ted) climate change in South Africa, Nigeria and Kenya: Re-imagining the public for engagement. Phd Thesis, Stellenbosch: Stellenbosch University.

Olteanu, A., Castillo, C., Diakopoulos, N., and Aberer, K. (2015). *Comparing events coverage in online news and social media: The case of climate change*. Proceedings of the Ninth International AAAI Conference on Web and Social Media.

Pilger, J. (2005). *Tell me no lies: Investigative journalism and its triumphs*. London: Vintage.

Pointer, R. and Matsiko, S. (2023). How are African talking about climate change and who is doing the talking. *Journal of African Media Studies*, Special Issue: Shifting, *15*(2), 247–271.

Polit, D.F. and Beck, C.T. (2004). *Nursing research: Principles and methods* (7th edn). Philadelphia: Lippincott Williams & Wilkins.

Ranger, T. (2005, October). The rise in patriotic journalism and its possible implications. *Westminster Papers in Communication and Culture*, Special Issue, pp. 8–17.

Sackey, R., Asiamah, E. O., and Osei-Mensah, B. (2022). Data journalism in Africa: A systematic review. *Advances in Social Sciences Research Journal*, 9(12), 248–259.

Schäfer, M.S. and Painter, J. (2021). Climate journalism in a changing media ecosystem: Assessing the production of climate change news around the world. *Wiley Interdisciplinary Review, 12*(1), https://ideas.repec.org/s/wly/wirecc.html

Schudson, M. (1989). The sociology of news production. *Media, Culture and Society*, 11, 263–282. https://doi.org/10.1177/016344389011003002

Streubert, H. and Carpenter, D. (1999). *Qualitative research in nursing: Advancing the humanistic perspective* (2nd edition). Philadelphia, PA: Lippincott Williams and Wilkins.

Wilcox, J. (2011). *Single sourcing is source of news*. Accessed 4 February 2025, from https://joewilcox.com

Yin, R.K. (2011). *Qualitative research from start to finish*. New York: Guilford.

14 Beyond the Headlines

Examining Media and Internet Discourse on Climate Change in Mozambique

Dércio Tsandzana

14.1 Introduction

UN Global Assessment Report on Disaster Risk Reduction, published in 2022, noted that Mozambique is Africa's third most exposed country to disaster risk (United Nations, 2022). Thus, it is essential to critically analyse the portrayal of climate change in Mozambique's media and Internet[1] channels to discern the potential impact on public awareness, perception, and action. The concept of climate change has expanded from its roots in natural science to include aspects of international politics, economics, popular culture, and religion. As a result, it has acquired new meanings and fulfils new functions (Hulme, 2009).

According to Attfield (2020), temperature increases pose challenges to human health and resilience, as well as the extension of deserts and increased problems for people living on the semi-arid fringes of deserts in their efforts to sustain themselves. "Climate change also increases the intensity and frequency of extreme weather events such as storms, hurricanes, floods, droughts, and wildfires, and these resulting changes force many human settlements to relocate to more hospitable areas" (Attfield, 2020: 1). States across the world are looking for the best ways to reduce disaster risks and to mitigate the impact of ongoing climate change. However, what hasn't been discussed is the role of the media or the Internet in this equation for less developed countries like Mozambique.

In regard to the literature, it is largely proven that "research in this tradition focuses on a wide range of historical and contemporary social and economic processes and practices that bring about vulnerability and seeks to understand how adaptation can be grounded in existing policies and regulations" (Artur and Hilhorst, 2012: 531), rather than the role of media and Internet in that process. There is a consensus that climate change has emerged as one of the most pressing challenges facing humanity in the 21st century, with its far-reaching impacts affecting ecosystems, economies, and societies worldwide.

As the world becomes increasingly interconnected through the power of media and the Internet, public discourse on climate change has gained unprecedented significance in shaping perceptions, policy decisions, and actions taken to address this complex issue. In recent years, the spread of information has been greatly facilitated by the rise of digital media and social networking platforms (Castells,

DOI: 10.4324/9781003660217-18

2012). The speed and ease with which information can be disseminated have given rise to a vibrant online discourse surrounding climate change, including debates on its causes, consequences, and possible mitigation strategies. The media, as a central conveyor of information, plays a crucial role in shaping public understanding and attitudes towards climate change. However, amid the wealth of information, there is also a risk of misinformation, sensationalism, and oversimplification of climate-related topics in the media and online platforms.

This chapter explores how climate change is represented and discussed in Mozambican media and online platforms, focusing on the narratives, actors, and impacts of such discourse. Key research questions guiding this analysis include: How do Mozambican media and Internet platforms frame the issue of climate change, and what dominant narratives emerge? Who are the primary actors shaping this discourse, and how are local communities, particularly those most affected, represented? These questions aim to uncover the interplay between media, technology, and societal engagement in addressing climate change in Mozambique.

The main argument unfolds as follows: following this introduction, Section 14.2 outlines the theoretical framework; Section 14.3 explores the climate change debate in Africa; Section 14.4 reviews the broader literature on media and climate change; Section 14.5 examines the media and Internet landscape in Mozambique; Section 14.6 presents the methodology; and Section 14.7 discusses the findings and their implications. The chapter concludes with a general reflection aimed at opening space for further research.

14.2 Theoretical Approach

This chapter examines how the media and Internet platforms affect public opinion and policy discussions around climate change in Mozambique. It accomplishes this by using an agenda-setting strategy for media material and by interpreting the construction of climate-related narratives using framing theory. These theoretical tools shed light on how the public becomes concerned about climate change and how different media outlets influence public knowledge and participation.

This approach is based on agenda-setting theory, which was first put forth by McCombs and Shaw (1972). According to the thesis, the media actively shapes reality by choosing and giving priority to the topics that are most frequently discussed, rather than merely reflecting it. The media creates a hierarchy of topics by influencing the public's opinion of what is most important through frequent coverage. This idea is essential to comprehending how environmental issues, particularly those pertaining to climate change, become more visible and make their way onto the public and policy agenda in Mozambique.

In addition to exposure, the framing of issues is important. This is the point at which Entman's (1993) framing theory becomes crucial. The way that information is arranged and presented to affect how audiences perceive and react to it is known as framing. By highlighting some perspectives—such as presenting climate change as a humanitarian disaster, a problem for policy, or an issue of international justice—and downplaying others, media frames have the power to influence public perception. Decision-makers' and citizens' perceptions of the problem's nature and importance are directly impacted by these narrative decisions.

The importance of media framing is further highlighted in Mozambique, where the effects of climate change are becoming more noticeable due to extreme weather occurrences, coastline erosion, and agricultural disturbance. Determining whether media coverage reflects lived realities and fosters open and knowledgeable public conversation can be aided by knowing which frames predominate in the climate discourse. Additionally, it shows which actors are allowed a say in how the discussion is shaped and whose viewpoints are ignored or marginalized.

This chapter aims to provide a thorough examination of the discourse surrounding climate change in Mozambican media by fusing agenda-setting and framing theory. It looks into how they are created and placed in context in addition to what subjects are prioritized. The chapter uses this dual perspective to evaluate how much media and online platforms help the creation of responsive climate policies and how much they contribute to an informed and engaged public debate. In the end, this analysis creates room for more investigation into how the media may promote environmental consciousness and motivate group action in Mozambique.

14.3 Understanding Climate Change in Africa

Climate change is a global threat that can impose strain on numerous industries and jeopardize their long-term viability. The agricultural sector is vulnerable worldwide because of insufficient production and supplies. Global feeding patterns are being challenged, particularly in African countries, where agriculture is an important economic component (Fields, 2005). Therefore, reducing the effects of climate change is critical and requires worldwide effort. Most African countries routinely deal with climate-related concerns such as droughts, which are common in semi-arid sub-Saharan Africa.

As noted by Guillaumont and Simonet (2011), these changes pose serious concerns in many geographic areas, particularly in developing countries. These countries are often regarded as being more sensitive to the consequences of climate change because of their lower adaptation potential. As a result, development agencies and governments are stepping up efforts to assist in the development of adequate adaptation measures. A thorough understanding of each country's vulnerability to climate change is required to guide adaptation (Guillaumont and Simonet, 2011).

Since the early 1970s, Africa became a special arena to emphasis on climate change (Glantz, 1992), given its vulnerability to the effects of climate change and urbanization patterns, which could have huge ramifications for climate change adaptation and mitigation. Despite the abundance of studies on climate change in Africa (Bedair et al., 2023), there is a dearth of review studies that highlight the broader research landscape (Tenaw and Hawitibo, 2021; Baninla et al., 2022). In Africa, the rise in temperatures across the continent, which exceeds the global average by 1.5 times over the last century, is a challenge (Diffenbaugh and Giorgi, 2012).

Some studies have also been conducted to estimate the impact of climate change on sub-Saharan Africa (Maino and Emrullahu, 2022). Okonjo-Iweala (2020), for example, estimated that climate change will severely diminish Africa's GDP through decreased crop yields, lower agricultural and labour productivity, and harm to human health using Kompas et al.'s (2018) modelling data. In general, Africa is facing exponential collateral damage, posing systemic threats to its economies, infrastructure, water and food systems, public health, agriculture, and livelihoods, threatening to reverse its development gains and push the continent back into deeper poverty.

In fact, it is argued that "despite having contributed the least to global warming and having the lowest emissions, the African region is the most vulnerable continent to climate change impacts" (Mavume et al., 2021: 1). According to various reports (Al-Zu'bi et al., 2022), nearly all African countries committed to increasing climate action by lowering greenhouse gas emissions and developing resilience after signing and ratifying the Paris Agreement. While adaptation to the negative effects of climate change is critical for the continent, many of these commitments are contingent on receiving appropriate financial, technical, and capacity-building support.

As noted by González and Sánchez (2022), although Africa is the region most vulnerable to climate change, few studies have focused on how Africans perceive climate change. For instance, despite a strong scientific consensus about the existence of anthropogenic climate change, only 51% of Africans believe that human activity is the main cause of climate change. It was also demonstrated just 56% of the continent's population has heard about climate change. "These gaps in climate change beliefs erode public support for environmental policies and limit individual behavioural changes that could reduce greenhouse gas emissions" (González and Sánchez, 2022: 1).

It was demonstrated that actual changes in local climate conditions are stronger predictors of climate change beliefs in Africa than access to information, political ideology, or demographics across different model specifications (González and Sánchez, 2022). Similarly, worsening perceptions of climate conditions for agriculture are a strong predictor of climate change beliefs for all five dimensions. In addition, not speaking French, English, or Portuguese is linked to a poor understanding of climate change (González and Sánchez, 2022). However, the question that remains to be answered is: Was the role of media in this discourse?

14.4 Media and Climate Change: A Brief Review

The relationship between media and climate change extends beyond national borders, with global media platforms playing a significant role in shaping public discourse, awareness, and action on this critical issue. As climate change becomes increasingly recognized as a global challenge, media outlets worldwide have assumed a crucial responsibility in communicating its complexities and implications to diverse audiences (Abbass et al., 2022).

Amplified by Paris Agreement, climate change coverage by media has experienced significant growth over the years (Santos et al., 2022). Major news agencies, television, newspapers, and digital media platforms have dedicated resources to report on climate-related developments, scientific findings, and environmental events around the world (Hase et al., 2021). This widespread coverage has played a crucial role in elevating climate change to the forefront of public consciousness, fostering a sense of global urgency and collective responsibility (Kunelius and Roosval, 2021).

Even if there is lack of consensus, media outlets can serve as crucial intermediaries between scientific research, policy decisions, and the public. They act as channels through which information about climate change is disseminated to diverse audiences (Williams, 2015). From traditional print and broadcast media to the expansive realm of digital platforms, media plays a fundamental role in sharing climate-related information, research findings, and environmental developments with society (Boykoff and Rajan, 2007).

Climate change has transformed into a compelling global news story, transcending geographical boundaries, and capturing international attention (Schäfer et al., 2021). From extreme weather events to climate-related policy decisions and climate activism, media outlets amplify climate narratives that resonate with audiences across different cultures and contexts (Salvesen, 2018). This global perspective has fostered a sense of interconnectedness and mutual concern for climate-related challenges faced by communities worldwide. The influence of media is not limited to raising public awareness; it also impacts international climate policy (Weingart et al., 2000).

Climate journalism often scrutinizes international agreements, emissions reduction targets, and the effectiveness of climate policies, thereby holding leaders accountable for their commitments. Journalists and news organizations are central players in the climate change communication process (Ganapathy, 2022). Reporting on complex scientific issues like climate change poses unique challenges for journalists. Factors such as editorial guidelines, media ownership, and resource constraints influence newsroom decisions, affecting the depth and accuracy of climate change reporting.

14.5 Climate Change and Media Discourse in Mozambique

In 2019, two strong tropical cyclones hit Mozambique during the same season—the first time since record. Tropical Cyclone Idai in March and Tropical Cyclone Kenneth, both in 2019, caused more than 600 deaths. In March 2022, Mozambique was hit by another cyclone and this time was Tropical Cyclone Gombe, which reportedly caused more than 60 deaths. In March 2023, Mozambique was hinted by a very intense tropical cyclone Freddy, an exceptionally long-lived, powerful, and deadly storm that traversed the southern Indian Ocean for more than five weeks in February and March 2023. Freddy is both the longest-lasting and highest-ACE-producing tropical cyclone ever recorded worldwide.

Little has been said in terms of research about the role of the media in the way climate effects are reported in Mozambique (Mandamule et al., 2024), even if some reports should be highlighted (IMS, 2023). The National Institute for Disaster Risk Management and Reduction (INGD) was the first to use models on a regional scale to predict future climate change in Mozambique in their study titled "Impact of climate change on disaster risk in Mozambique". It gave a perspective on the potential effects of climate change on investment and poverty reduction initiatives, as well as the vast stretches of coast that are home to investments and human habitation (Van Logchem and Queface, 2012).

In fact, it is observed that different studies have been carried out on Mozambique, most of which deal with the effects or causes of climate change, but little or almost nothing has been written from the point of view of the role of media. These studies do not discuss about the role of the discourse propagated by the Internet or other communication tools. For example, Almeida and Jacobs (2022) noted that with its long coastline and numerous rivers, Mozambique is very vulnerable to the effects of climate change and has been repeatedly hit by tropical storms, cyclones, and floods. The case studies in this study prove this.

Mavume et al. (2021) noted that Mozambique is one of the most vulnerable to problems associated with the climate change in Africa. Events that are incredibly dangerous and destructive are extraordinary and have been linked to the incidence of disasters with significant socio-economic effects (Mavume et al., 2021). Additionally, Mozambique is one of the hotspots in Southern Africa due to its high levels of poverty, extreme vulnerability to climate change, and heavy reliance on the sector of rain-fed agriculture to support its economy, provide jobs, and ensure food security. In other words, countless factors jeopardize the country's situation.

According to certain scholars (Almeida and Jacobs, 2022), the period following Cyclone Idai was difficult for many Mozambicans. It was proved that in the absence of a clear expropriation process, state officials had great leeway in determining their own method to expropriation, frequently to the prejudice of persons on the ground. Others (Artur and Hilhorst, 2012) illustrated that the successes and failures of climate change adaptation depend more on societal perceptions and the sensitivity of institutions than on the availability of appropriate policies, technologies, or funding, notwithstanding the importance of these last three aspects.

14.5.1 Media and Internet Landscape in Mozambique

The government of Mozambique has a significant impact on the mass media. The most common kind of media is radio. The autonomous Supreme Mass Media Council controls media outlets. Although the Council is formally independent, it is answerable to the government, since it is appointed by and reports to the President of the Republic. With the adoption of the 1990 Constitution (revised in 2023) and the 1991 Press Law, the concepts of media freedom first became apparent in Mozambique. Other laws cover topics connected to media freedom in an indirect manner.

The fight of a prominent and unified group of journalists on the eve of the 1990 Constitution's passage led to the principle of press freedom and freedom of expression being incorporated in the national Constitution (Bussotti et al., 2015). The Press Law (no. 18/1991) defines the concept of "right to information", including the right to disseminate information, opinions, and ideas by the press. However, the media sector faces challenges. As pointed by Ronning (2011), the low capacity of the press to collect advertising income and remain competitive and sustainable from an economic point of view is one of the major difficulties facing the private media sector in Mozambique.

The frequency spectrum is a good in the public interest and is therefore public property for addressing the group demands of the media or radio communications. The National Communications Institute (INCM), which is part of the Ministry of Transport and Communications, is where broadcasting license requests are made. This body recommends licenses to the Council of Ministers, who has the authority to approve them, and it can be renewed after ten years.

Regarding to the Internet, according to Kepios (2025), as of the beginning of 2025, there were approximately 6.96 million Internet users in Mozambique, representing an online penetration rate of 19.8%. This means that around 28.2 million people—roughly 80.2% of the population—were not using the Internet at the start of the year. In the same period, Mozambique recorded 3.70 million active social media user identities, amounting to just 10.5% of the total population. Additionally, there were 17.7 million cellular mobile connections, highlighting the widespread use of mobile technology. It is important to note, however, that many individuals use multiple mobile connections, so the number of mobile subscriptions often exceeds the total population (Kepios, 2025).

14.6 Methodology

This chapter adopts a qualitative research approach. I have conducted a comprehensive analysis of news articles, social media posts and blogs, from Mozambique and international media outlets covering climate change in Mozambique. This analysis reveals that the discourse on climate change in Mozambique is primarily focused on natural disasters, such as cyclones and floods, while the human impacts of climate change, such as displacement, food insecurity, and health risks, receive little attention. I have examined three cases that characterized the media and online discussion on climate change in Mozambique. In the first case, I looked at Cyclone Idai in 2019 (Beira); the second is Cyclone Kenneth also in 2019, which hit north of Mozambique (Cabo Delgado); and the third and final case was in Maputo province, particularly Boane, south of Mozambique in 2023 (see Table 14.1).

The analysis was made using reports news published on the three climatic phenomena that occurred in Mozambique. Technically, I have adopted a content analysis considering three categories: (i) the subject of the report—what was reported and how; (ii) the target audience—who are the actors mentioned in the news; (iii) the degree of sensitivity of the news—positive or negative bias.

Table 14.1 Case studies

Case	*Period*	*Impact*
Cyclone Idai	2019/Beira (central Mozambique)	Cyclone Idai was one of the worst tropical cyclones to ever hit the Southern Hemisphere and Africa. More than 1,500 people died because of the long-lasting storm's catastrophic damage and humanitarian disaster in Mozambique, Zimbabwe, and Malawi. In Mozambique, it caused more than 900 fatalities and approximately USD $2 billion of damage based on the impact to infrastructure.
Cyclone Kenneth	2019/Pemba (north Mozambique)	Cyclone Kenneth formed in the Mozambique Channel, near northern Madagascar and on the east of Aldabra Atoll, on 23 April 2019. Cyclone Kenneth was ranked as a Category 4 storm on the Saffir-Simpson scale, with gust wind speeds reaching a maximum of 270 kph and average wind speeds of 185 kph. Affected around 250,000 people and caused 45 casualties.
Floods	2023/Boane (south Mozambique)	Floods in Boane, northern Mozambique obliterated crops and swallowed two bridges that provide access to Maputo for commerce. Local media reported that the tragedy brought on by global warming puts food security in jeopardy. In addition, out of the more than 50,000 individuals who were reportedly affected by the water, 18,000 people had to be evacuated, according to the National Institute for Disaster Management (INGD).

Source: By the author.

The three cases analysed date from March 2019 to March 2023, a period that severely impacts climate phenomena in Mozambique. During that period the media, but especially the Internet, reported and reflected consistently on the effects of climate change in the country. I intentionally selected nine news reports, published during that period, but also two tweets shared on social networks, specifically on X/Twitter. These publications were obtained with the due authorization of its authors under the respect of anonymity and data privacy. I adopted to include online approach because social media and digital forums provide spaces for diverse voices to be heard, but they also harbour echo chambers that reinforce pre-existing beliefs and misinformation.

The news reports were identified through digital platforms, as all selected reports were available online. The selection process was intentionally curated to reduce the volume of news pieces analysed, aligning with the qualitative nature of this research. This deliberate approach reflects the context and aims of the study, utilizing purposive sampling commonly employed in qualitative methodologies. The results are presented and discussed thematically. The first theme examines the subject of the reports, focusing on the primary topics or issues addressed in the news pieces. The second theme explores the target audience, aiming to understand the intended readership or viewership of the content. Finally, the third theme assesses the degree of sensitivity, analysing both negative and positive biases in media coverage to provide a nuanced understanding of the reporting's tone and implications.

14.7 Results

(i) Subject of the report—what was reported and how: this section refers to the focus or topic covered in the news piece. This includes what was reported and how it was presented by the media outlet. Understanding the subject helps to grasp the core information being conveyed, while analysing how it is presented sheds light on the tone, style, and emphasis of the reporting about the climate events in Mozambique (see Table 14.2).

1. *The Guardian* reported that "And the biggest impact may be the least visible in one of Mozambique's most densely populated regions. After several years of drought in southern Africa, the subsistence farmers who populate the coastal plains have lost their crops at harvest time. They have no food or produce to take to market".
2. The Weather Network reported that "Cyclone Kenneth, a formidable storm expected to make landfall in Mozambique on Thursday, is drawing significant parallels to Cyclone Idai, which caused catastrophic damage across the region in March".
3. In its piece, *The East African* mentioned that "Mozambique is one of the poorest countries in the world, with most of its population living on less than $1 a day and ranked 181 on the UN Human Development Index out of 189 countries".

(ii) The target audience—who are the actors mentioned in the news: this section refers to the specific group or demographic that a news article or piece is intended to reach and engage. The actors mentioned in the news can be individuals, organizations, or entities that play a role in the events or issues being

Table 14.2 The subject of the report

Title	*News Report*	*Period*
No food, no shelter, no livelihood: families in Mozambique hit by Cyclone Idai	*The Guardian*	March 2019
A repeat disaster: Mozambique pushed to brink by cyclone	The Weather Network	March 2019
Mozambique closes 90 schools due to heavy flooding in Maputo	*The East African*	February 2023

Source: By the author.

Table 14.3 The target audience

Title	*News Report*	*Period*
Cyclone Idai: Mozambique survivors desperate for help	BBC	March 2019
Mozambique rescue teams struggle to save thousands	*The Guardian*	March 2019
Mozambique braces for deadly cyclone's landfall	*China Daily*	February 2023

Source: By the author.

reported. Identifying the target audience helps understand who the news is intended for, while recognizing the mentioned actors provides insight into the key players involved in the story (see Table 14.3).

1 BBC was one of the first international media reporting about climate changes events in Mozambique. In its peace, the UK reported that "Those who survived the disaster have had little reprieve to mourn the loss of their loved ones or salvage the little that is remaining of their belongings. They are in desperate need of food, shelter, and clothing, as the BBC's Pumza Fihlani reports from Beira".
2 *China Daily* tended to report quoting external actors, not necessarily for Mozambique. In of the statements, it says that "The United Nations said that at least 43,426 people had reportedly been affected, including 16,588 people displaced and 10 deaths. Mozambique's disaster relief authority has established accommodation centers and has coordinated government efforts to provide food and nonfood items to those affected".
3 *The Guardian* stated that "The magnitude of the disaster is almost unexplainable, with water as far as the eye can see, exhausted and hungry people trapped on exposed rooftops for days and mothers throwing their babies into rescue boats but having to stay behind in the trees they are holding on to, according to a rescue worker in the devastated port city of Beira".

(iii) The degree of sensitivity—positive or negative bias: this section refers to how strongly or intensely a particular viewpoint or perspective is presented in each context. A positive bias indicates favouritism or a favourable inclination towards a certain viewpoint, while a negative bias suggests a predisposition against it (see Table 14.4).

1 D+C/E+Z reported that "Cyclone Idai devastated Central Mozambique—particularly the port city of Beira—on 14 March. More than 600 people died and over 1,900 were injured. Idai made it clear that the country is not prepared for such a catastrophe".
2 News 24 reported that the National Institute for Disaster Management (INGD) announced that the deaths occurred in the capital Maputo, while 14,000 people

were evacuated out of more than 30,000 people had been affected by the flooding. In one of the statements, it can be read "(…) We still have many people to rescue, (but) we are using small boats. We think that if we had helicopters, we would be in a better position to respond to the emergency, but we don't …".

3 Vox reported that "Additionally, more than 58,660 homes have been destroyed by the storm. Beira, a city with a population of 400,000, continues to be mostly without power, and it's still very difficult to bring supplies into the city via roads".

14.7.1 Discussion

The Guardian's headline (Table 14.2) could be seen as offering a somewhat narrow perspective on the climate challenges facing Mozambique. While it rightly underscores the less apparent consequences of climate change, it may inadvertently overshadow other crucial facets of the issue. One could argue that the report lacks a broader contextualization of the climate crisis in Mozambique. It does not delve into the historical, systemic, and multifaceted factors that contribute to the current predicament. Such a comprehensive understanding is pivotal in formulating effective and sustainable solutions. Moreover, the attribution of the impacts primarily to climate change and drought might simplify a complex scenario.

Climate-related adversities are often intricately connected with various socio-economic and political factors, as well as land use practices. By not acknowledging these interconnections, the article might not offer a complete picture. The piece could carry implicit biases, whether they be political, economic, or environmental. Additionally, it might be noted a potentially excluding certain perspectives or voices that could provide a more rounded view of the issue. Furthermore, it may be seen as slightly one-sided due to the absence of a discussion on potential solutions or mitigation strategies.

The Weather Network (Table 14.2) draws attention to the impending arrival of Cyclone Kenneth in Mozambique, emphasizing parallels with the previous destructive event, Cyclone Idai. However, there are some aspects that might warrant scrutiny. First, its headline's focus on drawing parallels between Cyclone Kenneth and Cyclone Idai. This comparison is indeed relevant for understanding potential impacts, but it may overshadow unique characteristics and challenges posed by each cyclone. This could potentially lead to an oversimplified narrative. In addition to that, the use of the term "formidable storm" might be seen as potentially sensationalist. In addition to that, the mention of "catastrophic damage" caused by Cyclone Idai serves to highlight the severity of the previous event. However, do not provide specific impacts and lessons learned from Cyclone Idai, which could inform preparedness and response efforts for Cyclone Kenneth.

The report from *The East African* (Table 14.2) sheds light on Mozambique's economic condition and its global ranking on the UN Human Development Index.

While this information is vital for understanding the socio-economic context of the country, there are some points that merit examination. It is noted that it is important for reports to balance these statistics with narratives of resilience, innovation, and progress within the country. Moreover, it is worth noting that statistics can change over time due to various factors, including policy changes and global economic shifts. Therefore, the report could benefit from providing context on how these figures have evolved and the potential efforts being made to address these challenges.

On target audience, BBC report (Table 14.3) tends to focus on the immediate suffering experienced by the affected population. While this is crucial for mobilizing aid and generating empathy, it might sometimes overshadow the long-term impacts and systemic vulnerabilities that contribute to such suffering. In some cases, BCC might employ language or imagery that could be perceived as sensationalist. While it is important to convey the seriousness of the situation, care should be taken to avoid exploitative or overly dramatic portrayals.

In its report, *China Daily* (Table 14.2) appears to rely heavily on information from external sources, particularly the United Nations. While citing reputable organizations is important for accuracy, it is also crucial to seek and include local perspectives and insights to provide a more comprehensive understanding of the situation. In fact, the report does not include perspectives or voices from the affected communities or local experts. Incorporating these voices could provide valuable insights into the experiences and needs of those directly affected.

The Guardian's (Table 14.3) report employs vivid and emotive language to convey the magnitude of the disaster. While this is effective in evoking empathy and highlighting the severity of the situation, it's important to balance such descriptions with factual reporting. I understand that it is important to balance emotive language with factual reporting and to provide context about the source of information. Including statistical data and perspectives from affected communities and experts would further enrich the narrative.

On the degree of sensitivity, while the D+C/E+Z (Table 14.4) report mentions the impact of the disaster, it does not discuss the response efforts or relief initiatives that were implemented following the cyclone. Including this information would provide a more complete picture of the situation. In the case of News 24 (Table 14.4), it can be highlighted that the report focuses on ongoing rescue operations, but it could also mention the broader response efforts, relief

Table 14.4 The degree of sensitivity of the news

Title	*News Report*	*Period*
Mozambique faces enormous challenges	D+C/E+Z	May 2019
What Mozambique's unfolding flooding	Vox	March 2019
Mozambique floods kill four in capital area	News 24	February 2023

Source: By the author.

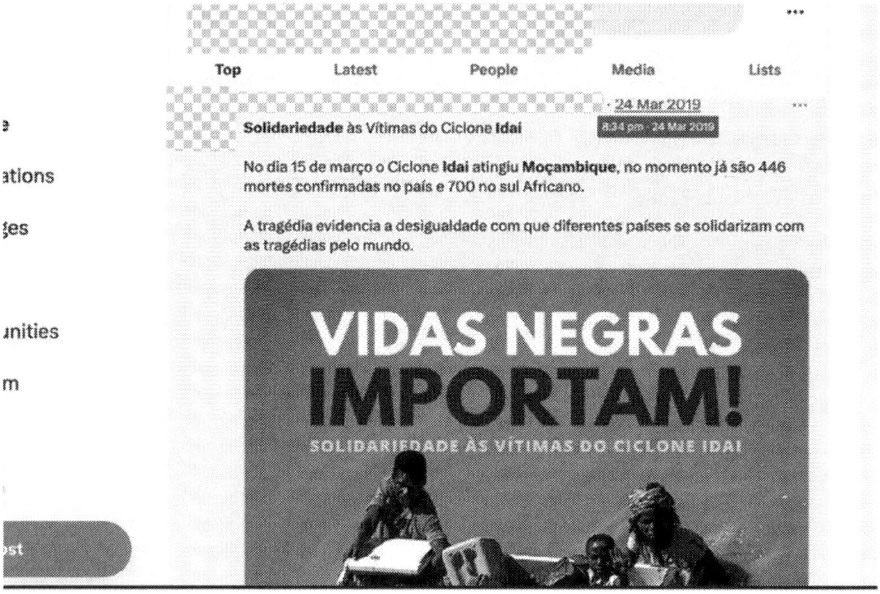

Figure 14.1 Online discussion of climate change case 1 (copyright granted, March 2019).

initiatives, and coordination with partners. Finally, Vox (Table 14.4) report predominantly focuses on the destruction of homes, which provides an important measure of the storm's impact on infrastructure. However, this focus may inadvertently overshadow other critical aspects of the disaster, such as impacts on livelihoods, healthcare, education, and overall community well-being. Focusing solely on the destruction of homes may lead to an oversimplified narrative of the disaster.

On social media networks, I have observed the emergence of discourses that run counter to what the traditional media (newspapers and television) have reported. For example, in one of the publications (24 March 2019), the slogan "Black Lives Matter" was used (see Figure 14.1)—talking about Cyclone Idai, a user emphasises the need to look at the case of Mozambique as no different with other regions across the world.

Figure 14.1 is a call for international solidarity, without looking at the colour or emergency dimension of the victims affected by the cyclone. In another context, there was the creation of a large broad solidarity movement to support the victims of the floods in Boane (Figure 14.2).

The last screenshot (below) illustrates the creation of a broad solidarity movement by Mozambicans locally, with a view to creating a sense of support for the victims. It should be noted, however, that WhatsApp was the platform chosen to launch this national campaign.

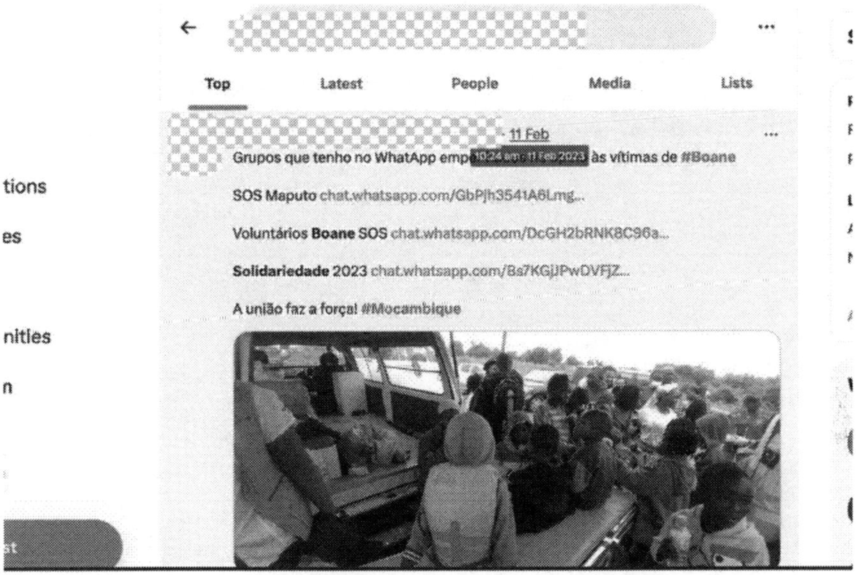

Figure 14.2 Online discussion of climate change case 2. Captured by the author with authorization (February 2023).

14.8 Conclusion

This chapter noted that media coverage of climate issues in Mozambique serves as a vital source of information, advocacy, and accountability. It plays a critical role in informing the public about the immediate and long-term consequences of climate change, as well as the efforts being made at local, national, and international levels to address these challenges. The role of media in fostering understanding and driving action remains essential. It was also observed that natural disasters are expected to become the new normal for Mozambique and the rest of the world, significantly affecting how governments approach and prioritize development challenges.

The main conclusion indicates that the discourse on climate change in Mozambique is primarily focused on natural disasters, such as cyclones and floods, while the human impacts of climate change, such as displacement, food insecurity, and health risks, receive little attention. Moreover, the study finds that the traditional media on climate change in Mozambique often adopts a "disaster narrative", portraying Mozambicans as passive victims of natural disasters and failing to acknowledge their resilience and adaptation strategies. This framing perpetuates a sense of helplessness and undermines the agency and empowerment of Mozambicans in responding to climate change, even if online discourse (social media networks) tends to be more about solidarity and hope.

The study concludes that media and online discourse on climate change in Mozambique needs to move beyond the disaster narrative and focus on the human impacts of climate change, including the experiences and perspectives of affected communities. By amplifying the voices of those most impacted by climate change, the media and online discourse can play a crucial role in raising awareness, generating support, and advocating for effective climate change policies and interventions.

Note

1 By personal choice and standardization, the initial of Internet will be in capital letter.

References

Abbass, K., Qasim, M., Song, H. et al. 2022. A review of the global climate change impacts, adaptation, and sustainable mitigation measures. *Environmental Science and Pollution Research* 12 (29): 42539–42559.

Al-Zu'bi, M. et al. (2022). African perspectives on climate change research. *Nature Climate Change* 12: 1078–1084.

Almeida, B. and Jacobs, C. (2022). Land expropriation – The hidden danger of climate change response in Mozambique. *Land Use Policy* 123(December): 106408.

Artur, L. and Hilhorst, D. (2012). Everyday realities of climate change adaptation in Mozambique. *Global Environmental Change* 22(2): 529–536.

Attfield, R. (2020). Africa and climate change. *Utafiti* 14(2): 281–294.

Baninla, Y., Sharifi, A., Allam, Z. et al. (2022). An overview of climate change adaptation and mitigation research in Africa. *Frontiers in Climate* 4: 976427.

Bedair, H., Alghariani, M. S., Omar, E. et al. (2023). Global warming status in the African continent: Sources, challenges, policies, and future direction. *International Journal of Environmental Research* 17(45): 1–23.

Boykoff, M. and Rajan, S. 2007. Signals and noise. Mass-media coverage of climate change in the USA and the UK. *EMBO Reports* 8(3): 207–211.

Bussotti, L., Barros, D. M. and Grätz, T. (2015). *Media Freedom and Right to Information in Africa*. Lisboa: Centro de Estudos Internacionais.

Castells, M. (2012). *Networks of Outrage and Hope: Social Movements in the Internet Age*. London: Polity.

Diffenbaugh, N. and Giorgi, F. (2012). Climate change hotspots in the CMIP5 global climate model ensemble. *Climatic Change* 114(4): 813–822.

Entman, R. M. (1993). Framing: Toward clarification of a fractured paradigm. *Journal of Communication* 43(4): 51–58.

Fields S. (2005). Continental divide: Why Africa's climate change burden is greater. *Environmental Health Perspectives* 113(8): A534–A537.

Ganapathy, D. 2022. *Media and Climate Change Making Sense of Press Narratives*. London: Routledge.

Glantz, M. (1992). Global warming and environmental change in sub-Saharan Africa. *Global Environmental Change* 2(3): 183–204.

González J. and Sánchez. A. (2022). Multilevel predictors of climate change beliefs in Africa. *PLOS ONE* 17(4): 1–14.

Guillaumont, P. and Simonet, C. (2011). To what extent are African countries vulnerable to climate change? Lessons from a new indicator of physical vulnerability to climate change. Fondation pour les études et recherches sur le développement international. Working Paper. No. 8.

Hase, V., Mahl, D., Schäfer, M. S. et al. (2021). Climate change in news media across the globe: An automated analysis of issue attention and themes in climate change coverage in 10 countries (2006–2018). *Global Environmental Change* September (70): 102353.

Hulme, M. (2009). *Why We Disagree About Climate Change. Understanding Controversy Inaction and Opportunity*. Cambridge: Cambridge University Press.

International Media Support – IMS. (2023). Reimagining the role of media in Mozambique in the era of terrorism and natural disasters: A media assessment report. IMS.

Kepios (2025). Digital in Mozambique, retrieved on https://datareportal.com/reports/digital-2025-mozambique, accessed on 20 May 2025.

Kompas, T., Pham, V. H., and Che, T. N. (2018). The effects of climate change on GDP by country and the global economic gains from complying with the Paris Climate accord. *Earth's Future* 6(8): 1153–1173.

Kunelius, R. and Roosvall, A. (2021). Media and the climate crisis. *Nordic Journal of Media Studies* 3(1): 1–19.

Maino, R. and Emrullahu, D. (2022). Climate change in sub-Saharan Africa fragile states: Evidence from panel estimations. *IMF Working Papers* 2022(54): A001.

Mandamule, U., Tsandzana, D., and Lasse, N. (2024). As redes sociais da Internet: um novo repertório na governação dos desastres "naturais" em Moçambique? *Destaque Rural*, 299. Maputo: Observatório do Meio Rural.

Mavume, A. F., Banze, B. E., Macie, O. A., and Queface, A. J. (2021). Analysis of climate change projections for Mozambique under the representative concentration pathways. *Atmosphere* 12(5): 1–51.

McCombs, M. E. and Shaw, D. L. (1972). The agenda-setting function of mass media. *Public Opinion Quarterly* 36(2): 176–187.

Okonjo-Iweala, N. (2020). *Combatting Climate Change. An Urgent Call for Comprehensive Global and Local Action*. Foresight Africa. Brookings Institute.

Ronning, R. (2011). *Mozambique: Financially Viable Media in Emerging and Developing Markets*. Swedish International Development Cooperation Agency. WANIFRA.

Salvesen, I. (2018). *Should Journalists Campaign on Climate Change?* Oxford: Reuters Institute for the Study of Journalism.

Santos, F., Paulo, L., and Pedersen, J. (2022). The climate change challenge: A review of the barriers and solutions to deliver a Paris solution. *Climate* 10(5): 75.

Schäfer, M. and Painter, J. 2021. Climate journalism in a changing media ecosystem: Assessing the production of climate change-related news around the world. *Wiley Interdisciplinary Reviews: Climate Change* 12 (1): 1–29.

Tenaw, D. and Hawitibo, L. (2021). Carbon decoupling and economic growth in Africa: Evidence from production and consumption-based carbon emissions. *Resources, Environment and Sustainability* 6: 100040.

United Nations. (2022). *Global assessment report on disaster risk reduction*. www.undrr.org/gar

Van Logchem, B. and Queface, A. J. (2012). *Responding to climate change in Mozambique: Synthesis report – INGC: Maputo*.

Weingart, P., Engels, A., and Pansegrau, P. (2000). Risks of communication: Discourses on climate change in science, politics, and the mass media. *Public Understanding of Science* 9: 261–283.

Williams, A. 2015. Environmental news journalism, public relations and news sources. In A. Hansen and R. Cox (Eds.), *The Routledge Handbook of Environment and Communication*. Oxon: Routledge. 197–206.

15 Podcasts, Sustainability, and Environmental Communication in the Middle East and North Africa

Thouraya Snoussi

Introduction

Environmental issues are increasingly encroaching into the news agenda in the Middle East and North Africa (MENA), rapidly becoming a subject matter of debate in the region's public sphere. International media events such the United Nations Climate Change Conference celebrated in both Egypt and the United Arab Emirates have brought needed attention to one of the most pressing issues of our lifetime. it does not go unnoticed that these events took place in a region that hosts some of the biggest producers and highest consumers per capita of fossil fuels (Al-Saad & Salman, 2012). In addition, these are countries also known for mostly hosting co-opted media systems that are not very familiar with addressing pollution issues (Farahat, 2016; Richter & Kozman, 2021; Sabbagh, 2021).

In this paradoxical setting, the question that arises is if there are venues in which we could see a deep public examination of environmental issues. This chapter argues that podcasts offer such platforms due to the fact that they have emerged as a medium for disseminating knowledge, engaging audiences, and fostering dialogue (Arab News, 2021; Jin, Segal, & Carroccio, 2020). With their accessible and immersive nature, podcasts have the potential to play a crucial role in raising environmental issues in the region (Al Moghadam, 2022; Al Sharif, 2022). Unlike traditional media, podcasts are not limited by geographical boundaries or time constraints as they can be accessed anytime, anywhere, and by anyone with an internet connection.

Consequently, this study asks if podcasts in the MENA region can serve as a critical media space for promoting environmental consciousness. The study examines the three most downloaded podcast series in the region and uses them as case studies to understand better the news agenda they promote, media representations of environmental issues and how journalists and producers frame these issues in the context of podcasting. The aim is to understand what the dominant themes are represented in the environmental podcasts in the MENA region. This is in addition to pinpointing how environmental podcasts in the MENA region portray the environmental matters and determining which rhetorical narratives are most prevalent. Overall, the study asked how podcasts in the MENA region frame and shape public

DOI: 10.4324/9781003660217-19

understanding of events and issues related to environmental matters and which framing strategies tend to be mostly adopted.

The literature highlights environmental journalism as a vital field that communicates complex environmental issues, such as climate change, pollution, and biodiversity loss, to the public in an accessible way (Kovarik, 2020; Abd Al-Elah & Al-Saraj, 2021; Qusien & Robbins, 2023; Kamboh & Ittefaq, 2023; Carlson, 2011; Kovarik, 2020; Comfort & Park, 2018; Fahy, 2018; Brüggemann et al., 2020; Bødker & Morris, 2022; Pintak, 2019; Pease & Dennis, 2018; Yee & Fassihi, 2021; Dennis et al., 2020). Despite challenges posed by constrained media systems, this form of journalism plays a crucial role in raising awareness, promoting accountability, and shaping policy debates. With the rise of digital media, radio and podcasts have emerged as key platforms for environmental reporting, offering space for in-depth storytelling, expert insights, and community engagement. These audio formats, especially in the MENA context, leverage the region's oral traditions and growing podcast culture to provide more independent, critical, and inclusive coverage of environmental topics (Snoussi et al., 2024). This chapter explores how podcasts, in particular, frame and develop distinct environmental news agendas.

Methodology

This study undertook an exploratory qualitative approach which entailed systematic interpretation of the content of the podcasts. It involved the categorization, coding, and qualitative examination of the content to identify patterns, themes, or trends fallowing established and tested methodologies (Elmasry, 2012; Krippendorff, 2004; Neuendorf, 2016). The author chose to do this analysis because it enables one to draw insights, make inferences, and understand the underlying messages, ideas, or representations within the analyzed content. The emphasis was placed on themes (topics raised, angles addressed), narrative strategies and framing strategies. This was done in order to highlight degrees of professional autonomy in the way the agenda set, themes addressed as well as the diversity of topics.

This analysis focused on how environmental issues are framed in media representations. Framing analysis involves identifying the dominant perspectives or "frames" used in media content, which help shape how audiences perceive issues and influence their support for related policies and actions. By examining podcast representations of environmental topics in the MENA region, the author gained deeper insights into both the nature of the content and the unique role these platforms play in creating spaces for critical and distinctive environmental discourse.

The sample included three of the most popular podcasts from the MENA region in the Arabic language, namely, Mad wajazr (High Tide and Low Tide) by Sowt Podcasts (Jordan), Azya' wa Lakin (Fashion, but ...) by Vice Media (UAE), and Akhdar (Green) by Independent Arabia (KSA). For this study the author took the final season from each podcast that contained information about environmental

Table 15.1 The number of podcasts

	Podcasts title	*Airing period*	*No. of episodes*	*%*
1	Mad wajazr (High Tide and Low Tide)	14-09-2021 to 15-12-2021	12	44.44
2	Akhdar (Green)	27-03-2022 to 21-08-2022	6	22.22
3	Azyaa Wakaken (Clothes but …)	05-08-2022 to 14-10-2022	9	33.33
Total			27	99.99

Source: Author.

issues raised in the region (11 episodes, 9 episodes, and 6 episodes, respectively). They are among the most popular and very few to discuss environmental issues (Table 15.1).

The author explored various frames to strategically present information, aiming to shape audience perception and interpretation. Message framing, involving deliberate choices, targeted communication goals such as attitude, belief, or behavior shaping. This analysis included Gain-framed and loss-framed messages, focusing on potential gains or losses to influence attitudes, beliefs, and intentions; Attribute framing, shaping attribute processing and evaluation; Goal framing messages, aligning attitudes, beliefs, and behaviors with desired outcomes; and Temporal framing messages, influencing attitudes, perceptions, and decision-making through time-related aspects.

The data collection involved systematic listening to episodes, noting delivery styles and presentation strategies, followed by a categorization based on predominant tone and informational tendencies. Episodes where transcribed and translated into English and then analyzed by the research team fallowing research strategies used in similar studies (Hasan & Hoon, 2013).

Findings and Discussion

Overall, the data indicates a rich content that is far more diverse and dynamic that the type of media encounter in other media outlets and channels in the region when dealing with environmental issues. A variety of genres, news styles, and reporting was also revealed in the podcasts far more than one tends to find in other media outlets in the region (Hasan & Hoon, 2013). Having said that, we also found dominant themes represented in environmental podcasts in the MENA region (Table 15.2).

The data demonstrates that podcasts of Mad Wajazr and Al Akhdar are comprehensive in terms of covering all topics of environmental programs. At the same time, Azyaa Walaken's podcast specializes in fashion and the latest trends in clothing in relation with environmental issues. Therefore, the content of specialized podcast

Table 15.2 Titles, timing, and duration of the podcast's series

Podcasts		*Diffusion date*	*Duration (min)*
Mad Wajazr (High Tide and Low Tide)			
1	Miyawaki: In the beginning was the tree	14-09-2021	17:11
2	Car alternatives	21-09-2021	15:41
3	Cooling houses or warming the planet?	07-10-2021	15:32
4	Shake (Take) your trash	12-10-2021	23:54
5	What planes do to our climate	19-10-2021	20:45
6	From war to serving the environment	27-10-2021	21:53
7	The Gas that will change the world	09-11-2021	18:40
8	How was construction before air conditioners?	16-11-2021	13:45
9	So we can avoid another virus	23-11-2021	15:40
10	Sorry, reducing plastic use won't save the planet	30-11-2021	12:52
11	Why do women get affected more by climate change	09-12-2021	17:35
12	The environment and Conference of the Parties COP	15-12-2021	17:11
Total			4 hours 15 min
Akhdar (Green)			
1	Our Lives without Trees?	27-03-2022	8:00
2	Are we experiencing a water crisis?	24-04-2022	7:40
3	The rivers of the Arabian Peninsula	22-05-2022	9:56
4	Reserves ...A necessity or luxury?	29-06-2022	9:27
5	Recycling Waste	24-07-2022	9:56
6	Climate Change	21-08-2022	10:46
Total			1 hour 26 min
Azyaa Wa laken (clothes but …)			
1	Our fashion says a lot about us	05-08-2022	12:51
2	Donating clothes is not always a good idea	12-08-2022	10:15
3	Greenwashing for Trademarks	19-08-2022	9:34
4	Luxury fashion and its impact on the environment	26-08-2022	14:04
5	Egyptian Cotton: History and complex politics	02-09-2022	10:44
6	3D fashion design with Marwan Al Hussien	09-09-2022	23:35
7	But I wore this before – with Lana Al Baik	16-09-2022	21:49
8	Made in our country with Ahmad Al-Shareef	07-10-2022	35:01
9	I'm sorry environment; I'll do a better job	14-10-2022	16:33
Total			3 hours 10 min

Source: Author.

Azyaa Walaken is analyzed separately (Tables 15.3 and 15.4). Consequently, the content of Mad wajazr and Al Akhdar will be analysed together.

In terms of topics raised during last season of the podcasts, the findings indicate richness in terms of themes and agenda, criticality in terms of how these topics are approached, and variety in relation to genres and reporting instruments used.

Table 15.3 Topics raised in Mad wajazr final season

Topics	*Description*
Planting trees (three episodes)	Three episodes dedicated to discussing the benefits of planting trees.
Micro mobility	The episode deals with micro mobility as a mode of transportation that involves lightweight, often electric-powered vehicles designed for short-distance travel within urban or suburban areas.
Ozone depleting substances	The main topic of this episode was ozone-depleting substances (ODS) as chemicals that contain chlorine or bromine atoms and are capable of depleting the ozone layer in the Earth's stratosphere.
Montreal Protocol + Conference of Parties (COP) (two episodes)	One episode delved into the Montreal Protocol, an international environmental treaty crafted to safeguard the Earth's ozone layer by gradually eliminating the production and usage of ozone-depleting substances (ODS). In the second episode, the focus shifted to discussing the preliminary arrangements leading up to COP 28 in the UAE, scheduled for December 2023.
Waste management + Garbage disposal (three episodes)	Three episodes focusing on topics related to waste management and garbage disposal.
Aviation and global warming + drones	The episode was focusing mainly in the aviation and its contribution to global warming primarily through the emission of greenhouse gases (GHGs), particularly carbon dioxide ($CO2$), as well as other pollutants such as nitrogen oxides (NOx), water vapor, and particulate matter. Also, a particular focus was on drones that have gained popularity in various sectors, including commercial, military, recreational, and environmental applications.
Renewable energy (two episodes)	Two episodes on renewable energy sources that are naturally replenished on a human timescale and are considered environmentally sustainable.
Protecting the environment	The episode was focusing on how to protect the environment via a wide range of actions and initiatives aimed at preserving and conserving Earth's natural resources, ecosystems, and biodiversity for current and future generations.
Hydrogen energy	The main theme was on hydrogen energy that refers to the use of hydrogen as a clean and renewable energy source.

(*Continued*)

Table 15.3 (Continued)

Topics	*Description*
Endemic/viruses and the environment	The episode explains the relationship between endemic viruses and the environment is complex and multifaceted, encompassing various factors such as climate change, habitat degradation, biodiversity loss, and human activities
Plastic use	The episode focuses on the plastic use that has become ubiquitous in modern society, with plastic products ranging from packaging materials and household items to electronics and medical devices.
Climate change impact on women (two episodes)	The episodes discuss the fact that climate change impacts women and men differently due to existing social, economic, and cultural inequalities. Women, particularly in developing countries and marginalized communities, often bear a disproportionate burden of climate change impacts and face unique challenges in adapting to and mitigating its effects.

Source: Author.

Table 15.4 Topics raised in Akhdar podcast final season

Topics	*Description*
Planting trees	The episode describes the planting of trees as a powerful and effective strategy for mitigating climate change, conserving biodiversity, restoring ecosystems, and improving the quality of life for both humans and wildlife.
Water consumption	The episode explains that water consumption can vary significantly depending on factors such as geographic location, climate, population density, economic activities, and water management practices.
Sustainable water resources	The episode raises sustainable water resources management that involves ensuring that water is used efficiently, equitably, and in a manner that preserves the quality and quantity of available water for future generations.
Role of valleys in Arabia	The episode highlights that valleys play several important roles in the Arabian Peninsula, both environmentally and culturally in water management and settlement.

Table 15.4 (Continued)

Topics	*Description*
Distinguished experiences in rehabilitating valleys	The episode continues with the role of valleys in Arabia, discussing distinguished experiences in rehabilitating valleys.
Role of natural reserves in our lives	In this episode, the focus was on nature reserves, national parks, wildlife sanctuaries, and conservation areas, playing crucial roles in our lives.
Global practices in operating reserves exceptionally and sustainably	The episode states a combination of best practices, innovative strategies, and effective governance mechanisms in operating reserves exceptionally and sustainably.
Waste management	The focus on this episode was on the collection, transportation, processing, recycling, and disposal of waste materials in a manner that minimizes environmental impact, conserves resources, and protects public health.
Waste recycling	The episode deals with converting waste materials into new products or materials to prevent waste from being disposed of in landfills or incinerators.

Source: Author.

This variety also applied to formats and timings. In Mad wajazr (High Tide and Low Tide) podcast, the data reveals 13 different topics while in Al Akhdar the number is 9. The timing for episodes of Mad wajazr (High Tide and Low Tide) podcast is between 12:52 minutes and 23:54 minutes; as for Al Akhdar episodes, the timing was shorter: between 7:40 minutes and 10:46 minutes. The periodicity of the podcasts is weekly for Mad wajazr (High Tide and Low Tide) and monthly for Al Akhdar (Table 15.5).

Data shows that both podcasts, Mad wajazr and Al Akhdar, stressed the waste management issues; the time allotted was 23:54 minutes in the episode # 4 of Mad wajazr and 9:56 minutes in the episode # 5 in Al Akhdar ($N = 33.50$ minutes). The time allotted to speak about planting trees was 17 minutes in Mad wajazr and 8 minutes in Al Akhdar ($N = 25$ minutes); an entire episode was devoted to talking about the topic and covering all its aspects. Lastly, the time allotted to discuss the climate change was 17:35 minutes in the episode #4 of Mad wajazr and 10:46 minutes in the episode #6 of Al Akhdar ($N = 28{:}21$ minutes).

It is worth adding that the podcast Mad wajazr emphasized some topics that were not addressed in Al-Akhdar, and devoted lengthy special episodes to them, such as the issues related to the aviation and global warming (episode 5), drones, renewable energy, and protecting the environment (episode 6). The allotted time was 20:45 and 21:53 minutes, respectively (Figures 15.1 and 15.2).

Table 15.5 Angles and themes addressed in the podcasts

Podcast	*Episodes*	*Topics covered*	*Common vs Distinctive themes*
Al Akhdar	3 and 4	Talking about best practices in operating reserves exceptionally and sustainably, as well as global experiences in rehabilitating valleys. This kind of content does not exist in Mad wajazr.	Distinctive
Al Akhdar	2 and 4	Raised issues related to water like water consumption, sustainable water resources, and the role of natural reserves.	Distinctive
Mad wajazr	3 and 12	These were dedicated to talk about international conferences related to environment Montreal Protocol and Conference of Parties (COP); this kind of coverage does not exist in the Al Akhdar sample.	Distinctive
Mad wajazr	5 and 6	Two episodes were devoted to issues related to planes like aviation and global warming, and drones. Al Akhdar podcast did not raise these issues.	Distinctive
Mad wajaz	6 and 7	This was the only to address the renewable energy.	Distinctive
Mad wajaz	10	The plastic use also was raised only in Mad wajazr as well as alternatives for cooling houses, micro mobility, and endemic/viruses and the environment.	Distinctive
Mad wajazr Al Akhdar	1 1	Planting trees	Common
Mad wajazr Al Akhdar	4 and 12 5	Waste management and recycling	Common
Mad wajazr Al Akhdar	11 5	Climate change	Common

Source: Author.

While variety does not necessarily translate to diversity in these cases, it does to a certain degree. Not only were journalists and podcast news producers using a variety of rhetorical strategies and journalistic genres, but also approached the different topics from distinctive activities (e.g., fashion and sustainability) within the news-beats environmental (Table 15.6).

To be sure, a fashion and branding specialist, podcaster Azyaa Walaken has produced most of his episodes on topics related to clothing trends from a special angle: the environment. Episodes were published weekly with a time slot ranging from 9:34 minutes while talking about "Greenwashing for Trademark" in episode 3 to 35:01 minutes dedicated to promoting local brands in episode 8 "Made in our country with Ahmed Al."

Figure 15.1 Common themes in podcasts.

Source: Author's data.

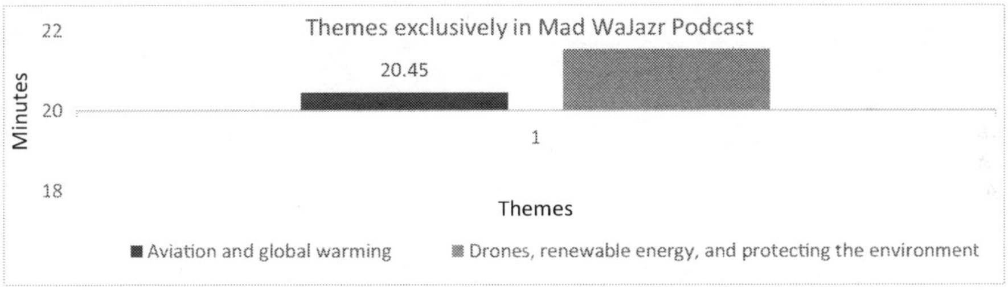

Figure 15.2 Themes exclusively in Mad Wajazr podcast.

Source: Author's data.

It should be noted that except for episode 7 "But I've Wore This Before – with Lana Albeik" and episode 2 "Donating Clothes Isn't Always a Good Idea," where the entire episodes were devoted to a single topic, two themes were raised in each one of the remaining seven episodes. Therefore, the time specified for each topic has been calculated based on the number of minutes allocated to it in each episode and not depending on the duration of the episode as a whole (Figure 15.3).

To sum, two distinct types of programming appear in the sample: generalized and specific. The Mad wajazr and Al Akhdar podcasts produce various eco-related topics, while Azyaa Wa Lakin is devoted to eco-related fashion styles. The dominant topics

Table 15.6 Content of Azyaa walaken

Topics	*Description*	*Duration*
Fashion	Six episodes out of nine were devoted to fashion in relation to concepts like identity and culture, sustainability, cotton industry, environment, fashion design on digital platforms, and fast fashion	N = 69 minutes
Measures of environmental protection	Two episodes dealt with this subject, which are greenwashing, steps that can be taken to reduce environmental damage, and ways to wear old clothes	N = 40 minutes
Sustainability	The topic sustainability was raised in four episodes while taking about fashion, and how brands can operate sustainably	N = 38 minutes
Environmental activities of branding	Three episodes were discussing issues related to misinformation in the environmental activities of trademarks and ways to support local production. The issue of labor rights has been raised in this area	N = 27 minutes

Source: Author.

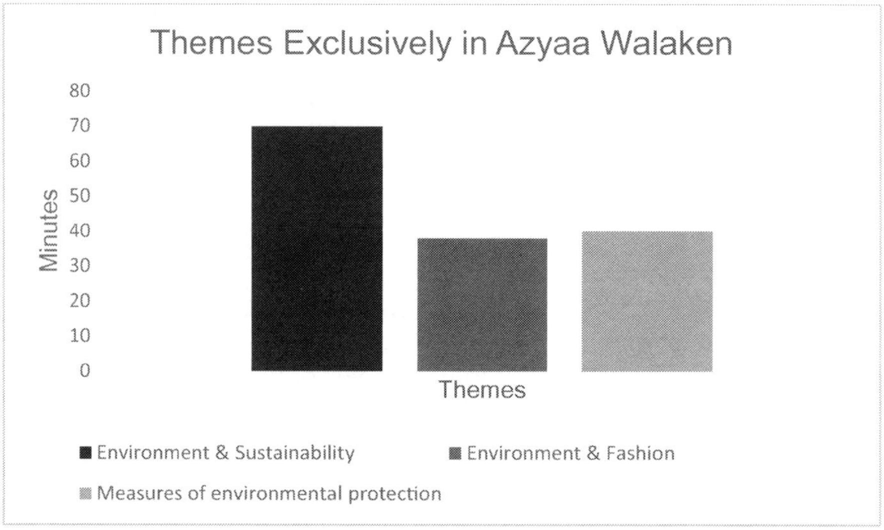

Figure 15.3 Dominant themes in the specialized podcast (Azyaa Walaken).

Source: Author's data.

produced by generalized podcasts are waste management, tree planting, and climate change. In the fashion podcast, Greenwashing for Brands was the dominant issue.

In relation to narrative strategies in the environmental podcasts. Journalists tend to diverse narrative strategies to engage listeners and effectively convey their intended message. They include storytelling that refers to the art and practice of transmitting narratives, stories, or information through radio. This includes developing engaging characters, building a narrative arc with a beginning, middle, and end, and incorporating elements such as suspense, conflict, and resolution. In the studied sample, storytelling was used in two episodes of Mad wajazr podcast as follows: episode 8 "How was construction before air conditioners?"; and episode 10 "Sorry, reducing plastic use won't save the planet".

On the other hand, the storytelling strategy was used in Al Akhdar podcast only in episode 3 labeled "The rivers of the Arabian Peninsula"; besides, this technique is combined with monologues in the following episodes: episode 5 "Recycling Waste" and episode 6 "Climate Change". In the case of Azyaa Walaken, the storytelling is tracked in episode 1 "Our fashion says a lot about us" and episode 4 "Luxury fashion and its impact on the environment".

Another important element found in the podcast space was the use of interviews. Something that comes across very distinctive from other types of media channels. The use of interviews as a strategy allows for a more personal and interactive narrative, providing different perspectives and insights. Mad wajazr podcast relies almost entirely on interviews as it is used in 10 episodes out of 12, while Azyaa Walaken podcast had only a few (Table 15.7).

Table 15.7 Themes and topics of the podcasts

Mad Wajazr (High Tide and Low Tide)	*Azyaa Walaken (Clothes but ...)*
Ep. 1 – *Miyawaki: In the beginning was the tree*	Ep. 6 – *3D Fashion Design with Marwan Al Hussien*
Ep. 2 – *Car alternatives*	Ep. 7 – *But I wore this before – with Lana Al Baik*
Ep. 3 – *Cooling houses or warming the planet?*	Ep. 8 – *Made in our country – with Ahmad Al-Shareef*
Ep. 4 – *Shake (Take) your trash*	
Ep. 5 – *What planes do to our climate*	
Ep. 6 – *From war to serving the environment*	
Ep. 7 – *The Gas that will change the world*	
Ep. 9 – *So we can avoid another virus*	
Ep. 11 – *Why do women get affected more by climate change*	
Ep. 12 – *The environment and Conference of the Parties (COP)*	

Source: Author.

Another genre is that of "formal talks," which refer to formal presentations or speeches delivered in an audio format to engage the audience through spoken words or to provide information or opinions. Talk shows can be presented by experts, public figures, or the radio hosts themselves. Podcasts often have hosts or guests giving talks on specific topics, sharing ideas, experiences, or opinions with the audience, but what the author mean by formal conversations in this study is formal speech that uses a serious tone away from spontaneity and humor. The podcast Mad wajazr includes formal talks along with interviews in the episode 12 dealing with the environmental and conference of the Parties COP (in newsreel style).

One of the most interesting findings was encountering documentaries and investigative journalism pieces that are not often seen in other types of media spaces. In the case of Mad wajazr podcast this included seven episodes furnished with documentaries and investigative journalism along with the interviews as follows: episode 1 "Miyawaki: In the beginning was the tree"; episode 2 "Car alternatives"; episode 3 "Cooling houses or warming the planet?"; episode 4 "Shake (Take) your trash"; episode 5 "What planes do to our climate"; episode 6 "From war to serving the environment"; and episode 7 "The Gas that will change the world." The fourth episode of Al Akhdar podcast was shown in the style of documentary and investigative journalism. The episode is titled "Reserves...Necessity or Luxury?" This strategy was also adopted in the fifth episode of Azia Walken podcast, entitled "Egyptian Cotton: History and Complex Politics."

Another narrative strategy used was that of monologues and personal stories in the form of speech or dramatic performance in which one character speaks alone, expressing her thoughts, feelings, or thoughts. While the monologue and personal stories were absent in Mad wajazr podcast, Al Akhdar podcast relied heavily on it, in four episodes as follows: episode 1 "Our Lives without Trees?"; episode 2 "Are we experiencing a water crisis?"; episode 5 "Recycling Waste"; episode 6 "Climate Change." Meanwhile, Azyaa Walaken podcast adopted the same strategy in the following three episodes: episode 2 "Donating clothes is not always a good idea"; episode 3 "Greenwashing for Trademarks"; and episode 9 "I'm sorry environment; I'll do a better job."

It is important to note that these strategies sometimes overlap in the sample, as many podcasts combine techniques to create a unique narrative style. Examples of this shared style are on the Mad wajazr podcast, where the first seven episodes of last season featured documentaries and interviews. The author believe this blend fits with the essence of documentaries, which often incorporate interviews, archival footage, reenactments, and expert analysis to provide a balanced and thought-provoking perspective on the chosen topic.

The podcast Mad wajazr also combined interviews formal talks in the episode 12 dealing with the environmental and conference of the Parties COP. This refers to the method of providing breaking news for environmental-related events such as COP. Moreover, in episodes 5 and 6 of the Al Akhdar podcast, monologue and storytelling were used together while discussing the issues of waste recycling and climate change.

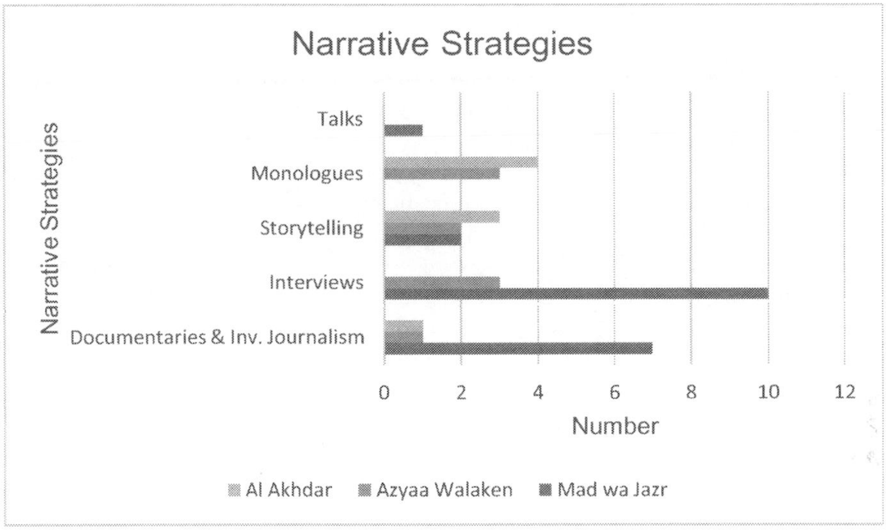

Figure 15.4 Narrative strategies.

Source: Author's data.

In fact, monologues and storytelling can be used together to enhance the richness and depth of the narrative (Figure 15.4).

While in Mad Wajazr podcast, documentaries, investigative journalism, and interviews were the narrative strategies most employed, storytelling and monologue are the techniques Al Akhdar use in their episodes; Azyaa Walakin podcasts are based on interviews and monologues. Lastly, Mad Wajazr is the only podcast that uses the talks and discussions. Thus, Mad Wajazr appears as a journalistically inclined podcast, while Al-Akhdar adopts the style of narration and anecdotes.

The power of interviews has to be enphasized, as the podcast provides a platform for interdisciplinary conversations, bringing together experts from diver countries in the region. These talks promote a holistic understanding of environmental challenges, encourage listeners to view environmental issues through a broader lens, and learn about the interconnections between environmental health, human well-being, and social justice. Indeed, podcasts offer a unique platform for experts, scientists, and environmental advocates to share their insights, research findings, and experiences with a wide audience. Through interviews, discussions, and storytelling, podcasts can effectively educate listeners about pressing environmental issues. Expert voices bring credibility, depth, and nuance to the conversation, helping listeners develop a comprehensive understanding of complex environmental challenges.

Framing Analysis

In terms of frames, a special focus was put on the strategic presentation of information in a way that emphasizes certain aspects or perspectives of a message to influence how it is perceived and interpreted by the audience.

Message framing: Involves making choices about how to frame information to achieve specific communication goals, such as promoting certain attitudes, beliefs, or behaviors. In this sense, we looked at a series of framing categories that include: Gain-framed and loss-framed messages: They involve framing information in terms of either potential gains or potential losses to influence individuals' attitudes, beliefs, and behavioral intentions. Gain-framed messages emphasize the benefits, positive outcomes, or rewards associated with a particular action or decision.

Attribute framing: Involves presenting information about a decision or option in terms of different attributes, such as gains versus losses, positive versus negative descriptions, or percentages versus frequencies. This framing technique can influence individuals' perceptions, attitudes, and choices by shaping how they process and evaluate the presented attributes.

Attribute framing messages: Refers to a persuasive communication strategy that involves emphasizing different attributes or characteristics of a decision, option, or behavior to influence how it is perceived and evaluated. This framing technique focuses on presenting information in a way that highlights specific features, benefits, or consequences associated with the target attribute.

Goal framing messages: Refers to the persuasive techniques that focus on the presentation and framing of goals to influence individuals' attitudes, beliefs, and behaviors. These strategies aim to align individuals' goals with desired outcomes or actions, emphasizing the benefits or motivations associated with pursuing those goals.

Temporal framing messages: Involves the manipulation of time-related aspects in persuasive messages to influence individuals' attitudes, perceptions, and decision-making. By emphasizing different time frames, such as the present, the future, or the past, communicators can shape how individuals interpret information and evaluate potential outcomes. Present-focused framing emphasizes immediate benefits, rewards, or consequences of a decision or action.

Fallowing these categories, the author found that they are present in all of the podcasts as indicated in Table 15.8.

However, this presence is not homogenous and instead varies according to the nature of the topic and content in the podcast, as indicated also in Table 15.9.

It is to point out that podcasts can convey poignant stories and experiences, making environmental issues relevant and personal using a time framing strategy. The framing analysis suggests that the podcasts tend to humanize the impact of environmental degradation on communities and ecosystems. Besides sharing personal narratives, real-life experiences, and the voices of those affected, podcasts tend to cultivate empathy and emotional connection and motivate listeners to act and make positive changes in their lives as indicated in Figure 15.5.

Table 15.8 Framing strategies in the environmental podcasts in the MENA region

Framing strategies podcast	*Gain-framed*	*Loss-framed*	*Attribute framing*	*Goal framing*	*Temporal framing*	*Total*	*%*
Mad wajazr	2	6	6	2	8	24	52.17
Akhdar	1	1	3	1	5	11	23.91
Azya Wa Laken	1	1	5	1	3	11	23.91
Total	4	8	14	4	16	46	99.99
%	8.69	17.39	30.43	8.69	34.78	99.99	

Source: Author.

Table 15.9 Framing in podcasts in terms of episodes

Type of framing	*Podcast series*	*Frequency*
Gain-framed messages	Mad wajazr	2
Loss-framed messages	Mad wajazr	6
Attribute framing	Mad wajazr	6
	Azyaa Walakin	5
	AlAkhdar podcast	3
Goal framing	Mad wajazr	2
Temporal framing	Mad wajazr	5
	AlAkhdar	5
	Azyaa Walakin	3

Source: Author.

It should be added that the timeframe or branding strategies used may differ in specific segments of the podcast mentioned in this study as they were combined in some episodes, mainly combining attribute with goal framing (Mad wajazr, episode 4 and 7; Al Akhdar, episode 2; Azyaa Walakin, episodes 6, 7, and 8).

Content analysis of podcasts in the MENA region reveals two distinct types of programming: generalized and specific. While Mad wajazr and Al Akhdar podcasts produce all kinds of eco-related topics, Azyaa Walakin is dedicated to talking only about eco-related clothing and fashion styles. The dominant topics produced by the public podcast are issues of waste management, tree planting, and climate change. On the niche podcast, topics like "Greenwashing for brands and Made in our County" were related to garments from a specific angle, that is, the environment.

It is worth to emphasize the fact that waste management has emerged as a critical topic for environmental awareness in the podcast, highlighting the need for sustainable waste practices to address the growing global waste crisis. The podcast has delved into various aspects of waste management, including waste generation, recycling, landfilling, and waste reduction strategies.

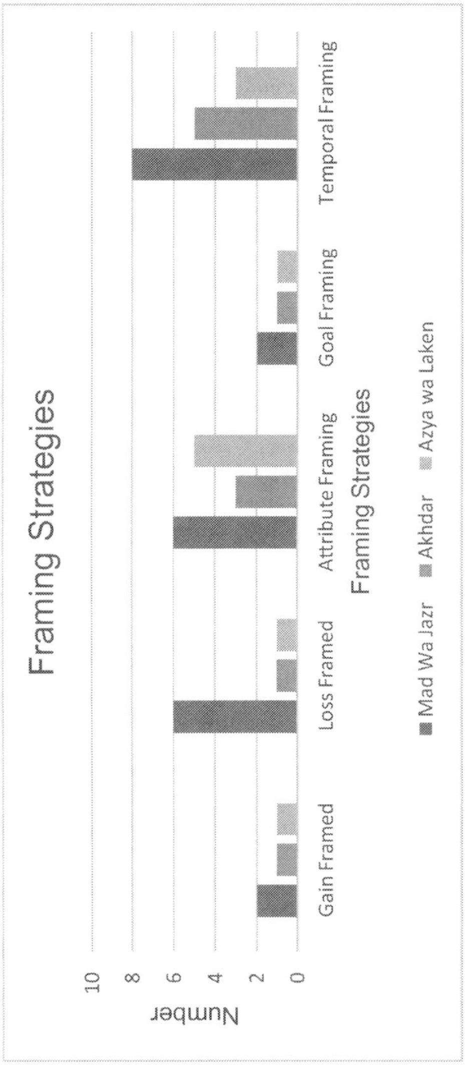

Figure 15.5 Framing strategies.

Source: Author's data.

In addition, the podcast explores innovative approaches to waste management, such as circular economy models, composting, and zero-waste lifestyles, promoting the idea of reducing, reusing, and recycling waste. The "Zero Waste Countdown" podcast explores the zero-waste movement and offers practical tips and advice for reducing waste in everyday life. They cover composting, plastic-free living, and sustainable fashion reminiscent of Azia walakin (the fashion podcast included in this study). Another example is the "Living Zero Waste" podcast that shares stories and strategies for living a zero-waste lifestyle. It discusses waste reduction, sustainable consumer choices, and the challenges and successes of individuals and communities striving for zero waste. The "Plastic Pickup" podcast focuses on reducing plastic waste and preserving the oceans. It discusses plastic pollution issues, interviews experts, and shares the stories of individuals and organizations working toward plastic-free solutions.

Planting trees as a topic of awareness in podcasts has also gained significant attention in recent years due to its potential to address multiple environmental challenges. Tree planting initiatives have been featured in various environmental podcasts as a tangible and accessible action that individuals and communities can take to make a positive impact on the planet. These podcasts highlight the benefits of tree planting in sequestering carbon dioxide, restoring degraded ecosystems, enhancing biodiversity, improving air and water quality, and providing socioeconomic benefits to local communities. For instance, the podcast episode "The Power of Tree Planting" by the "Climate Connections" podcast explores the role of tree planting in climate change mitigation and the stories of individuals and organizations engaged in tree planting efforts. Another podcast, "The Environment: Tree Planting" by "Sustainability Defined," discusses the benefits and considerations of tree planting projects in urban areas. These podcasts provide insights into the science, practical aspects, and success stories of tree planting initiatives, inspiring listeners to take part in tree planting campaigns or support organizations involved in reforestation efforts.

On Climate Change

In the meantime, climate change has been a prevalent topic in environmental podcasts, focusing on its causes, impacts, mitigation strategies, and adaptation measures. Discussions often cover topics such as rising temperatures, extreme weather events, sea-level rise, and the importance of transitioning to renewable energy sources. Several popular podcasts focus on climate change, such as Minnesota Public Radio's "Climate Cast"; this podcast features interviews with scientists, policy experts, and environmental advocates to explore different aspects of climate change and its impacts. Global Optimism's "Outrage + Optimism" podcast offers a mix of analysis and discussion to inspire action and highlight positive solutions in the face of the climate crisis. Furthermore, Eric Holthus' "Warm Greetings" podcast explores the intersection of climate change, culture, and society while delving into climate impacts, politics, activism, and the human side of the climate crisis.

Finally, Doc Society's "Mothers of Invention" podcast features the stories of the women leading the charge against climate change and explores feminist perspectives, climate justice, and innovative solutions to address the climate crisis. In all these cases, several environmental topics seemed to be underreported in the studied sample such as pollution, which was only addressed in the episode about car alternatives in the Mad wajazr podcast. Besides, sustainability was shown while dealing with sustainable brands at Azyaa Walakin.

The narrative strategies adopted in the sample differ from one podcast to another, but the dominant narrative method used in most of the episodes' sampled for this study was referring to the serious news character, as it used investigative journalism techniques and documentaries. Also, storytelling and monologues were techniques used by Al-Akhdar podcast, while Azyaa Walakin's is based on interviews and monologues. As for the framing strategies, data shows that the studied podcasts mostly present their information based on temporal messages, which is a common framing strategy as several environmental podcasts in the world include time framing approaches.

Conclusion

The prominent narrative technique in podcasts considered in this study reminds us not only of the importance of journalistic news styles and audio storytelling, but also of its possibilities in all types of societies. Moreover, not only a variety and rich set of approaches were used here, such as interviews and discussions, but they often were integrated into a variety of genres and formats that underscore a type of richness and variety in the narrative strategies that is hardly found in other places; something that applied not only from one podcast to another but also from episode to episode in which multiple narratives are sometimes incorporated in the same episode, indicating a higher level of professionalism.

As for the framing strategies, temporal techniques, as well as branding techniques, were prevalent. The fear-mongering approach of presenting the ecological content was minimal while the monologue, the storytelling, and the enticement style toward some "green" practices were abundant. In the meantime, the author intends to study sample speech from the episodes to determine the trend and depth of environmental messages, which is indicative of the level of expertise of environmental podcasts in the region.

To conclude, podcasts can play a pivotal role in raising environmental awareness by combining educational content, engaging storytelling, diverse perspectives, and global reach. Through immersive and intimate format, as well as its framing strategies, podcasts seem can connect listeners with the natural world, foster empathy, and inspire action. As podcasts continue to grow in popularity and influence, it is imperative that this medium be harnessed to promote environmental awareness, amplify diverse voices, and drive positive change for a sustainable future.

By harnessing the power of podcasts, journalists operating in the region, despite existing limitations, can create a more informed, inclusive, and environmentally aware society. Given this potential, future research on the role of podcasts in raising environmental awareness and the way news audiences engage with these podcasts could provide valuable insights into its impact on audience perceptions and behaviors. These are issues that urgently need the attention of researchers and experts, mostly to push for new spaces where we can all become more aware about the environment and the challenges that society faces.

Data Availability

Data will be made available on request.

Declaration of Competing Interest

The author declares that she has no known competing financial interests or personal relationships that could have appeared to influence the work reported in this chapter.

References

Abd Al-Elah, H. R. H., & Al-Saraj, S. K. (2021). Science journalism within the framework of media richness theory. *PalArch's Journal of Archaeology of Egypt/Egyptology*, 18(7), 2835–2844.

Al Moghadam, L. (2022). *A podcast is a golden opportunity for the Arab press to squander*. Al Jazeera Media Institute.

Al Sharif, H. (2022). *Podcast initiatives in the Arab world*. Maharat Foundation.

Al-Saad, H. T., & Salman, N. A. (2012). Status of oil pollution in the Arabian Gulf and Shatt Al-Arab Estuary: A review. In V. Subramanian (Ed.), *Coastal environments: Focus on Asian regions* (pp. 36–46). Springer.

Arab News. (2021). New report explores state of podcasting in MENA in 2020. Arab News. www.arabnews.com/node/1809551/%7B%7B

Bødker, H., & Morris, H. (2022). *Climate change and journalism*. Routledge.

Brüggemann, M., Lörcher, I., & Walter, S. (2020). Post-normal science communication: Exploring the blurring boundaries of science and journalism. *Journal of Science Communication*, 19(3), A02. https://doi.org/10.22323/2.19030202

Carlson, M. (2011). Whither anonymity? Journalism and unnamed sources in a changing media environment. In B. Franklin & M. Carlson (Eds.), *Journalists, sources, and credibility: New perspectives* (pp. 37–48). Routledge.

Comfort, S. E., & Park, Y. E. (2018). On the field of environmental communication: A systematic review of the peer-reviewed literature. *Environmental Communication*, 12(7), 862–875. https://doi.org/10.1080/17524032.2018.1514315.

Dennis, E., Martin, J., & Allagui, I. (2020). Media use in the Middle East. Northwestern University in Qatar. Retrieved from www.mideastmedia.org/survey

Elmasry, M. H. (2012). Journalism with restraint: A comparative content analysis of independent, government, and opposition newspapers in pre-revolution Egypt. *Journal of Middle East Media*, 8(1), 1–34).

Fahy, D. (2018). Objectivity as trained judgment: How environmental reporters pioneered journalism for a "post-truth" era. *Environmental Communication*, 12(7), 855–861. https://doi.org/10.1080/17524032.2018.1495093

Farahat, A. (2016). Air pollution in the Arabian Peninsula (Saudi Arabia, the United Arab Emirates, Kuwait, Qatar, Bahrain, and Oman): Causes, effects, and aerosol categorization. *Arabian Journal of Geosciences*, 9, 1–17. https://doi.org/DOI10.1007/s12517-015-2203-y

Hasan, Md. M., & Hoon, T. B. (2013). Podcast applications in language learning: A review of recent studies. *English Language Teaching,* 6(2), 128–135. http://dx.doi.org/10.5539/elt.v6n2p128

Jin, L., Segal, A., & Carroccio, B. (2020). Investing in the podcast ecosystem in 2019 [Information]. Andreessen Horowitz. https://a16z.com/2019/05/23/podcast-ecosystem-investing-2019/

Kamboh, S. A., & Ittefaq, M. (2023). Advocacy journalism and climate justice in a Global Southern country. *Local Environment*, 28(4), 412–432. https://doi.org/10.1080/13549839.2022.2155937

Kovarik, B. (2020). The rise of environmental journalism in Asia, Africa, and Latin America. In D. B. Sachsman & J. Myer Valenti (Eds.), *Routledge handbook of environmental journalism* (pp. 52–69). Routledge.

Krippendorff, K. (2004). *Content analysis: An introduction to its methodology*. Sage.

Neuendorf, K. A. (2016). *The content analysis guidebook*. Sage Publications.

Pease, E. C., & Dennis, E. E. (2018). *Radio-the forgotten medium*. Routledge.

Pintak, L. (2019). Middle Eastern and North African journalism. In T. P. Vos, F. Hanusch, D. Dimitrakopoulou, M. Geertsema-Sligh, & A. Sehl (Eds.), *The international encyclopedia of journalism studies* (pp. 1–12). John Wiley & Sons.

Qusien, R., & Robbins, D. (2023). Science journalism in Pakistan: The challenges faced by environmental reporters. *Journalism Studies*, 25(5), 459–479. https://doi.org/10.1080/1461670X.2023.2201854

Richter, C., & Kozman, C. (2021). *Arab media systems*. Open Book Publishers.

Sabbagh, R. (2021). There is a long road ahead for Arab journalists and free speech [News]. Al-Jazeera. www.aljazeera.com/opinions/2021/3/15/the-long-road-ahead-for-arab-journalists-and-free-speech

Snoussi, T., Hatamleh, I. H. M., Barkho, L., & Abusamra, N. (2024). Political listening and podcasting: The case of Tunisia. *Journalism and Media*, 5(4), 1433–1451.

Yee, V., & Fassihi, F. (2021). Clubhouse app creates space for open talk in Middle East. *New York Times*. www.nytimes.com/2021/05/02/world/middleeast/clubhouse-iran-egypt-mideast.html

Part IV

Platforms and Climate Change Communication

16 From Platformisation to Infomediation

Trends and Imaginaries of Participation in #GreenNigeria

Jendele Hungbo and Didunloluwa Tella

Introduction

Research on climate change communication has continued to mutate in different forms. While technological developments play a very crucial role in this mutation, competing media forms arising from a fast-changing technological architecture have also led to the reimagination of strategies and approaches to the climate communication enterprise. Climate change communication, in other words, has been pervasive on different media spaces moving from traditional mainstream media to new media platforms with changing approaches to modes of presentation for messages directed at different kinds of audiences. In the current environment dominated by numerous social media platforms like Facebook, Instagram, YouTube, TikTok etc., all forms of communication (including climate change communication) have had to evolve in different ways in order to be relevant to the new generation of actors whose understanding of behaviour change or change action generally should be a significant consideration for environmental sustainability. The popularity of social media platforms among young demographics further reinforces the need to pay closer attention to any form of communication calling for deliberate action or behaviour change among the public. This coupled with the constant availability of youth on digital spaces imply a pressing need to examine specific communication designs that get to them directly (Castro 2024; Fernandez et al. 2016; Gómez-Casillas & Márquez 2023; Leon et al. 2023). In specifically referencing climate change communication, Leon et al. (2023) harp on the primacy of interaction in making climate change communication through social media platforms effective.

Different types of changes have occurred over the years in the articulation of climate change communication. Among these changes is what Artieri et al. (2021) describe as 'platform-driven changes' which significantly alter 'emerging practices of everyday-life content production' (2021: 1). It is pertinent to note however that these changes have also resonated in the circulation and consumption of cultural products. The new patterns of information production and consumption which have trailed the ascendancy of social media platforms point to a season of significance in communicative practices that cannot be ignored in the general understanding of any targeted advocacy especially when the success of

DOI: 10.4324/9781003660217-21

such advocacy rests on public participation and engagement. Platformisation and platform studies have become some kind of interesting phenomena engaging the attention of scholars across scientific disciplines. As a concept, platformisation generally refers to the penetration of digital infrastructures as well as the reorganisation of cultural practices occasioned by digital revolution (see Poell et al. 2019). Platformisation of media forms is a process by which the architecture of digital platforms transform the communicative dynamics of public policy and the public sphere in general (Helmond 2015; Magaudda & Solaroli 2020; Nieborg & Poell 2018; van Dijck et al. 2018). Such transformation is driven by technology and the impatience of an energetic audience found mainly in the growing global youth population. Yet, platformisation debates have to be approached in a cautious manner as scholars have emerged to assume either techno-optimistic or techno-pessimistic positions based on their projections of what the future might hold for platformised communication processes (see Carpentier & Hroch 2024; Dutton 2023; Makananise 2025). Implicated in this debate is the undying contention over the heavy privileges accruing to the Global North over the Global South in things digital. Approaching the debate from an indigenous African languages perspective, for instance, Makananise and Madima (2025) have argued that a decolonisation of digital platforms is capable of redressing the historical and recurrent injustices that constantly disadvantage many Global South indigenous communities and contribute significantly to their marginalisation.

As Helmond contends, platformisation 'entails the extension of social media platforms into the rest of the web and their drive to make external web data "platform ready"'(2015: 1). After such readiness, information on the platforms have to be handled in particular ways in order to achieve desired objectives. This curating of information and data in consonance with set objectives transcends mere platform readiness, it involves giving consideration to many other factors in granting access and creating robust participation. The mediation of information on platforms may then offer another means of further engagement which throws more light on content and behaviour. This kind of approach, in a way, tends to take into account the discursive peculiarities and sociality that characterise participation in online engagements. Platformisation creates an automated media environment and encourages diversity and a multiplicity of views while infomediation on the other hand is expected to broaden participation. This is often seen in reactions which show participant input in form of reflections or contributions suggesting acceptance or rejection of proposals and ideas generated by initiators of messages or campaigns for specific purposes. As an alternative to mainstream outlets, digital platforms create avenues for alternative ideologies and views which might otherwise be muzzled by official considerations in traditional media forms which are generally submerged by power structures in society. The overall objective of this chapter therefore is in describing and gaining greater understanding of how the transition from platformisation to infomediation may contribute to the enhancement of participation in communicating climate change and environmental sustainability messages in Nigeria using quantitative content analysis as analytical tool. A better understanding of this transition has the potential for a more critical appraisal and execution of climate change communication strategies by relevant agencies interested in environmental sustainability. While platformisation can be taken for granted as a technological shift, the ways in which information exchanges are mediated by users of this technology become important for a better understanding of the resonance of the content noticeable in the expected response form the public.

This chapter focuses on the *#GreenNigeria*, a communication campaign by the Federal Ministry of Environment in Nigeria via social media especially X (formerly Twitter). #GreenNigeria is a slogan hashtag on X introduced by the Federal Ministry of Environment in Nigeria to signpost the activities of the ministry aimed at achieving environmental suitability. The launch of this hashtag continues to reflect in the reporting and postings about events hosted by the ministry as well as individuals and agencies associated with it. The hashtag therefore dominated posts by such agencies and individuals in reporting events and activities and as such became one of the interesting ways in which users 'attention was drawn to climate change and other environmental concerns'. As a slogan, #GreenNigeria resonates directly as an unambiguous advocacy instrument in the hands of the users as it speaks directly to the issue at hand. Communication campaigns like #GreenNigeria play a prominent role in determining what people think or perceive, how they feel, and what actions they take in relation to specific issues or phenomena. In specific climate change issues, communication campaigns may influence local and global agenda as people become more exposed to more information through online platforms especially social media (Segerberg 2017). X in this regard exemplifies an infomediation platform which provides opportunities for content creation, use and engagement (see Arias 2024; Fitzgerald & Winseck 2018; Sebbah et al. 2020; Siapera 2013).

Infomediation is a more media-centric perception of the power of the internet to alter the ecological and meaning-making dimensions of information in the digital age. While it recognises the key role that platformisation plays in the new era, it further captures, in greater detail, the technological and socio-cultural dimensions of the political economy of the internet-enabled communication environment of the new age with interest in user preferences as a major determinant of the distribution of access. As Smyrnaios and Rebillard argue, 'infomediation consists in computer-assisted matching between third-party content and users' individualised demands (including a set of socio-technical distribution and editorial packaging operations)' (2019: 32). In addition, the content so generated often constitutes an archive or a beehive of opinion, knowledge and contestations among new (social) imagined communities of belonging and participation who meet virtually.

By focusing on participation dynamics in terms of reactions and perceptions reflected by posts championing the *#GreenNigeria* campaign, the chapter extends the frontiers of research beyond questions of the mere adoption of technology in climate change communication to a more nuanced consideration of social and peculiar cultural contexts in unpacking communication messages. While such perspectives had been missing in climate change communication research, a focus on ways of participation also has a potential to constitute 'a deeper examination of how publics imagine climate change and its future role in social life' (Pearce et al. 2019) as information content and directions of discourse may yield a considerable amount of knowledge in trying to understand the ways X users react to this particular campaign.

Participation is a bit more nuanced than mere engagement as it is capable of implying different things in various contexts (Carpentier 2011; Cornwall 2008; Oliveira 2020). In the context of this chapter, participation is viewed as that practice of engagement that creates a broad-based opportunity for the inclusion of as many stakeholders as possible and by so doing holds a promise for the democratisation of the public space. Participation, as Villi and Matikainen (2016) argue, can be either implicit or explicit. While explicit participation may dwell on mere things allowed by automation and can be seen clearly from what is posted, implicit participation deals more with a greater sense of commitment to socio-cultural peculiarities and deeper objectives behind an advocacy project. Though interaction and engagement may be regarded as crucial conditions of participation, the idea of a participatory encounter requires more immersion in a process. This kind of participation is what Kelty and Erickson describe as 'an experiential-affective mode that emphasises the experience of being or becoming part of a collective, and the affective, communicational, and educational features of that experience' (2018: 71).

In bringing the social media dimension into climate change communication scholarship, different perspectives have emerged which continue to render the field prolific and provocative (see Evanglista & Garcia 2024; Tyagi 2021; Buthelezi 2024). While examining strategies for climate change communication through social media, for instance, Leon et al. call attention to how institutional communicators' perception influence the strategies and approaches used in climate change communication. They also point out how such strategies and approaches vary from one clime to another (Leon et al. 2023). Though online communication about climate change began with global NGOs, governments and their agencies across the world have keyed into the use of online platforms as a major strategy aimed at reaching a wider and remote audience. The current landscape of digital communication has therefore produced different forms of transformation which need to be factored into strategic corporate communication designs by organisations seeking to make impact through constant or sustained interactions or engagement with the public. It is indeed pertinent to note that in the current dispensation social media have become so normalised that ignoring them might become difficult in research and practice when dealing with new advances and approaches to any form of advocacy. In other words, it is indeed difficult to talk about any form of communication that has not been impacted significantly by social media one way or the other.

Theoretical Framework

This chapter is framed around digital intermediation theory (Gielens & Steenkamp 2019; Hutchinson 2021, 2023). Digital intermediation theory can be described as a successor to technological determinism and social construction of technology or social constructivism (Bijker et al. 2012; Cao 2024). While technological determinism emphasised algorithmical primacy and social construction of technology theory privileged the epistemological social context of stakeholders in understanding the functionality of technological innovations in society, digital intermediation focuses on the cultural characteristics of technologies, agencies and automation processes. Such cultural capital is quite useful in further understanding the ways in which participants relate on media platforms. In other words, cultural practices which define the identity of the people and at times reflect their material conditions may determine their discursive approaches to issues raised on a digital platform. Digital intermediation, to a large extent, problematises intermediation by critically examining user agency and how it is navigated within the context of owner control. As a clear departure from the theory of technological determinism, social construction of technology theory gives credence to the significance of social context in understanding the deployment of technological innovations (see Callon 2012; Douglas, Gislason et al. 2021; Hughes & Pinch 2012; Olohan 2017). Intermediation holds a great attraction as a theoretical entry point for this study in the sense that it speaks to the questions of acceptance or otherwise of technological novelties and argues that a better understanding of such acceptance or rejection often resides in happenings in the larger society as well as participatory culture dynamics defining each context (see Berglez & Al-Saqaf 2020; Wang & Yecies 2023). We attempt a critique of #GreenNigeria against the backdrop of Hutchinson's concept of 'microplatformisation' which is a form of 'fragmented platformisation' focusing on content production and circulation strategies that those individuals and organisations dealing with public issues find useful (Hutchinson 2023).

Methodological Considerations

This study focuses strictly on posts on X by agencies and individuals who posture as champions of climate change action in Nigeria by signing off their posts with #GreenNigeria. X is still regarded as one of the most effective social media platforms for the engagement of a wide range of interest groups in society. The brevity of posts is one major advantage. Such brevity makes participation exciting and this may also be part of the reason the platform attracts a huge youth demography (see Hungbo 2021; Leon et al. 2022). Original posts and replies using #GreenNigeria were manually and purposively selected from posts made in June 2024. The month of June coincides with the global United Nations annual commemoration of the World Environment Day which comes up on the 5th of June every year. The significance of including this day in the dataset is an anticipation of prolificity in posts on issues related to the environment which may also resonate with the #GreenNigeria slogan. The content of the selected posts was analysed for level of engagement and the kinds of social realities reflected in the posts. Some key questions asked in the cause of interrogating the posts include who are the people or institutions that used the hashtag? How frequently did they use the hashtag during the month of June 2024 and how much engagement has accompanied posts using #GreenNigeria during the said period? From a general perspective, engagements could come in form of views (the number of times a post was presumably read), quotes (when a post or a part of it is referenced directly or verbatim in a reply or another post), reposts or even likes. Being a precondition to participation, it is pertinent to consider the level of engagement with posts in trying to decipher how much participation has taken place on a platform.

Imaginaries of Participation in #GreenNigeria

It is important therefore to unpack a few of the posts that used #GreenNigeria during the month of June 2024.

Ecosmart Initiative is an X user with a handle @ecosmart_I. The user is also a recent joiner of the platform having joined only in March 2024. The major preoccupation of posts from this handle seems to revolve around tree planting. A cursory look at two short posts from the handle seems to validate this trend. The posts about tree planting by the handle speak directly to environmental concerns. All the posts, like many others from the handle, express concern about tree planting. The first post shows concern about 'myths' around tree planting:

> Let's bust the myths around tree planting! What's one misconception you've heard about it?
>
> (@ecosmart_I, June)

Though the post had been viewed 110 times, interestingly there had been no further engagement with it two weeks after it was dropped on the site. This signals lack of interest on the part of participants to further engage with the subject matter of the post after viewing it:

> How many trees have you planted this year? Let's challenge ourselves to make Nigeria greener, one tree at a time! #GreenNigeria #PlantATree
>
> (@ecosmart_I June 2, 2024, X)

This post had been viewed 16 times and had also attracted no further engagement.

The 12th of June is celebrated as Democracy Day in Nigeria. There was an environmental advocacy post from a user who identified as Safer Space Initiative with the handle @SaferInitiative to mark the day:

> Environmental degradation, pollution, and climate change threaten our health, resources, and social stability. These challenges impact our democratic values and the quality of life for all Nigerians. Let's work together to promote sustainability and protect our future. Join Safer Space Initiative in tackling these issues through advocacy, education, and community action. Together, we can make a difference. #ClimateAction #SaferSpaceInitiative #GreenNigeria
>
> (@SaferInitiative Jun 12, 2024)

This post was viewed seven times with only on reply.

Similarly an agency of the ministry NESREA posted:

> The NESREA family joins fellow Nigerians in celebrating this milestone—the 25th Democracy Day. As we march on, let us join hands to ensure a cleaner and healthier environment for all Nigerians Happy Democracy Day Nigeria. #June12DemocracyDay #GreenNigeria
>
> (@NESREANigeria June 12, 2024)

Only eight people had viewed this post without any further action.

The lack of engagement is not indicative of a lack of popularity or a complete lack of interest in the handled entirely. This is as a result of the fact that more engagements occurred on posts that did not use the hashtag. For instance, the following posts received different reactions from followers:

> NESREA has sealed 4 construction sites in Jos, Plateau State, in continuation of it's*[sic]* nationwide crackdown on environmental offenders. The facilities were shut down for failing to adhere to environmental laws and standards.
>
> (@NESREANigeria June 29, 2024)

The above post was viewed 600 times. There were five reposts, sixteen likes and one reply. In a similar vein the same handle posted on the 4th of June and had a more active engagement from followers:

> NESREA DEVELOPS CODE OF CONDUCT AND ETHICS FOR STAFF The National Environmental Standards and Regulations Enforcement Agency, (NESREA) with the support of the United Nations Office on Drugs and Crime, (UNODC), is developing a Code of Conduct to guide the behavior staff.
>
> (@NESREANigeria June 4, 2024)

The above post was viewed 878 times. It was reposted five times with one reply and twelve likes.

In a post by a user, (@NgugiEvuti), who has close ties with the Ministry of environment we see an encapsulation of the main objective of #GreenNigeria and similar campaigns:

> The ultimate goal is to keep Nigeria cool, protect her coastline, strengthen food security and halt the encroaching Sahara desert—permanently. #GreenNigeria
>
> (@NgugiEvuti, 19 June, 2024)

This post appears to be one of the most popular with on 1,833 views, 2 replies and 54 reposts and 52 likes.

Government agencies and officials, especially those of the Federal Ministry of Environment, fall among the most conspicuous users of #GreenNigeria in their X posts. The Minister for Environment for instance posts a weekly review of his activities on his verified X handle under the caption 'Minister's Week in Review' or 'Highlights of Last Week' when the post is delayed. Each time these weekly posts appear, they often end with '#GreenNigeria'. The same trend is observed in posts emanating from the institutional X handle of the Environment Ministry, @FMEnvng. One of such posts declared:

> Dr. @SalakoIziaq mphasizd at #WorldEnvironmentDay that nature-based solutions (NBS) are key to tackling climate change and biodiversity loss. NBS harness nature to address environmental challenges and secure a sustainable future. #ReviveOurLands
>
> #GreenNigeria (X, June 5, 2024)

The post above referenced the activity of a junior minister in the ministry. While the handle parades over 96,000 followers, the participatory engagement of the post is minimal. After a week and a record of 719 views, there were 37 likes, 15 reposts and only 2 replies. It is also instructive to note that the two replies are self-posted repeats of the same message.

Another post from the same handle on June 5 calls attention to biodiversity:

> This #WorldEnvironmentDay, let's focus on "Reviving Lands for a Brighter Future." Together, we can restore degraded lands, improve biodiversity, and secure a sustainable future for generations to come. #ReviveOurLands #GreenNigeria

The above post attracted a good measure of engagement with 1,400 views, 63 likes, 28 reposts and 1 reply.

A great number of those who used the hashtag in their posts on X are limited to officials of the ministry or their followers whose commitment to the ideals of environmental principles cannot be established. It is observed that there is a limited co-occurrence of use of the hashtag on X except for the commemoration of the global World Environment Day on the 5th of June. Such limitation indicates a limited articulation of #GreenNigeria as an advocacy slogan aimed towards behaviour change or greater imbibing of a new culture of protection of the environment through climate change communication. Evidence from the handles and discussions on the different posts suggest that articulation of the climate change advocacy discourse using the #GreenNigeria slogan is very weak. This stems from a low level of engagement and participation.

Conclusion

The internet, as Dutton contends, does not merely alter the technological affordances of communication. It 'changes how we do things and the outcome of those activities' (2023: 1). The way we design and use technology, which could also be a function of several other attenuating factors tied to our cultural orientation, is quite significant in the outcome we get at the end of the day. Social media may have become a 'fifth estate' because of their pervasiveness thus making them a very attractive platforms for information dissemination for specific targeted campaigns, yet the need still exists for a critical consideration of the socio-political realities which define stakeholders and participants in designing, evaluating and understanding such activities. Social subject positions, lived realities and orientation matter in the ways people mediate information, campaigns or slogans on social media as such information displayed on social media should be seen generally as socially constructed phenomena which require rigorous examination for the understanding of their import and impact. The discursive power held by the producer of a social media message is often fragile and as such greater participation through infomediation which incorporates the social realities of participants may be important in giving resonance to climate change communication campaigns such as #GreenNigeria.

References

Arias, M. (2024). Framing Digital Identities through Social Media in Museums *Internet Histories,* 8: 1–2, 169–187.

Artieri, G., Brilli, S., & Zurovac, E. (2021). Below the Radar: Private Groups, Locked Platforms, and Ephemeral Content—Introduction to the Special Issue. *Social Media + Society,* 7: 1–7.

Berglez, P., & Al-Saqaf, W. (2020). Extreme Weather and Climate Change: Social Media Results, 2008–2017. *Environmental Hazards,* 20: 4, 382–399.

Bijker, W., Hughes, T. & Pinch, T. (eds.) (2012). *The Social Construction of Technological Systems: New Directions in the Sociology and History of Technology*. London: The MIT Press.

Buthelezi, N. (2024). Probing Climate Change Perceptions in Vulnerable Enclaves: Resilient Realities in the Buffalo City Municipality, South Africa. *Lusophone Journal of Cultural Studies*, 11: 1–27.

Callon, M. (2012). Society in the Making: The Study of Technology as a Tool for Sociological Analysis. In Douglas, D., Hughes, T., & Pinch, T. (eds.), *The Social Construction of Technological Systems: New Directions in the Sociology and History of Technology*. Cambridge: MIT Press, pp. 83–103.

Cao, T. (2024). Rethinking Openness: A Social Constructivist Approach to the Promises of the New Museology. *Internet Histories*, 8: 1–2, 114–135.

Carpentier, N. (2011). *Media and Participation: A Site of Ideological-democratic Struggle*. Bristol: Intellect.

Carpentier, N., & Hroch, M. (2024). Introduction: The Construction of the Future of Platforms. *Central European Journal of Communication,* 35: 1, 1–16.

Castro, M. (2024). Climate Emergency and Youth Activism: A Case Study in Lisbon *Lusophone Journal of Cultural Studies*, 11: 1–19.

Cornwall, A. (2008). Unpacking Participation: Models, Meanings and Practices. *Community Development Journal,* 43: 3, 269–283.

Douglas, D., Hughes T., & Pinch, T.(eds) (2012). *The Social Construction of Technological Systems: New Directions in the Sociology and History of Technology*. Cambridge: MIT Press.

Dutton, W. (2023). *The Fifth Estate: The Power Shift of the Digital Age*. Oxford: Oxford University Press.

Evanglista, S., & Garcia, M. (2024). Climate Change Narratives in TikTok Brazil: From Diagnosis to Despair. *Lusophone Journal of Cultural Studies*, 11: 1–22.

Fernandez, M., Piccolo, L. S., Maynard, D., Wippoo, M., Meili, C., & Alani, H. (2016). Talking Climate Change via Social Media: Communication, Engagement and Behaviour. *Proceedings of the 8th ACM Conference on Web Science*, pp. 85–94.

Fitzgerald, S., & Winseck, D. (2018). Media Economics: Missed Opportunities, Mischaracterizations. *The Political Economy of Communication,* 6: 1, 83–97.

Gielens, K., & Steenkamp, J. (2019). Branding in the Era of Digital (Dis)intermediation. *International Journal of Research in Marketing,* 36: 3, 367–384.

Gislason, M. K., Galway, L., Buse, C., Parkes, M., & Rees, E. (2021). Place-based Climate Change Communication and Engagement in Canada's Provincial North: Lessons Learned from Climate Champions. *Environmental Communication,* 15: 4, 530–545.

Gómez-Casillas, A., & Gómez Márquez, V. (2023). The Effect of Social Network Sites Usage in Climate Change Awareness in Latin America. *Population Environment*, 45: 2, 6–28.

Helmond, A. (2015). The Platformization of the Web: Making Web Data Platform Ready. *Social Media + Society,* 1: 2, 1–11.

Hungbo, J. (2021). Twitter, Youth Agency and New Narratives of Power in #RhodesMustFall. In Ugor, P. (ed.), *Young People and Popular Culture in Africa.* Rochester: University of Rochester Press, pp. 235–254.

Hutchinson, J. (2021) Digital Intermediation: Unseen Infrastructures for Cultural Production. *New Media & Society,* 25: 12, 3289–3307.

Hutchinson, J. (2023). *Digital Intermediation: Unseen Infrastructures for Cultural Production*. London: Routledge.

Kelty, C., & Erickson, S. (2018). Two Modes of Participation: A Conceptual Analysis of 102 Cases of Internet and Social Media Participation from 2005–2015. *The Information Society,* 34: 2, 71–87.

Leon, B., Bourk, M., Finkler, W., Boykoff, M., & Davis, L. (2023). Strategies for Climate Change Communication through Social Media: Objectives, Approach and Interaction. *Media International Australia*, 188: 1, 112–127.

León, B., Negredo, S., & Erviti, M. C. (2022). Social Engagement with Climate Change: Principles for Effective Visual Representation on Social Media. *Climate Policy*, 22: 8, 976–992. https://Doi.Org/10.1080/14693062.2022.2077292

Magaudda, P., & Solaroli, M. (2020). Platform Studies and Digital Culture Industries. *Sociologica*, 14: 3, 267–293.

Makananise, F. (2025). Decolonial Thinking of Digital Media Inequalities and Indigenous Language Marginalisation of the Global South from the South African Context. In Makananise, F. & Madima, S. (eds.), *Decolonising Digital Media and Indigenisation of Participatory epistemologies: Languages of the Global South.* New York: Routledge, pp. 1–14.

Makananise, F., & Madima, S. (2025). A Decolonial Study of Digital Media and the Epistemic Indigenisation of the Indigenous Future. In Makananise, F. & Madima, S. (eds.), *Decolonising Digital Media and Indigenisation of Participatory Epistemologies: Languages of the Global South.* New York: Routledge, pp. 1–14.

Nieborg, D., & Poell, T. (2018). The Platformization of Cultural Production: Theorizing the Contingent Cultural Commodity. *New Media & Society*, 20: 11, 4275–4292.

Oliveira, C. (2020). Proposed Solutions to Citizen Engagement in Virtual Environments of Social Participation: A Systematic Review. *International Journal of Electronic Governance,* 12: 1, 76–91.

Olohan, M (2017). Technology, Translation and Society: A Constructivist, Critical Theory Approach. *Target,* 29: 2, 264–283.

Pearce, W., Niederer, S., Ozkula, M., & Querubin, N. (2019). The Social Media Life of Climate Change: Platforms, Publics and Future Imaginaries. *Wiley Interdisciplinary Reviews: Climate Change,* 10: 2, 1–13.

Poell, T., Nieborg, D., & van Djick, J. (2019). Platformisation Internet Policy Review, 8: 4, 1–13.

Sebbah, B., Sire, G., & Smyrnaios, N. (2020) Journalism and Platforms: From Symbiosis to Dependency. *About Journalism* 9: 1, 12–17

Segerberg, A. (2017). Online and Social Media Campaigns for Climate Change Engagement. *Oxford Research Encyclopedia of Climate Science.* https://oxfordre.com/climatescience/view/10.1093/acrefore/9780190228620.001.0001/acrefore-9780190228620-e-398. Retrieved 21/7/2024.

Siapera, E. (2013). Platform Intermediation and Journalism. *Culture Machine* 14. culturemachine.net/wp-content/uploads/2019/05/509-1144-1-PB.pdf. Retrieved 21/7/2024.

Smyrnaios, N., & Rebillard, F. (2019). How Infomediation Platforms Took Over the News: A Longitudinal Perspective. *The Political Economy of Communication,* 7: 1, 30–50.

Tyagi, A. (2021). Challenges in Climate Change Communication on Social Media. Unpublished PhD Thesis, Pittsburgh: Carnegie Mellon University.

van Dijck, J., Poell, T., & de Waal, M. (2018). *The Platform Society: Public Values in a Connective World.* Oxford: Oxford University Press.

Villi, M., & Matikainen, J. (2016). Participation in Social Media: Studying Explicit and Implicit Forms of Participation in Communicative Social Networks. *Media and Communication*, 4: 4, 109–117.

Wang, D., & Yecies, B. (2023). Collaborative Cultural Intermediation and Communitainment Value in the Creator Economy: The Expanding Case of Webtoons. *Emerging Media,* 1: 2, 218–243.

17 Communicating climate change on TikTok

Actors, cultures and practices in Africa

Gopolang Ditlhokwa and Lyton Ncube

Introduction

Reflecting on some landmark events such as the 1992 Rio de Janeiro conference, advocacy efforts have pushed for pragmatic agendas and frameworks for addressing climate change. To achieve their agenda, advocacy groups have instrumentalised digital platforms. These media platforms, particularly in the digital age, have also been instrumental in disseminating information and defining public perceptions of climate change. Platforms like TikTok have amplified these efforts by activists, advocacy groups and individuals, providing interactive digital communication spaces and accessible channels for sharing climate-related content. Scholarly research on the role of digital platforms like TikTok as avenues of activism on issues such as climate change tends to be more prominent outside of Africa (see, for example, Basch, Yalamanchili & Fera, 2022; Hautea, Parks, Takahashi & Zeng, 2021), despite the Global South often bearing the brunt of climate change impact. Research that examines climate change communication on digital platforms like TikTok is still emerging in the Global South. This literature tends to be growing in contexts like China (see, for example, Sun, Jia, Razzaq & Bao, 2024). This chapter fills this gap by examining how climate change activists, be they individuals or organisations, disseminate climate change messages on TikTok. The intention is to understand, among other issues, the actors and dominant narratives of climate change on TikTok, how the platform has democratised climate change communication and the potential for misinformation and disinformation on climate change.

TikToking cultures are growing across the world, and as they grow exponentially, they are shaping in revolutionary ways the digital media landscape. This growing culture of TikToking is more evident in countries that have advanced technologies and strong digital infrastructures. As TikTok expands, it has driven new ways of creating content (Cho, 2020). And in the same vein, fostered new ways of public engagement with issues (Shafie, 2021). This short-form video content platform has experienced a seamless integration into global digital ecosystems (Abidin & Lee, 2023). As it integrates into global cultures, TikTok facilitates the export of cultural products and their proliferation in different contexts through different cultural products like memes, dance etc. It has also become a preferred platform for influencers to influence various public debates and exchange information on

DOI: 10.4324/9781003660217-22

various subjects. Amongst the youth, TikTok is appropriated as a space for knowledge production and often, for humour (Vazquez-Calvo, 2023).

The platform has, furthermore, morphed into a space on which the commodification and fetishisation of culture are prevalent (Madarang, 2020). Despite all these challenges and accusations of fetishisation and commodification, TikTok remains, nonetheless, a dynamic platform that allows different publics to discuss and debate topics, both sensitive and non-sensitive, tension-building and divisive topics like climate change. Its unique positionality as a short messaging platform allows it to provide a digital arena for different forms of advocacy.

In this chapter, we examine how TikTok has been used for climate change advocacy. Through a content analysis of climate change-related material on its platform, we examine how the platform has been used for climate change communication. We, furthermore, examine the kind of climate change content available on this platform. Our focus is more on climate change discourses relating to Africa. We use digital archival research to gather our data. For our data, we make use of climate change-related videos drawn from TikTok, specifically relating to the African continent. There are thousands of climate change videos on TikTok. Because of this, we purposefully sample our videos using an 'episodic' approach. By this we mean that we focused on climate change videos that were posted in relation to certain climate change-induced events. Some were made during major climate change meetings like COP27. We also sampled videos that related to cyclones – Idai, Freddy etc. We also focused on climate change-induced events across the rest of Africa in order to widen our sample. Thus, we added events like droughts in the Sahel and the Sudan that are often debated on TikTok as climate change-induced events. Our period of sampling started in 2019. To further narrow the sample, we eliminated any videos that did not specifically mention climate change in this period. Thus, 'climate change' was our key search term. This still left us with about a hundred videos. We finally eliminated those that were in languages other than English. Our aim was to understand the kind of climate change information the videos had, what it was generally meant to achieve in our own view and finally, the authenticity of the message. However, in some instances, we refer to some videos which are made secondary references by some other videos. The reason being that they had been deleted for violating community standards. For example, climate change videos containing misleading and false information.

The chapter is organised as follows. In the next section we simultaneously provide a brief literature review and theoretical framework on which our discussion is anchored. After this section, we then present our findings and conclude with a discussion of our findings.

Literature review and theoretical framework

Literature on TikTok and its role in climate change communication is still emerging, especially in Africa. The majority of the extant literature is situated in research contexts of China and in parts of the Global North. TikTok as a social media platform is mostly used by the millennials and 'Generation Z' (Cervi, 2021).

In countries like China, young people are beginning to worry more about climate change (Sun, Jia, Razzaq & Bao, 2024). As such, youth activism on climate change has been growing on this platform (Basch, Yalamanchili & Fera, 2022). TikTok has quickly become a constructive outlet for the free expression of dissent on climate change policies (Hautea, Parks, Takahashi & Zeng, 2021). For instance, as Basch and Ethan (2024) note, 'Generation Z' and millennials voice their concerns about climate change and oppose fossil fuels that drive climate change. Nguyen (2023) agrees with Hautea, Parks, Takahashi and Zeng (2021) on the role of TikTok in climate change activism, noting that TikTok, much more recently, has led to increased activism, rallying support for climate change movements and disseminating climate change-event information. Sun, Jia, Razzaq and Bao (2024) note that TikTok is central to raising awareness; in addition to this, it is often used by 'environmental influencers', opinion leaders who have significant power and control over climate change discourses on the TikTok platform and any other social media platform. Researchers like Cody et al. (2015) argue that TikTok has become more popular because of its climate communication change discourses. In addition to this, some other researchers agree that climate change is a subject, on TikTok, more popular with men than women (Nieto-Sandoval & Ferré-Pavia, 2024).

In Africa, there is also a growing scholarly interest in TikTok and everyday discourses. Existing literature indicates its popularity, particularly among urban youth, as well as its emerging potential in areas such as political communication, advocacy and representation. Nwafor et al. (2024) emphasise TikTok's value in Nigeria for disseminating climate change information, positioning it as a tool for raising awareness about critical issues. Similarly, Morah and Nwafor (2024) argue that TikTok's participatory features have enabled it to play a growing role in political communication, allowing users to engage with civic issues in new ways. These studies indicate a shift in how TikTok is being used in Africa, moving beyond entertainment to more serious applications. However, the picture is far from uniform, with structural challenges limiting the platform's broader impact.

Research by Nkomo and Matli (2022) explores TikTok's potential in education, specifically among rural learners in Zimbabwe and South Africa. Their study shows how TikTok facilitates the sharing of reading materials and improves accessibility to educational content. Yet, these benefits are not evenly distributed. Barriers such as poor Internet connectivity, limited access to mobile devices and a cultural focus on academic reading over leisure constrain its use. These findings reflect the broader digital divide that characterises much of Africa, where rural and underserved communities often struggle to access digital platforms effectively. This raises important questions about whether TikTok's potential can be fully realised in contexts where infrastructural inequalities persist (Hesmondhalgh et al., 2023). The question of whether TikTok is taken seriously as a platform for advocacy in Africa is another area of scholarly debate. Critics such as Haslem (2022) argue that the platform's emphasis on fleeting trends and playful content undermines its capacity to sustain serious discussions. This critique is relevant in contexts like climate activism, where sustained engagement is critical for long-term effect. However, other studies suggest that TikTok can be an effective medium for charting serious discourse when creatively framed. Machirori (2023) provides a detailed analysis of how young Africans used TikTok during the COVID-19 pandemic to challenge Afropessimism. The study reveals that these users disrupted global stereotypes of Africa as a crisis-ridden and underdeveloped continent by creating humorous and relatable content, especially the dance challenges. Machirori's work proves TikTok's potential to breed positive narratives and engage audiences through creative approaches in addressing issues of representation.

TikTok's role in political engagement further shows its growing utility in Africa. Agabo (2024) demonstrates how TikTok has been used in Uganda to hold leaders accountable through satire and other forms of civic engagement, expanding the digital space for public debate. Similarly, Ureke (2024) indicates that TikTok was employed in Zimbabwe to encourage voter registration and visually critique political parties. This blending of entertainment and civic messaging shows that the platform has the ability to connect with younger, tech-savvy audiences. However, concerns remain about the risks of trivialisation. Critics warn that the focus on visually engaging, viral content may reduce serious issues to fleeting trends, diminishing their impact, especially that social media participants are shifting from being passive users to active creators of content (Wang, 2024; Nowacki, 2024). Nonetheless, proponents like Machirori (2023) and Ureke (2024) argue that TikTok's accessibility allows it to engage audiences beyond traditional media's purview.

Furthermore, TikTok, in this regard, becomes a valuable arena for gauging public sentiments on critical climate change concerns like global warming (Sun, Jia, Razzaq and Bao, 2024). While this may be the case, issues around information reliability on TikTok remain topical (Cody et al., 2015). Some of the climate information on TikTok requires verification (Sun, Jia, Razzaq and Bao, 2024). The tragedy, as stated by Sun, Jia, Razzaq and Bao (2024), is that TikTok can be a dangerous platform for climate change disinformation. Nieto-Sandoval and Ferré-Pavia (2024) argue that climate change communication is often event-driven, for example, during COP26 conferences. Basch and Ethan (2024) agree with the latter view, noting that TikTok offers networked communication on climate change events. Pereira and Ha (2024) argue, in agreement, that short videos on TikTok about climate have an impact on the discourses how it is viewed, even though there can be potential for misleading information. Misleading information on social media platforms emanates from the fact that some influential people on TikTok lack scientific facts, and some drive conspiratorial beliefs. The general agreement in this literature is that TikTok offers a platform for climate communication. Furthermore, this extant literature agrees that this platform allows the subaltern to speak. However, it is not without its danger, as it can be a purveyor of climate disinformation and misinformation.

This chapter is conceptualised on the concept of digital activism. Digital activism has a long history around which scholars have conceptualised it (Treré & Kaun, 2021). Simply expressed, digital activism or Internet activism involves the use of electronic-communication technologies such as social media, e-mail and podcasts for various forms of activism to enable faster and more effective communication by citizen movements and the delivery of particular information to large and specific audiences, as well as coordination (Howey, Tectonica & Weronika, 2024; Dun, 2024). Understood this way, digital activism becomes an articulation of resistance via digital platforms (Fotopoulou, 2016). Studies in digital activism became popular during the Arab Springs, and the #blacklives matter and #Occupy Wall Street movements, amongst others (Fotopoulou, 2016). Often digital activism is a preferred weapon for organisations struggling for social justice, as it offers new forms of interaction as well as new forms of resisting dominant forces in society.

Digital activism constitutes forms of continuities and resistance within social movements (Özkula, 2021). There are four distinct forms of digital activism that have come to define the symbolic interactionism tradition afforded by digital activism (Özkula, 2021). The first phase is the technological narrative phase, which sees technology as the main driver of digital activism. The second one is the communicative narrative, which sees activism as the basis of any activism. Then there is the online–offline narrative, which sees activism as based on the dichotomy of the offline and online. Lastly, there is the engagement narrative, which treats activism as based on the affordances for public engagement (Özkula, 2021). In this chapter, we draw on digital activism to examine the forms of climate change activism that emerge on TikTok. We find digital activism theory relevant to understanding climate change TikToking because it captures how digital platforms like TikTok enable new forms of environmental advocacy. TikTok's participatory culture allows users to engage in climate action by producing short-form video content that raises awareness, mobilises support and connects with global audiences. This aligns with the engagement narrative of digital activism, where public participation is enhanced through platform-specific affordances (Özkula, 2021). For example, TikTok users often employ hashtags, challenges and viral trends to make climate change messages accessible and engaging (Nieto-Sandoval & Ferré-Pavia, 2024). This signifies how digital activism operates within the affordances of TikTok's unique format. In the next section, we present our findings.

Findings

Our findings show that TikTok has been central to climate change communication amongst users in Africa, as elsewhere. Yet, it assumes an even more important role during major environmental catastrophes like cyclones, droughts, famines, floods, typhoons etc. We noticed that these events were often discussed as manifestations of climate change in Africa. We noticed that most of the climate change communicating videos on TikTok were often contextualised around these events and strenuously linked them to climate change on the continent. Furthermore, we noticed that climate change communication on TikTok was dominated by non-experts, who often communicated activists and awareness material. However, the fact that there was very little expert-driven material on climate change on TikTok drove climate change communication into the realm of misinformation and disinformation.

Non-expert climate change communication on TikTok: Activism and awareness

Empirical data indicates that due to its interactive nature, among other affordances, TikTok is transforming the production, distribution and consumption of climate change discourses in Africa. Belair-Gagnon and Holton (2018) argue that the contours of journalistic practice have changed substantially in the digital age. This is also applicable in the area of climate change communication, where non-traditional actors have taken centre stage, thanks to TikTok and other digital social media platforms. We observed that TikTok offers non-traditional climate change actors opportunities to participate and share experiences, fears, aspirations and concerns about climate change to a global audience.

We note that climate change communication has now been democratised in various ways, albeit with serious implications on the quality and credibility of the news or content. To begin with, non-expert communicators including ordinary men and women, activists and influencers play a critical role in sharing information about climate change incidence ranging from cyclones, devastating floods, heatwaves and droughts, among others. A video posted on 8 April 2024, about 'storms in South Africa[1]' illustrates our observation. The video shows how storms destroyed valuable property such as buildings, roads and cars, among others, in South Africa.

Despite being peripheral actors in climate change communication, non-climate change experts play a critical informative role. For instance, the TikTok video titled 'Cyclone Hidaya: The storm on Kenya's cost[2]', shared on 3 May 2024, demonstrates our observation. Following reports that Cyclone Hidaya could hit Kenya, a TikTok video was shared warning people to stay on the alert and take precautions to ensure their safety. The video also makes an attempt to provide 'thick descriptions' (see Geertz, 1973) of what Cyclone Hidaya is all about, its origins and potential destruction effects. Graphic illustrations are also provided to make the explanation more elaborate. However, most of the information in the video is not referenced or does not provide scientific evidence to support the claims. The same can be said about the video 'Cyclone Freddy between Mozambique and Malawi more than 70 people dead[3]', posted on 14 March 2023. The video footage shows houses destroyed by rains and voices of people crying in the background. The TikTok video also claims that 70 people lost lives, but there are no sources to support the claims. Of course, this is not something unusual for non-professional climate change communicators.

Our study notes that TikTok has also allowed marginalised voices to represent their own stories in digital public spheres. For example, in the Sahel region in Chad, climate change effects have far-reaching consequences. A video, 'In Chad climate change is already a reality[4]', chronicles the suffering of ordinary people in the Sahel region as a result of climate change. In the video, a woman explains from an experiential perspective what it means to live in the Sahel region in the era of disruptive climate change. The woman explains how water levels in Lake Chad have been decreasing at a decreasing rate, yet over 40 million people and animals depend on this lake. The decrease in water levels creates a conflict between communities who survive on this lake.

TikTok also allows climate change activists to take the mainstream media to task on their coverage of climate change discourses. For instance, in the wake of the 2022 devastating floods which hit South Africa's KwaZulu-Natal province, some of the influencers-cum-activists took their frustration to TikTok, asking the media to account for failure to provide timely and accurate information prior to the floods. For example, a video titled 'How is the heatwave? 😂 #southafrica #climatechange #naturaldisasters #wws[5]', posted on 24 October 2022, demonstrates our observation. The video raises provocative questions on whether the South African media are doing enough to cover climate change in the wake of disasters such as ever-recurring floods. The video provides statistical evidence claiming that 40,000 people were left devastated, 400 people died and 12,000 houses demolished, albeit without acknowledging sources of this information, which throws the issue of credibility into question.

We also established that TikTok climate change activism also targets world leaders to act and mitigate the multi-layered climate change crisis. A video posted on 8 November 2022, titled 'Youth climate activist from #Uganda, Leah Namugerwa's speech to world leaders at the UN Climate Change Conference #COP27 #ClimateAction #SharmElSheikh', demonstrates our point. In the video, a young lady laments how the climate change crisis is destroying the future of young people in Africa in multi-dimensional ways. Critically, we observed that there is a growing concern amongst citizens and activists that countries from the Global South are not responsible for climate change but pay the highest price of this vice. To gain legitimacy and traction, some of the videos utilise voices of authoritative elites such as presidents. For example, a video titled 'Mitigation the need for inclusion exclusive chat with President Hakainde Hichilema! Unpacking Zambia's role in global climate action & why & Africa's voice matters. Let's make change together! 💪 #ClimateChange #AfricaRising[6]' confirms our observation. Hichilema is quoted bemoaning that when decisions are made about climate damage and mitigation, Africans are excluded. Therefore, African voices are calling for inclusion and participation in decision-making processes on climate change.

TikTok climate change videos are also characterised by humour. For example, the video 'The heat in Ghana now is overwhelming what's causing the climate change … #fyp #heat #climatechange #africantiktok #ghanatiktokGH[7]' shared on 15 February 2024 confirms our observation. A commentator mimicking a sports commentary said, '… I don't know if it's the whole of Ghana or Accra, but there is something different about the weather, we all know that Africa is hot but the heat in the system now feels like an updated version, it feels like the earth is sitting on one place and the sun is rather going around the earth, because it doesn't matter whether it's a day or night, you can really feel the heat. It's not even about having air cones, the moment you step out of your room it's like angels are doing a barbecue in heaven …' The deployment of humour in climate change is something worth unpacking. Mbembe (2001) contends that humour must not only be taken as a form of crude and primitive culture, but rather it demonstrates power contestations in societies. In this case this humour should be read as an attempt by subalterns to speak against climate change injustices. Humour is also critical in popularising critical discourses.

There are many instances where climate change awareness and activism were raised through this platform. For example, Kobeboujee's video, posted on 15 February 2024, with a high engagement of 79.7K likes and 1,835 shares, gives a localised context on the heatwave in Ghana, illustrating how personal storytelling can drive climate activism and awareness. There are also, in some instances, videos about the World Economic Forum of 11 February 2020, which showcase it as a relevant educational platform, providing an in-depth look at how climate change affects Chad, a country facing severe environmental challenges. This video is an example of how climate education can be effectively communicated through digital platforms.

The UN Climate Change video featuring Leah Namugerwa, posted on 8 November 2022, exemplifies the blend of activism and education in climate advocacy. Leah's speech at the 27th session of the Conference of the Parties to the UNFCCC (COP27) in Sharm El-Sheikh, Egypt, brings to light the severe concerns of climate change, including landslides that have resulted in significant loss of life. Her address is paramount to the role of young leaders in both advocating for climate action and educating others about its importance. Equally, African stream's coverage of the Africa Climate Summit, posted on 5 September 2023, gives an account of the regional climate strategies and international cooperation. The video presents the actual footage recorded during the summit which was held from 3 to 6 September 2023, addressing climate change and the role of African nations in global climate discussions. Some videos also focus on practical strategies for mitigating climate change, educating the public and encouraging proactive responses. For instance, Just digit's video on 22 February 2023 showcases their regreening initiatives and the importance of sustainable practices. The video captures how minimal labour can contribute to large-scale environmental benefits, making an argument for regreening efforts as a viable solution to climate challenges. Similarly, the World Food Programme's video, posted on 31 August 2023, features agricultural technology in Mozambique. Its content shows how technology helps farmers cope with drought and other climate-related challenges.

TikTok and climate change disinformation and misinformation

The general consensus now is that climate change discourses are becoming more prevalent on TikTok. Famous climate change influencers like Wawa Gatheru, the Kenyan-born and US-based climate change activists, have made a huge following on TikTok through their environmental justice activism.[8] Several other climate change accounts have emerged over the years. Generally, climate change accounts on TikTok focus on raising awareness and activism. Studies[9] have shown that younger adults are more likely to engage with climate change content on TikTok and other social media platforms. This, according to these studies, has helped lessen climate change anxiety and turn into empowerment, helping next generations engage in informed conversations about climate change. The use of TikTok for climate change activism and awareness-building has increased during major events like COP27 and #ClimateAction global campaign, and other climate change-focused events.

But this has come with a caveat – the growing content on climate change denialism and outright misinformation and disinformation on TikTok. Thousands of TikTok videos have sprouted out that deny climate change. So, some of the videos outrightly deny that climate change is happening or will ever happen. However, it is difficult to reference such videos, as they are pulled down by TikTok for violating community regulations on the platform that seek to empower informed climate discussions. Every time such videos are uploaded, they generate several million views before they are taken down. What this means is that they are still viewed, even if it means for the short time of their existence, and they still have an impact on climate change communication. Misleading videos and texts, by, for example, a Kenyan farmer, Jusper Machogu, garnered thousands of views saying Africa will not be impacted by climate change.[10] The videos were still taken down because they violate community standards, but some were reported on; for instance, footnote 3 is an example of such a report.

Some climate change TikTok pages go beyond misinformation to outright disinformation. Climate change disinformation often took three major approaches. The first one is when users played down the effects of climate change. For example, the claim that Africa would not be impacted on by climate change. Some videos were pulled off TikTok because this still constitutes disinformation. The second type of climate change disinformation on TikTok was repurposing climate change-related disasters. For example, some TikTokers used destructive images of Cyclone Idai[11] as evidence of yet another disaster like Cyclone Freddy. The purpose of this kind of disinformation was to overemphasise the point being made. Sometimes users would repurpose the whole climate news cycle.[12] Under repurposing, events were mismatched in order to emphasise the urgency of the matter. The third way through which disinformation was spread was through backgrounding other issues and foregrounding certain events that suit the ideology of the TikTokers. For example, the damage that Storm Fillipo caused in Mozambique was often backgrounded because it did not sync neatly with the near-total destruction and wipe-out narrative TikTokers wanted, which aligned with the total lack of awareness on climate change that they wanted.[13] In the Sudan, old clips of arid land were used to drive the point that the Sudan drought crisis was a climate change crisis the world has seemingly forgotten.[14]

There are several dangers associated with climate change disinformation and misinformation. If these distortions and, in some instances, conspiracy theories are allowed to spread, they become an obstacle to meaningful social action against climate change. Secondly, climate change disinformation and misinformation may trigger disagreements amongst actors, and fossil fuel producers, who are the major targets of climate change awareness, may seize on this disagreement to question the validity of climate change research and question climate change facts. TikTok is peculiar in that it can lead to three types of biases identified by researchers at

Indiana University.[15] TikTok can, firstly, lead to cognitive biases on climate change. 'Cognitive biases are systematic errors in thinking that stem from our propensity to use mental shortcuts'. Furthermore, it can lead to social bias. This is the tendency to trust information that comes from people you know (or with whom you identify) over information from other sources. And, lastly, TikTok can promote algorithmic bias that comes from social media platforms themselves: Sites like Facebook and Twitter tailor the content that we see every day, based on what they predict you'll react to most passionately.

Discussion: TikTok, climate change communication and the emergence of multiple regimes of truths

As the climate crisis unfolds in Africa and beyond, a number of actors have come on board to share experiences and 'truths' on the subject. Focusing on critical incidences of climate change from Africa, we show that contrary to traditional practice, where climate change communication was a preserve for science experts, in the era of mass self-communication (Castells, 2012), non-expert climate change communicators are utilising TikTok and other technologies to participate in the production, sharing and consumption of climate change news. Through the lenses of platform activism, we show that platforms like TikTok offer avenues for collective and individual action by challenging the dominance of the gatekept traditional media, albeit with ethical implications. Though TikTok audiences provide insightful comments on climate change videos, they were not part of our unit of analysis.

Non-science communication experts have flooded and disrupted the climate change field, capitalising on their access to abundant interactive social media technologies such as TikTok. The involvement of non-traditional science actors, such as grassroots activists, community leaders and ordinary citizens in climate change communication, has the potential to decenter traditional narratives and amplify marginalised voices, which is commendable. Traditionally, science stories in the media have relied primarily on scientists, physicians and government and industry officials as expert sources (Kruvand, 2009). However, from the critical incidents of TikTok climate change communication analysed, the legacy media no longer enjoy a monopoly over climate change discourses. This is because TikTok, Facebook and YouTube, among other social media platforms, have opened up to 'peripheral' climate change communication actors to share their experiences. This democratisation of communication challenges elite discourses and specialised communicators, opening up new avenues for participation, activism and representation. In essence, TikTok increasingly contributes to the demotic turn (see Turner, 2010). The appropriation of TikTok by ordinary people for climate change communication purposes shows the agency of these people.

We demonstrate that TikTok has become central in climate change activism and awareness in Africa. This is because the Global South, Africa in particular, continues to be at the receiving end of climate change-induced problems such as drought, hunger, floods and displacements, among others. Most of the analysed videos expressed that Africa emits the least fraction of greenhouse gases compared to more industrialised economies in the European Union, America and China, but it is heavily impacted and disrupted by climate change. Some of the TikTok videos[16] frame climate change disasters in Africa in the context of colonial rapine and plunder relations between Western countries and Africa. Such TikTok videos underscore that Western countries got rich by looting African resources and became industrial powerhouses and now 'industrial-scale polluters', forcing Africa to pay for the price of environmental recklessness done by these countries. Importantly, activists also demonise hypocritical policies of some Western countries which encourage African countries to adopt green policies which they hardly stick to. For instance, Western countries discourage Africa from using coal, yet when there is an energy crisis in the West they revert to the same coal. Thus, activists call for climate change justice and retribution. Importantly, climate change activists also advocate for a better financing of global climate change programs. While this study does not focus on TikTok audiences themselves, it is important to recognise the degree to which climate change activists on the platform are able to capture attention and drive engagement around these issues. A notable example is Kobeboujee's video, as presented in the findings, which garnered 79.7K likes and 1,835 shares. This high level of engagement shows the potential for African TikTokers to build connections between themselves and their audiences, even in the face of structural barriers such as limited Internet access and high data costs in many African contexts. However, our study does not rule out the implications of content formats towards engagement, as observed by Duan et al. (2024), that short-form climate change videos yielded low audience engagement (in comments) compared to long-form videos.

We also note that the democratisation of climate change communication aided by interactive digital contributes to multiple regimes of truths (see Foucault, 1977) on climate change discourses. Peripheral actors in climate change communication are increasingly becoming integral in the social construction of the 'truth' about climate change. However, such 'truths' might not have been verified by climate change experts or through scientific ways of knowing. Every society has discourses and mechanisms which legitimise that truth, distinguishing it from false statements (Foucault, 1977).

We also show that the 'truths' we hold about climate change are products of the complex interrelatedness of power, knowledge and discourse (Foucault, 1977). This is demonstrated by how indigenous climate change 'experts' also utilise TikTok to clamour for space and recognition. These people engage in a fight for recognition in a space which delegitimises them. For instance, a video titled 'African rain making: Facts of fiction[17]' posted on 7 September 2022 validates our argument. In the video, Godwin Onasedu, a self-proclaimed African rainmaker, claims to have powers to command rain to fall or stop. Indigenous climate experts have traditionally been excluded from the mainstream public sphere because their ways of knowing generally lack 'scientific evidence'. Foucault (1972) notes that discourse authorises those who can speak on a topic and those who cannot at any given moment. Every regime of truth has to be understood in the context of existing power relations (Foucault, 1978).

The emergence of multiple TikTok climate change communication actors has got serious ethical implications in Africa and beyond. Trust issues emerge. Trust is a critical issue, especially when 'science' and 'facts' alone are not adequate to inspire public confidence, or alternatively when 'science' and 'facts' alone are under-developed and not adequate to present convincing explanations and solutions for social problems (Mhiripiri & Midzi, 2020). The media are supposed to furnish the public with credible information on climate change timely, clearly and accurately (Tshabangu & Salawu, 2021). As already alluded to, we witness the emergence of multiple regimes of truths around climate change which might not be trustworthy and fact based. There are ethical dangers that such truth regimes could constitute or perpetuate disinformation, misinformation and fake news. Mutasa (2024) asserts that the spread of climate change misinformation and disinformation has become a significant challenge. We argue that there is a need for audiences to be sensitised about fact-checking so that they do not fall victim to various forms of disinformation spread by TikTok videos. This is because both climate change misinformation and disinformation can have serious consequences when it comes to public understanding and engagement on this critical and technical issue (Mutasa, 2024). Journalism in general valorises fact-checking, a component which is somehow missing in the digital social media age (Ncube, Mare & Muzondo, 2023).

Our findings also indicate how TikTok serves as a platform where multiple regimes of truth surrounding climate change emerge and coexist. Drawing on the principles of digital activism theory, we affirm that TikTok's participatory culture enables users to craft and disseminate localised stories (whether factual or speculative). As we have observed earlier, such content as Kobeboujee's video, and others, may not be relevant to some communities, but others would find them very useful, hence the high-level engagement and metrics. However, the fragmented and transient nature of content on TikTok also raises questions about the long-term sustainability of such truths in influencing global climate agendas (see Foucault, 1977).

Conclusion

In this chapter, we examined the use of TikTok for climate change communication in Africa. Our chapter established that climate change communication has taken a 'demotic turn' on TikTok, allowing the 'subalterns', the non-expert citizens, to engage in climate change activism and awareness. Therefore, climate change public discourses are not restricted to the mainstream, with its agenda, and gatekeeping practices that have the potential to silence ordinary citizens' voices and their experiences. TikTok climate change communication proves that peripheral actors, in the mould of ordinary citizens, have moved to the centre of the climate change debate. There are now very limited institutional constraints to climate change communication on TikTok. However, the constraints imposed by the institutions themselves on TikTok can be felt through massive removal of disruptive climate change discourses that include inaccuracies like misinformation, disinformation and outright fake news. Suffice to mention that the disruptive practices of misinformation and disinformation remain a major threat to TikTok's climate change communication practices. Our chapter focused on only climate change videos generated during climate change-induced specific episodes like conferences, floods, cyclones etc. Future research should, perhaps, focus general climate change communication on TikTok outside these events. Future research can, in addition, focus on how audiences engage with climate change discourses, how influencers affect the circulation of particular types and topics of videos, and how this has a multiplier effect on climate change communication on TikTok.

Ethics statement

This chapter utilised data mined from comments that came from publicly trending and accessible TikTok content. The researchers ensured that all identifiable metadata and location data have been removed.

Notes

1 A TikTok video showing the aftermath of a tornado in the province of KwaZulu Natal can be accessed here: www.tiktok.com/@luzwide_/video/7378247960330636550
2 A video showcasing Cyclone Hidaya on Kenya's Coast can be accessed here: www.tiktok.com/@upstartkenya/video/7364640009397243141
3 See a video on Cyclone Freddy between Mozambique and Malawi here: https://vm.tiktok.com/ZGevtehvq/.
4 A video showing Hindou Oumarou Ibrahim, a Chadian environmental activist and geographer, explaining the reality of climate change in Chad can be accessed here: www.tiktok.com/@worldeconomicforum/video/6792231787226467590?is_from_webapp=1&sender_device=pc.
5 A video posted on TikTok by the World Economic Forum on Chad's climate change situation can be accessed here: www.tiktok.com/@mma_mablomo/video/7158022158151699718?is_from_webapp=1&sender_device=pc.
6 Watch the full video on Zambia's role in global climate action and why Africa's voice matters here: www.tiktok.com/@zambiansservezambia/video/7334583074321124640?is_from_webapp=1&sender_device=pc.
7 A video on the heat in Ghana can be accessed here: www.tiktok.com/@kobeboujee/video/7335829710435339526?is_from_webapp=1&sender_device=pc.
8 Check Gatheru on TikTok here: www.tiktok.com/@wawagatheru/video/7195339000113073451.
9 For example, a Pew Research Centre publication that is accessible here: www.pewresearch.org/short-reads/2021/06/21/on-social-media-gen-z-and-millennial-adults-interact-more-with-climate-change-content-than-older-generations/.
10 Follow the full story here about climate change denialism on TikTok here: www.bbc.com/news/articles/c133r4gyx1no.
11 For instance, here: www.tiktok.com/@jadechanty/video/7203043953002712326?is_from_webapp=1&sender_device=pc&web_id=7396334061718275590.
12 For example, here: Disinformation by repurposing climate change-related news; www.tiktok.com/@malume306/video/7203934823533006086?is_from_webapp=1&sender_device=pc&web_id=7396334061718275590Tropical.

13 One example of a video on Storm Fillipo can be followed here on TikTok: www.tiktok.com/@climate_kaputt/video/7346736021180075296?is_from_webapp=1&sender_device=pc&web_id=7396334061718275590.

14 This can be followed here: www.tiktok.com/@vicenews/video/7117971550027943174?is_from_webapp=1&sender_device=pc&web_id=7396334061718275590.

15 Follow the University of Indiana research here; www.nrdc.org/stories/climate-misinformation-social-media-undermining-climate-action.

16 Some of the climate justice online debates are found here: www.tiktok.com/@humannoresource/video/7364129807010925830?is_from_webapp=1&sender_device=pc

17 A video of video, Godwin Onasedu, a self-proclaimed African rainmaker, who claims to have powers to command rain to fall or stop can be accessed here: https://vm.tiktok.com/ZGevWGGbf/.

References

Abidin, C. & Lee, J. (2023). K-pop TikTok: TikTok's expansion into South Korea, TikTok Stage, and platformed glocalization. *Media International Australia*, 188(1), 86–111.

Agabo, S. (2024). *Digital activism through humour and satire: An analysis of the use of TikTok and X for political accountability in Uganda* (Unpublished master's thesis). Aga Khan University, East Africa – Graduate School of Media and Communications.

Basch, C. H., Yalamanchili, B. & Fera, J. (2022). # Climate change on TikTok: A content analysis of videos. *Journal of Community Health*, *47*(1), 163–167. https://doi.org/10.1007/s10900-021-01031-x.

Basch, E. D., & Ethan, D. (2023). Using TikTok for political action: The Willow Project, global climate change, and health. *Intersect: The Stanford Journal of Science, Technology, and Society*, *17*(1). https://orcid.org/0009-0004-4437-3236.

Belair-Gagnon, V. & Holton, A. E. (2018). Strangers to the game? Interlopers, intralopers, and shifting news production. *Media and Communication*, 6(4), 70–78. https://doi.org/10.17645/mac.v6i4.1490

Castells, M. (2012). *Networks of outrage and hope: Social movements in the Internet age*. Polity Press.

Cervi, L. (2021). Tik Tok and generation Z. *Theatre, Dance and Performance Training*, 12(2), 198–204.

Cho, M. (2022). Anonymous, QAnon, Tik-tok teens, K-pop fans. *NECSUS_European Journal of Media Studies*, 11(1), 169–193.

Cody, E. M., Reagan, A. J., Mitchell, L., Dodds, P. S., & Danforth, C. M. (2015). Climate change sentiment on Twitter: An unsolicited public opinion poll. *PLoS ONE 10*(8): e0136092. https://doi.org/10.1371/journal.pone.0136092

Duan, Y., Khoury, C., Joh, U., Smith, A. O., Cousin, C. & Hemsley, J. (2024). Comparing climate change content and comments across Instagram Reels, TikTok, and YouTube Shorts and long videos. *Proceedings of the Association for Information Science and Technology*, 61(1), 103–114.

Dun, A. (2024-01-10). Digital campaigning: Start here. *The Commons Social Change Library*. Retrieved 2024-04-19.

Fotopoulou, A. (2016). Introduction: Conceptualising feminist activism and digital networks. In *Feminist activism and digital networks. Palgrave studies in communication for social change*. Palgrave Macmillan. https://doi.org/10.1057/978-1-137-50471-5_1

Foucault, M. (2010). *Power/knowledge: Selected interviews & other writings 1972–1977.* Edited by Colin Gordon, translations by Gordon, Marshall, Mepham and Soper. Vintage Books. (Original work published 1977).

Foucault, M. (1972). *The archaeology of knowledge*. Tavistock.

Foucault, M. (1978). *The history of sexuality*. Pantheon Books.

Geertz, C. (1973) *The interpretation of cultures: Selected essays*. Basic Books.

Haslem, B. (2022). *TikTok as a digital activism space: Social justice under algorithmic control*. Old Dominion University.

Hautea, S., Parks, P., Takahashi, B. & Zeng, J. (2021). Showing they care (or don't): Affective publics and ambivalent climate activism on TikTok. *Social Media+ Society*, 7(2), 20563051211012344.

Hesmondhalgh, D., Campos Valverde, R., Kaye, D. B. V., & Li, Z. (2023). Digital platforms and infrastructure in the realm of culture. *Media and Communication, 11*(2), 296–306. https://doi.org/10.17645/mac.v11i2.6422.

Howey, N., Tectonica, & Paszewska, W. (2024-03-18). Movements, organising and transformational politics in the Digital Age. *The Commons Social Change Library.* Retrieved 2024-04-19.

Kruvand, M. (2009) Bioethicists as expert sources in science/medical reporting. *Newspaper Research Journal*, 30(3), 26–41.

Machirori, F. (2023). This is Africa: How young African TikTok trends challenged Afropessimism during COVID-19. *Journal of African Media Studies*, 15(2), 161–177.

Madarang, C.R. (2020, 10 September). *A deep dive into #CancelKorea trend and why it could be problematic. Interaksyon*. https://interaksyon.philstar.com/trends-spotlights/2020/09/10/176665/a-deep-dive-into-cancelkorea-trend-and-why-it-could-be-problematic/

Mbembe, A (2001) *On the postcolony*. University of California Press.

Mhiripiri, N. A. & Midzi, R. (2020) Fighting for survival: Persons with disabilities' activism for the mediatisation of COVID-19 information. *Media International Australia, 178*(1), 151–167. https://doi.org/10.1177/1329878X20967712

Morah, D. N. & Nwafor, O. A. (2024). Beyond tribal politics for e-participation and development: Social media influence on Nigeria's 2023 presidential general election. *Journal of Innovative Digital Transformation, 1*(1), 32–47.

Mutasa, B (2024, August 11). *The media's role in fighting climate change. The Standard.* www.thestandard.co.zw/standard-people/article/200030806/the-medias-role-in-fighting-climate-change

Ncube, L., Mare, A. & Muzondo, I. (2023). Social media and sports journalism in Zimbabwe. *Journalism Practice 19*(7), 1637–1657. https://doi.org/10.1080/17512786.2023.2279334

Nguyen, H. (2023). TikTok as learning analytics data: Framing climate change and data practices. In *LAK23: 13th international learning analytics and knowledge conference* (pp. 273–279). Association for Computing Machinery. https://doi.org/10.1145/3576050.3576055

Nieto-Sandoval, A. G. & Ferré-Pavia, C. (2024). Communicating climate change on TikTok during the climate summits: From the environmental issues to the politicization of discourse. *Environmental Communication, 18*(5), 569–588.

Nkomo, S. & Matli, W. (2022). A framework for embracing mobile gadgets to nurture the reading habits of rural learners in Southern Africa. *Africa Education Review*, 19(3), 36–54. https://doi.org/10.1080/18146627.2023.2248547

Nowacki, L. (2024). The old king is dead, long live the algorithmic king – the decline of Facebook and the rise of TikTok – comparative study of algorithmic design of social media platforms. *International Social Science Journal*, 74(254), 1325–1338.

Nwafor, G. U., Aghaebe, S. E., Bartholomew, C. E. & Umuze, A. N. (2024) Investigating the effectiveness of TikTok in promoting public awareness and engagement on climate change adaptation and mitigation measures in Nigeria. *Asian Journal of Education and Social Studies*, 50(12), 279–291. https://doi.org/10.9734/ajess/2024/v50i121696

Pereira, B. B., & Ha, S. (2024). Environmental issues on TikTok: Topics and claims of misleading information. *Journal of Baltic Science Education*, *23*(1), 131–150.

Shafie, T. (2021). Beyond Slacktivism: The cases of K-pop fans and TikTok teens. *International Journal of Social Science Research*, 9(2), 147.

Sun, Y., Jia, R., Razzaq, A. & Bao, Q. (2024). Social network platforms and climate change in China: Evidence from TikTok. *Technological Forecasting and Social Change*, 200, 123197.

Treré, E. & Kaun, A., 2021. Digital media activism. In F. Musiani & V. Schafer (eds), *Historicizing media and communication concepts of the Digital Age: Global governance*. De Gruyter.

Tshabangu, T. & Salawu, A. (2021). An evaluation of constructive journalism in Zimbabwe: A case study of *The Herald*'s coverage of the coronavirus pandemic. *Journal of African Media Studies*, 13(3), 477–490. https://doi.org/10.1386/jams_00060_1

Turner, G. (2010). *Ordinary people and the media: The demotic turn*. London: Sage.

Ureke, O. (2024). Politics at play: TikTok and digital Persuasion in Zimbabwe's 2023 General Elections. *Africa Spectrum*, 59(2), 254–278. https://doi.org/10.1177/00020397241241335

Vazquez-Calvo, B., Duarte-Marti, S. & Zhang, L. T. (2023). Commenting on learning Korean on TikTok and YouTube. *Interactive Learning Environments*, *32*(10), 6063–6080.

Wang, X. (2024). *Collision of beauty: Exploring users' perceptions of beauty trends generated by TikTok influencers versus Douyin influencers*. Syracuse University.

18 Social Media and Climate Change Activism

When Champions Meet Sceptics on X

Tendai Chari and Mmakgotso Theko

Introduction

Social media is increasingly playing an important role as a platform for climate change activism. Its immediacy, virality, accessibility, social networking and participatory approach to communication enable individuals to share their perspectives about climate change (Tanveer, 2024). The open architecture of social media enables climate change activists to leverage hashtags, tags, viral videos, comments, text, clickable hyperlinks, visuals and personal anecdotes to engage global audiences (Tanveer, 2024). Scholarship on the social media climate change nexus amplifies the utility of social media in creating awareness, mobilisation, coordination, engagement and formation of virtual networks for climate change action as well as holding government and corporations to account (Nisbet, 2009; Sultana et al, 2024). However, this literature downplays enduring discursive climate change struggles between believers and non-believers of climate change online (Treen et al, 2022; Bloomfield & Tillery, 2019). Consequently, the polarised and partisan exchanges which take place online between these groups are jettisoned from scholarship. As a result, there is a knowledge lacuna on how competing discourses about climate change interact on social media platforms. Underpinned by the framing theory and drawing from netnography – anchored on X posts, videos, pictures – and archived textual materials and in-depth interviews, this qualitative study fills this knowledge gap through an exploration of how two opposing climate change activists, namely, Jusper Machogu (Kenya) and Leah Namugerwa (Uganda), use X (formerly Twitter) to construct meanings about climate change. How these social media activists frame climate change on X and the possible implications of such framing are questions germane to this chapter. The rest of this chapter proceeds as follows: After this introduction, section two surveys literature on social media and climate change activism. This is followed by the research methodology in section three. Section four presents and discusses the study findings emanating from the netnography and in-depth interview. Section five provides concluding remarks, offering some critical reflections on the contribution of the study to the existing body of knowledge, discussing limitations of the study and offering directions for future studies.

DOI: 10.4324/9781003660217-23

Social Media and Climate Change Nexus

Climate change is one of the most pressing global challenges in contemporary times, with human activities being attributable to greenhouse emissions (IPCC, 2023). There is a burgeoning literature on the role of social media in climate change (Mavrodien, 2019; Abubakar et al, 2025). Due to its technological affordances, social media is increasingly becoming a key channel for ventilating climate change debates (Schaefer, 2025; Gaupp & Eker, 2024). It has been argued that owing to its open architecture, social media fosters engagement on climate change by enabling individuals to share information without the kind of gatekeeping found in traditional media (Tanveer, 2024). Consequently, social media brings about a vibrant online ecosystem which enables diverse voices to cross-pollinate, thus fostering a sense of community (Tanveer, 2024). Gaupp and Eker (2024) assert that social media has created new avenues for public debates, enabling individuals to raise their voices and exchange information about climate change. Because climate change is a contested topic, virtual sharing of information about climate change enables individuals to share their emotions, assert their identity and manage their self-concept (Gaupp & Eker, 2024). Thus, "people are able to increase their feeling of individual and collective agency when they share ideas, educate themselves and build relationships and feel inspired" (Gaupp & Eker, 2024: 8). Sultana et al (2024) found that social media is an essential source of information about climate change for many people with great potential to affect how the public feels and the issue. Sultana et al's study observed a link between social media use and increased awareness about climate change. Diehl et al (2021) found that people who got their news from social media were less sceptical about climate change, and using social media correlated with a decrease in climate scepticism. This suggests that social media has great potential to foster climate change agendas of individuals and groups (Gilardi et al, 2022). By disrupting the gatekeeping power of traditional media, diverse perspectives on climate change are possible through social media. While existing studies shed light on social media's potential to enlarge the range of voices and actors engaging in climate change discourses, none of them have specifically focused on competing climate change discourses on social media, particularly the struggle between believers and non-believers of climate change science.

Social Media Activism and Climate Change

Social media has transformed climate change activism by enabling activists to engage in radical politics, organising events and building global networks (Askanius & Uldam, 2011). The mobilisation role of social media in activism has been observed during social movements such as the Arab Spring, Black Lives Matter and #METOO movements, where platforms such as Twitter (now X), Facebook and Instagram were used to disseminate information rapidly. Hart et al's (2015) concur noting that climate change and environmental organisations are increasingly using social media to inform and mobilise publics. Such decisions are often underpinned by theories such as the advocacy coalition framework, which posits that policy change depends on the ability of advocacy organisations to effectively mobilise activists at opportune moments. Studies show that social media enables activists and protest movements to "self-mediate and distribute movement goals and frames" effectively, as activists can bypass the state, the market and mainstream media gatekeeping (Cammaerts, 2015: 5). The literature on social media and activism further identifies "core logics for the use of social media by protest movements" as (1) organising internally/recruiting/networking; (2) mobilising for and coordinating direct action; (3) dissemination of movement frames independently of the mainstream media; (4) discussing, debating/deliberating/deciding; (5) attacking ideological enemies; (6) surveilling the surveillers and (7) preserve protest artefacts. Underpinning these core logics is the recognition that social media provides mechanisms for engaging with content such as the absence of gatekeepers, virality, liking, sharing of content, dialogic communication, networking and commenting, all of which signal public attention and popularity of content (Hart et al, 2025).

Hart et al (2025: 387) contend that engagement contributes to further dissemination through shares and comments as well as through platform algorithms that amplify content that the audience would have previously engaged with before. In addition, social media "metrics provide social endorsement cues, which can affect others' perceptions of message content as well as their attitudes and behaviours". This means that engagement with social media content such as climate change links with broader forms of political activism with potential for galvanising collective action. While this literature sheds light on how social media can be leveraged for climate change activism, it is silent on the potential of social media to promote ambiguous meanings about climate change and the multiple versions of reality about climate change. While social media can promote visibility and engagement about climate change through activism, there are obstacles that stand in the way (Uldman, 2018), and the level of engagement is contingent on how the issue is communicated. Consequently, a growing strand of scholarship explores framing as a strategy for enhancing engagement with climate change on social media (Nisbet, 2009; Li & Su, 2018; Badullovich et al, 2020). Framing, rather than the traditional fact-based way of communicating information, has been found to be more suitable for communicating climate change because framing enables communicators to place emphasis on certain attributes over others, thereby shaping how the issue is understood in the public domain (Badullovich et al, 2020). Nisbet (2009) argues that climate change communication messages should be tailored to a specific audience and medium to break through communication barriers of human nature, partisan identity and media fragmentation. To achieve this, climate change communicators should use "carefully researched metaphors, allusions, and examples that trigger a new way of thinking about the personal relevance of climate change" (Nisbet, 2009: 15). This is because audiences rely on frames to make sense of an issue as much as journalists use frames to craft appealing stories and policymakers use frames to define "policy options and reach decisions" (Nisbet, 2009: 15). Extant literature shows that despite the negative consequences of climate change efforts, the efforts to curb climate change have made little headway, among other factors, due to its unobtrusiveness, widespread misunderstanding, denialism/scepticism, polarisation and politicisation of the issue. While framing has been identified as one potential technique of effective climate change communication (Badullovich et al, 2020), its effectiveness in framing climate change is marred by slacktivism (engaging with no real or little involvement or commitment), fake news, misinformation, disinformation and eco-chambers, resulting in little constructive engagement and polarisation of public opinion, thus complicating the translation of online activism into tangible real-world outcomes (Sathisha, 2018).

Nisbet (2009) demonstrates how different frames were employed in the USA to perpetuate the divide between Democrats and Republicans. For instance, several pro-Republican think tanks, political leaders and commentators continued downplaying the urgency of climate change even when some of their leaders, such as Senator John McCain and others, urged action against global warming and framed climate change as "uncertain". This is contrasted to Democrats who tried to counter the "uncertainty" frame, emphasising a "Pandora's Box of looming climate crisis" (Nisbet, 2009: 19) and other discursive tropes painting "catastrophic and apocalyptic climate change scenarios" (De Graff et al, 2024) through the use of "fear appeals" whereby climate change was depicted as an existential threat to humanity (Bilfinger et al, 2024). This strand of literature throws insights into the contradictory discursive constructions of climate change, but unlike this study, which specifically focuses on social media (X), their focus is on traditional media and offline public discourses. Dinan & Miller (2023) note that the battleground between climate change champions and denialists has shifted from offline to the cybersphere, where stringent editorial standards of the legacy media are not applicable. To the best of our knowledge, there are no studies that explore the contestation between climate change supporters and sceptics on social media. As a result, there is limited understanding of contrarian climate change framing practices, particularly from a Global South perspective. We therefore build on existing studies by demonstrating how climate change sceptics and supporters harness social media to frame their climate change agendas and, more precisely, how these competing discourses interact on social media.

Framing Climate Change

Framing is a useful lens for understanding how climate change is communicated on social media platforms such as X, not least because of the cacophonous voices and diversity of agendas circulated online about climate change. Guran and Ozarslan (2022) note that framing theory and frame analysis are suitable for researching social media communications with the potential to illuminate the research of dynamic communication processes "in the fast lives of virtual communities and vast effects of online interaction of social, political, commercial institutions" (Guran & Ozarslan, 2022: 446). Although framing has a much longer history, symbolic interactionist Erwing Goffman is credited for developing the concept into a constructionist theory in his book *Framing Theory* (1974), wherein he defined frames as "schemata of interpretation" that enabled individuals to "locate, perceive, identify, and label" events and situations in their social context (cited in Guran & Ozarslan, 2022: 446). Entman (1993: 52) views framing as essentially entailing selection and salience, wherein to frame is

> … to *select some aspects of a perceived reality and make them more salient in a communication text, in such a way as to promote a particular problem definition, causal interpretation, moral evaluation, and/or treatment recommendation* for the item described.
>
> (Entman, 1993: 52; emphasis original)

In the context of this study, framing implies selecting some aspects of climate change and making them more salient to engender certain perceptions, actions, behaviours and actions by individuals, activists or policy outcomes. Nisbet (2009) contends that in order to slice through the numerous communication barriers in climate change communication, messages need to be tailored to a specific audience and medium. He defines frames as "interpretive storylines that set a specific train of thought in motion, communicating why an issue might be a problem, who or what might be responsible for it, and what should be done about it" (Nisbet, 2009: 15). Nisbet argues that framing is "an unavoidable reality of the communication process", adding that "there is no such thing unframed information" (Nisbet, 2009: 15).

Framing is important in understanding climate change communication given the divergent moral evaluations and political polarisation associated with the issue (Song et al, 2025). Studies have found a correlation between message framing and individuals' engagement with climate change as well as behavioural intentions and support for climate policy. Several studies report both a positive and negative link between social media use and changing opinions about the climate change issue (Mavrodieva et al, 2019). That social media platforms can be used for both raising awareness and nurturing climate scepticism about climate change speaks to the explanatory power of framing theory. The power accorded to climate change activists to select information based on their preferences and tailor it for specific audiences demonstrates the extent to which individuals may appropriate social media power to promote their climate change agendas. The implications of this framing power are far-reaching on individual attitudes, behaviour, action and policy and are discussed in our findings below.

Methodological Discussion

This chapter explores opposing frames on climate change by two climate change activists and their implications on climate change action. It examines how two African climate activists, namely, Jusper Machogu (Kenya) and Leah Namugerwa frame climate change on X and the possible implications of such framing on climate action. The study addresses the following questions:

- How do the identified social media activists frame climate change on X?
- What are the implications of such framings on climate change action?

To address these questions, the study deployed netnography (virtual ethnography) – a qualitative research approach that applies and adapts ethnographic research techniques to gain insights into the culture of online communities and the relationship of actors in these communities (Juliadi et al, 2021).

In keeping with netnography's concern with studying people's beliefs, values, customs and behaviours, the data collected for this study comprised X posts and archival textual material such as pictures, videos and graphical materials posted on the two activists' X handles, @JusperMachogu and @NamugerwaLeah, respectively. The textual and audio-visual data were complemented by a Zoom interview with Machogu.[1] While the netnographic data collection took place between November 2024 and March 2025, the Zoom interview with Machogu took place on 4 April 2025 and lasted for about 35 minutes. The naturalistic orientation of netnography and the immersion of the authors into the online enclaves of the two climate change activists enabled the authors to generate rich and thick descriptions and grounded interpretations about the phenomena under consideration (Kozinets et al, 2024). X was suited for this exploration because it is by far the most popular social platform on which climate change discussions are held (Sultana et al, 2024). The choice of X is also inspired by scholars who advise that when choosing netnographic field sites, researchers must go for sites that generate "higher traffic of postings" and/or have "more detailed or rich descriptively rich data" (Kozinets et al, 2014: 265).

Data obtained from netnography and the interview were thematically coded and subjected to explicate emerging climate change frames. The unit of analysis therefore was the X post, the sentence in textual data, the pictures, and the themes/tone and language in audio-visual texts. To this end, Kuypers's (2009) criteria for identifying frames, namely, a text that (1) defines the problem, (2) diagnoses the problem's cause, (3) makes a moral judgement and (4) suggests a remedy, were adopted in this study (cited in Olteamu et al, 2015: 290). The emerging frames, which are categorised into pro-climate change frames and anti-climate change frames, are discussed below.

Pro-Climate Change Frames

The pro-climate frame emerges from X posts by 21-year-old Ugandan climate change activist Leah Namugerwa. Namugerwa who was inspired by the Swedish climate change activist, Greta Thunberg, regularly posts climate-related content on her X handle @NamugerwaLeah which commands 23,500 followers, making her one of the most influential female voices on climate change in Africa. Her tweets and retweets are anchored on the call for action to mitigate the effects of climate change. She is well known for leading tree-planting campaigns and for advocating the banning of plastic bags in Uganda. As an integral member of the Fridays for Future Uganda, Namugerwa leads school strikes where learners skip school on Fridays to protest and demand urgent action on climate change in the country in what is popularly known as #Schools Strike4Climate.

Climate Action Frame

The Climate Action Frame is pivoted on content that elicits engagement such as extreme weather conditions and purported inaction by policymakers. In her tweets and retweets, Namugerwa emphasises the need for urgent attention to the climate change problem, urging all hands on deck. For instance, one of her posts, which attracted 2,000 likes and 523 retweets, stated:

> @NamugerwaLeah. The effects of locusts, COVID-19 and floods happening now in Uganda are hitting children harder. I don't see any NGO coming to help starving children. I've today donated some food parcels. This is my first stop over. I'll continue to give what I can. #Action More than Words.

The reference to locusts, COVID-19 and floods underscores the multiple crises confronting Uganda and, by extension, humanity, illustrating how different crises intersect with climate change. By highlighting the plight of "starving children", the post demonstrates how vulnerable groups are disproportionately affected by climate change. The call for action frame is anchored on remedying climate change through activities such as banning plastic bags, planting trees and putting in place the appropriate environmental policies. By foregrounding the vulnerability of children and implicating the triple crisis of locusts, COVID-19 and climate change, Namugerwa uses selectivity as a framing technique (Entman, 1993) to make salient the named crises. References to "starving children" and extreme weather conditions like floods demonstrate how climate change is sustained through fear appeals (De Graff et al, 2024; Bilfinger, 2024).

To demonstrate the urgency of climate change action, Namugerwa's tweets and retweets employ a significant level of crisis framing such as the hashtag #Climate Change. In one of her posts, Namugerwa wrote:

> @NamugerwaLeah: Floods have led to landslides that have left a 13-year-old girl dead, and two injured. The injured were ferried to hospital. More floods in several districts. #ClimateEmergency#Climate Crisis (18 September 2021).

In a retweet from Fridays For Future Uganda's post, Namugerwa accentuates the urgency of climate change action:

> @NamugerwaLeah: Climate crisis is happening. Climate solution is everyone's business. Youth are important agents of change in raising awareness and taking action. Yesterday we were in Hoima (South Western Uganda). Our next destination is Gulu in Northern Uganda.

Yet another retweet from Fridays for Future Uganda on 29 September 2021 read:

> @NamugerwaLeah: FLOODS FORCE A PRIMARY SCHOOL TO RELOCATE IN UGANDA

The running thread in these posts is how climate change is framed as an emergency which needs urgent action. To jolt her audience into action, Namugerwa spotlights extreme weather events such as floods to underscore the urgency of climate change in Uganda. The reference to the death of a 13-year-old child is an appeal to foster increased engagement of policymakers. The power of emotions is demonstrated by the increased audience engagement by her followers who express their sympathy and urge her to continue doing the good work of helping the poor. Accentuating climate change as a crisis and the use of emotional tropes heightens engagement and spurs stakeholders to look for solutions to the climate change problem. Such framing is well suited to social media platforms such as X because of their immediacy, virality and shareability of content, which means that messages have a wider reach and high engagement (Hart et al, 2025). However, the fixation on adverse weather conditions narrows the climate change frame as the audience is likely to view climate change through the lens of disaster journalism, where overemphasis on catastrophes may overshadow cross-cutting dimensions of climate change. Emphasis on the crisis frame is further demonstrated through posts supported by videos of flooded rivers, collapsed dam walls and houses submerged in water, as well as photographs with victims of floods displaced from their homes. The videos and photographs are underpinned by text calling for action, such as donations to those affected by the floods. To achieve this, she uses a testimonial approach wherein her own posts and retweets from third parties galvanise action through emotional appeals. The emotional appeal is illustrated through a retweet of climate scientist and activist Peter Kalmus's tweet (@ClimateHuman) thus:

> @NamugerwaLeah: If you aren't terrified by climate breakdown you don't understand its full irreversible implications. This is just the truth and there is a good reason to say it. So far, we are not collectively scared enough to tear down the fossil industry and that needs to change (29 August 2021).

By using the voice of an influential climate scientist Namugerwa presents climate change as irrefutable, thereby enhancing the credibility of her messages. Given the unobtrusive nature of climate change, visuals of extreme weather events such as floods and co-opting voices perceived to be credible become crucial.

Equally important is the use of effective rhetorical communication strategies to galvanise climate change action. Such words as "irreversible implications" and "truth" evoke a sense of finality, projecting climate change as a phenomenon beyond contestation. Suggesting that if one is not "terrified" by climate change, they do not understand its "irreversible implications" heightens the urgency of the climate crisis, underscoring the way in which pro-climate change is pivoted on "doomsday" scenarios (De Graff et al, 2024) and "fear-mongering" discourses whereby climate change is projected as "an existential threat or existential danger" (Saab, 2023: 114). Saab (2023: 114) argues that representing climate change as "an impending catastrophe, a looming apocalypse and a path to irreversible damage that may even result in human extinction" is problematic. Similarly, De Graff et al (2024) argue that the use of doom framings of climate change is not always effective in communicating climate change. Williams argues that in order to compete for attention and to avoid sceptical counterclaims climate change messages should be packaged in a way that makes the situation a matter of common sense (cited in Russill & Nyssa, 2009).

The action-centred approach to climate change necessitates that Namugerwa occasionally uses a testimonial approach whereby influential voices in climate change are conscripted to emphasise the urgency of climate change. For instance, a Fridays for Future Uganda tweet reposted by Namugerwa on her X handle which features a video of Swedish climate activist Greta Thunberg being interviewed by a Channel 4 journalist. The short video clip underscores the necessity of collective action because climate change is no longer a matter of speculation "but a fact".

> Greta Thunberg: Many people think that climate change is an opinion, but it's not an opinion, it's a fact. You can't say "I don't think that the heart is pumping blood", I don't think so, it's a fact. Scientists often talk about tipping points. That means that once we pass a certain point, it means that the earth starts warming itself … we still have time. But that time will not last long. So, we need to do something now and we need to do it quickly and we need to do it drastically …
>
> (@NamugerwaLeah, 31 August 2021)

The urgency in Thunberg's message resonates with Namugerwa's action-centred approach to climate change. The reference to "tipping points" is suited for initiating action in a context where policymakers are lethargic because the tipping point discourse invokes irreversible shifts in the Earth systems (Russill & Nyssa, 2009). Such discursive constructions imply that humanity has reached the precipice of the climate system, beyond which there will be no redemption (Russill & Nyssa, 2009).

An extreme form of tipping point discourse is found in Namugerwa's retweet of a #COP26 message wherein Sir David Attenborough, "the People's Advocate on Climate Change" at #COP26 was quoted in a video saying:

> To date the biosphere has been our greatest ally, absorbing much of our impact. But now we are on the *verge* of passing the *tipping points boundaries* that, *once passed,* will *unleash irreversible* and *self-amplifying change.* Then all the innovation, wealth and political will in the world *will not be enough to save our civilisations.* Our natural world will go from being our biggest ally to our *biggest foe.* As the Greenland *icesheets tip into an irreversible thaw* and the Amazon enters a *self-perpetuating* period of drying, the *Earth will stop masking our warming and start to accelerate it.* There is strong evidence that we are within a *decade of passing these tipping points* – and committing hundreds of millions of people to floods, famine, fires, and forced migration. *Staying below 1.5 degrees* is the *only change* we have of *avoiding these tipping points and stabilising our world again.*
>
> (@NemugerwaLeah, 30 August 2021; emphasis by the authors)

Climate change is depicted as irreversible, and the solution prescribed is to ensure that we do not exceed the threshold of 1.5 degrees in global warming. That the phrase "tipping point" is mentioned three times in a message comprising 148 words demonstrates the extent to which climate action is leveraged on emotional fear. The blazing fires, melting glaciers, receding sea shows and dry rivers backgrounding the video concretise the existential threat posed by climate change to humanity and are also expected to jolt the audience to act immediately. However, the doomsday framing of climate change and the conscription of highly influential personalities have their own limits on social media platforms such as X, as demonstrated by a string of negative comments in response to Sir David Attenborough's message. Some commenters questioned who had bestowed "The People's Advocate on climate change at #COP26" title on Sir David Attenborough;

> @Clim8resistance: "Which people appointed him the "People's Advocate? I don't remember a vote. He doesn't speak for me". (30 August 2021)
>
> @Energy_Peoples: "I'm sorry. This is scaremongering propaganda nonsense laced with tired old tropes about 'tipping points'. What does the 'people's' advocate have to say about human flourishing, human adaptability and raising millions out of poverty by access to cheap abundant reliable fuel?" (30 August 2021)
>
> @Phil_Invest
>
> "Absolute nonsense by a TV presenter …" (30 August 2021).

These responses to Sir David Attenborough's message, underscore the limits of the tipping point frame and its boomerang effect on platforms such as X, where diverse opinions about climate change intermingle. While this kind of dissonance warrants a separate investigation, it is the vulnerability of the tipping point frame on social media that is worth noting. As demonstrated in the next section, social media platforms with open architectures such as X are becoming fertile grounds for gaining insight into the competing discourses around climate change.

Anti-Climate Change Frame

The sceptical frame finds expression through 30-year-old Kenyan agricultural engineer, farmer and founder of the company Fossil Fuels for Africa, Jusper Machogu. Through his X handle, @JusperMachogu, Machogu disseminates information that contradicts the perspectives of climate change champions. With 28,900 followers on X and 41,500 posts to date, Machogu has been described as a climate change denialist and "a flagbearer for fossil fuels in Africa" (Silva, 2024). His views on climate change are underpinned by the following themes: campaigning for fossil fuels in Africa, denial of the climate change emergency/crisis, Western nations' hypocrisy and critique of global power relations. His X cover photograph features an image of a forest with the words "WE Love CO2 and You" overlaid on the photo, reflecting rejection of the dominant climate change narrative and campaigns to ban fossil fuels. The "We Love CO2" text underscores his belief that CO2 is beneficial to Africa's development, as increased CO2 levels are beneficial to plant growth and agricultural productivity.

Scepticism Frame: A Counter-narrative

Unlike Namugerwa who views climate change as an emergency, Jusper Machogu believes that climate change is not an issue in Africa.

> @JusperMachogu: Nobody over here is worried about climate change. We have 99 problems, but climate change isn't one. This is the best time to be alive but how we wish we concentrated on solving our biggest problems such as hunger, clean water and cooking, education access, proper clothing, access to electricity etc. Fossil Fuels for Africa. (28 November 2024).

The post highlights the polarity between climate change champions and sceptics, and the way in which climate change is mediated by contextual, historical, social, economic and political factors. It counters Namugerwa's messages which are inspired by the dominant narrative where climate change is viewed through a scientific lens. Whereas dominant narratives that align with Namugerwa's activism consider climate change action a priority, Machogu views climate change as a remote concern for Africa and therefore subservient to the plethora of socio-economic challenges confronting the African continent. He describes climate change as "alarmism" and "brainwashing", and holds the notion that extreme weather conditions and events, such as sea-level changes, droughts, heatwaves etc., have nothing to do with climate change (Zoom Interview, 8 April 2025).

> There are a few people in the world who are doing the brainwashing, the alarmism, starting with the UN, the big organisations, the World Bank, IMF, all of those, but the UN, it's like they are the top. And so, we have very few people being brainwashed that there is this existential threat of climate change and that we should ensure that it doesn't become bigger and bigger, and the only way to do is limiting the burning of fossil fuels. And that notion is not … It's not correct because when you look at the data, the data says something different and that is why I believe we have to do a lot of work, especially … and that is why we have to really jump on to the ship and tell, especially the Western organisations, those, like the IMF and World Bank and the UN that we don't buy the climate alarmism.
>
> (Zoom Interview, 8 April 2025)

The reference to "brainwashing", "alarmism" and attribution of these to the "West" and big organisations such as the IMF and the World Bank shows how sceptics construct climate change as a discourse between polar worlds. This polarity demonstrates that despite claims of near-unanimous consensus about climate change within the scientific community, the polarity around climate change persists and is underpinned by two opposing logics, namely, "sceptics" and the "convinced" (Treen et al, 2022: 68). The technological affordances of X enable sceptics to circulate information that counters dominant narratives about climate change (Treen et al, 2022).

A recurrent theme in Machogu's anti-climate change discourses is the campaign for fossil fuels in Africa. He argues, "Fossil Fuels developed the West, and it kept up it there. They will develop the South" (Zoom Interview, 8 April 2025). Using the hashtag #Fossil Fuels for Africa, Machogu regularly posts messages, retweets, videos and pictures which validate his perspectives on fossil fuels. An example is a retweet of one Linnea Lueken who argued that "Alarmists will claim the planet is hurtling towards disaster …" (@JusperMachogu, 15 April 2025), an allusion to the tipping points discussed earlier. Like Namugerwa, Machogu validates his views on climate change through retweeting posts from people whose views align with his in relation to climate change. The refrain on the need for fossil fuels in Africa is distinct in different posts and interviews with mainstream media organisations and podcasters. A case in point is a retweet of the CO2 Coalition's post postulating the advantages of fossil fuels in the developing countries.

> @JusperMachogu: Without the advantage of fossil fuel use, we in the developed nations of the world would be subjected to a life of poverty, disease, and death. If you don't think so, look no further than the poignant example set by East Africa where Grace, a Kenyan woman uses excrement and wood simply to cook her food. "If I continue like this, I'll go blind" says Grace (12 April 2025).

Another CO2 Coalition tweet features a video clip in which Machogu is interviewed by a journalist, stating:

> @JusperMachogu: I think it's pretty obvious that the West has got what it has because of fossil fuels. When people say Africa does not need fuels, I wonder, I don't think they want what's best for us (31 May 2025).

By framing the banning of fossil fuels through the prism of double standards and conspiracy, Machogu views coloniality as embedded in climate change discourses. This demonstrates how climate change intersects with politics and economics in the global arena. The words "development" and "neo-colonialism" appear not far from "fossil fuels" in his X posts. In a video titled "A MESSAGE from a Small-Scale Farmer in rural Kisii to the UNCOP28 in UAE", Machogu advocates for fossil fuels in Africa to address poverty and lack of electricity and clean water as well as asks African leaders to rebuff the neo-colonial climate policies of Western countries.

> My message is Africa needs fossil fuels. We need fossil fuels to better our lives. We need fossil fuels for economic development. You guys preaching to us, the

> Western governments and Non-Governmental Organisations preaching to us about climate change. Climate change is not a big problem to Africans. We have far bigger problems. We have people sleeping hungry. We have people who are very poor. We have people without electricity. We have people who don't have access to clean water. Those are our big problems. We have people cooking with firewood and you are telling us we should care about the climate change. Anyway, what is climate change? So, my message to these leaders, to African leaders, don't be bribed, stand firm, don't even attend the COP28 meeting. If those people want to make decisions for us, policies for us, let them do make those policies. But we are going to say no to neo-colonialism in the name of climate change.
>
> (@JusperMachogu, 29 November 2023)

The sentiment about neo-colonial machinations of the West is bolstered thus:

> But that is a debate (banning of fossil fuels) that is happening in Africa. It is not happening elsewhere. Those environmental people, they are not going to China telling the Chinese to stop burning coal, but they are telling Africans not to start burning our fossil fuels. So, people need to understand that our way to prosperity and our way to develop ourselves and have industries, and have proper transportation, and have refrigerators and beat poverty and hunger is by using fossil fuels and we should never … A hungry man should never have to worry about climate change, about something going to happen 30 years from now …
>
> (Zoom Interview, 8 April 2025)

The rhetorical strategy in the sceptical frame is a rebuttal of the universalistic approach to climate change by foregrounding contextual and ideological motivations of champions of climate change. Framing climate change as an issue between "them" and "us" exposes its polarising dimension. By framing the fossil fuel debate through a neo-colonial and conspiratorial lens, Machogu constructs climate change as an alibi for perpetuating Western hegemony. This aligns with the thinking of decolonial scholars who question the perpetuation of unequal political and economic power relations between Africa and the West (Maldonado-Torres, 2007). By evoking the discourse of coloniality in climate change, Africans are constructed as pawns who pander to the whims of the West on matters of climate change, thus obfuscating calls for climate action (Treen et al, 2022). Machogu mobilises the language of hypocrisy (Gunster et al, 2018) to weaken the moral credibility of supporters of climate change while raising his own standing by adopting an anti-colonial frame. By representing climate change as a remote problem for Africa, Machogu's message, unlike that of Namugerwa, who treats climate change as an emergency, has the potential to demobilise efforts against climate change, particularly given the networked nature of X (Bloomfield & Tillery, 2019). Treen et al (2022) note that the sceptical discourse on climate change has the potential to "confuse" the climate change discourse and increase political polarisation.

Bridging the Dichotomy: Climate Justice Frame

Despite the existence of opposing framings of climate change, it would be an oversimplification to argue that these frames are mutually exclusive. Apart from the fact that both Namugerwa and Machogu are from the youth sub-group, there are several commonalities regarding their style of activism which include co-opting the voices of third parties considered influential in the climate discourse. In addition, the antagonism between the pro-climate change and anti-climate change is, however, moderated by a climate change justice frame. For instance, both activists critique power inequalities and the disproportionate impacts of climate change on marginalised groups/Africa/the developing world.

In one of her X tweets, Namugerwa stated:

> @NamugerwaLeah: Fact: Those who contributed the least to #ClimateCrisis are suffering the most, but they are also listened to the least. They have the lowest representation, lowest coverage, lowest funding, the list can go on … They're even struggling to attend@COP26. What a World. (30 August 2021).

Namugerwa also reposted a Fridays for Future tweet which opined that:

> @NamugerwaLeah: Time is running out for millions of people who are already losing their lives, their homes and their livelihoods to #ClimateCrisis. These people have contributed least to the global climate emergency, yet they are being hit the most hardest. @UNOCHA (30 August 2021)

These two posts highlight the disproportionate effects of climate change on marginalised countries, representation in climate policies, lack of funding, exclusion and marginality induced by climate change. These concerns intersect with some of the injustices pointed out in Machogu's complaints about rich countries making climate-related policies and decisions on behalf of poor nations and the neo-colonial machinations informing climate change policies. Machogu's conspiracy theory about the West not wanting Africa to use fossil fuels because they want Africa to remain poor and his assertions that "there is a cabal of organisations who feel threatened because once Africa develops" (Zoom Interview, 8 April 2025) demonstrates how the climate change issue intersects with coloniality (Maldonado-Torres, 2007). This perspective resonates with studies that highlight climate injustices in Africa whereby the continent is least responsible for climate suffers the most devastating consequences of climate change (BBC World Trust, 2009).

Like Machogu, Namugerwa posted opinion articles from third parties to validate her perspectives on climate change injustices. Two notable opinion articles shared by Namugerwa originated from the Fridays for Futures Uganda X posts which she reposted on her handle. One of the opinion articles was authored by Jason Hickel, Dylan Sullivan and Huzaifa Zoomkawala and titled "Rich countries drained $152 tn from the global South since 1960: Imperialism never ended it just changed from"

(20 September 2021). The article's critique of Western nations' extractivism in the Global South echoes Machogu's indictment on Western nations who impose their climate change policies on Africa because "they feel threatened because once Africa develops we are going to start consuming our own resources …" (Zoom Interview with Jusper Machogu, 8 April 2025). Hickel et al's contention that "[w]e have long known that the industrial rise of rich countries depended on extraction from the global South during the colonial era" resonates with Machogu's assertions that "Fossil Fuels are what Africa needs to develop, to become more productive and to feed its people and the wider world" (Zoom Interview with Jusper Machogu, 8 April 2025). Similarly, the article "The Climate Crisis Inequality All the Parties Are Ignoring" by Gamage Michelle (The Tyee, 17 September 2021), shared by Namugerwa on her X handle (18 September 2021), highlights the unequal and disproportionate impact of climate change on marginalised communities, thus demonstrating how climate change conspires with colonialism, capitalistic extractivism and racism to perpetuate inequalities, a notion that recalls Namugerwa and Machogu's X posts about the effects of climate change on impoverished communities in their countries. Despite their contrary views on climate change broadly, Namugerwa and Machogu find common ground on climate change justice, thereby moderating the polarity between the pro-climate change frame and the anti-climate change frames discussed in this chapter. This demonstrates how the polarities between climate change believers and non-believers are ultimately mediated and moderated by context.

Conclusion

This chapter examined the framing of climate change on X using a climate change advocate, Leah Namugerwa from Uganda and a climate change sceptic, Jusper Machogi as lenses to illuminate the way in which competing views on climate change interact on social media in the context of the Global South. The study reveals that the competing frames on climate change, categorised as "pro-climate change" and "anti-climate change", have broader implications on climate change action, adaptation behaviour and lifestyle change. However, despite the existence of two competing frames, the two activists find common ground through a climate change justice frame through wherein both are able to critique unequal power relations and the disproportionate effects of climate change on marginalised groups in the Global South. This means that within the Global South context, particularly in Africa, the dichotomy between climate change advocates and sceptics needs more nuancing as it is mediated by a complex set of factors rooted in colonialism and neo-colonialism. While the study opens critical vistas for understanding the constellation of contrarian views on climate change on social media, its scope is limited because it is based on two activists who use the same platform. Future studies could focus on more activists using different social media platforms with different technological affordances to obtain broader perspectives on the interaction between climate change believers and non-believers in the cybersphere.

Note

1 The authors could not get hold of Namugerwa to have an interview.

References

Abubakar, M., Maharri, Z., Satter, T. and Ahmad, S.R. (2025) Role of Social Media in Promoting Environmental Advocacy and Climate Change Awareness. *NUST Journal of Natural Sciences*, 10(1).

Askanius, T. and Uldam, J. (2011). Online Social Media for Radical Politics: Climate Change Activism on YouTube. *International Journal of Electronic Governance*, 4(1/2): 69–84.

Bilfinger, L., Brumernhenrich, B. and Jucks, R. (2024). The Effects of Fear Appeal on Reactance in Climate Change Communication. *Journal of Experimental Social Psychology*, 115: 1–9.

Bloomfiled, E.F. and Tillery, D. (2019). The Circulation of Climate Change Denial Online: Rhetorical and Networking Strategies on Facebook. *Environmental Communication*, 13(1): 23–34.

Cammaerts, B. (2015). Social Media Activism. In R. Mansell and P. Hwa (Eds) *The International Encyclopaedia of Digital Communication and Society* (pp. 1027–1034). Oxford: Wiley-Blackwell.

De Graff, J., Bal, M., de Wit, J. and Stok, M. (2024). Climate Change Doom Communication From a Fear Appeal Perspective. *European Journal of Health Communication*, 5(4): 26–50.

Diehl, T., Huber, B., Gil de Zuniga, H. and Liu, J. (2021). Social Media and Beliefs about Climate Change: A Cross-National Analysis of News Use: Political Ideology and Trust in Science. *International Journal of Opinion, Research*, 33: 197–213.

Dinan, W. and Miller, D. (2023). Managing the Climate Apocalypse: Think Tanks, Policy Planning Groups, and The Corporate Capture of Sustainable Development. In A. Hansen and R. Cox (Eds) *The Routledge Handbook of Environmental Communication* (pp. 132–145). London: Routledge.

Entman, R.M. (1993). Framing: Toward Clarification of a Fractured Paradigm. *Journal of Communication*, 43(4): 51–58.

Gaupp, F. and Eker, S. (2024). *Climate Activism, Social Media and Behavioural Change.* A Literature Review. Working Paper 24-007. International Institute for Applied Systems Analysis. Retrieved from https://pure.iiasa.ac.at/id/eprint/19639/1/WP-24-007.pdf. Accessed 26 May 2025.

Gilardi, F., Gessler, T., Kubli, M. and Muller, S. (2022). Social Media and Political Agenda Setting. *Political Communication*, 39(1): 39–60.

Gunster, S., Fleet, D., Paterson, M., and Saurette, P. (2018). "Why Don't You Act Like You Believe It?": Competing Visions of Climate Change Hypocrisy. *Fronters in Communication*, 3(49): 1–14.

Guran, M.S., and Orzaslan, H. (2022). Framing Theory in the Age of Social Media. *Selçuk University Journal of Social Sciences Institute*, 2022(48): 446 457.

Hart, P.S., Feldman, L. and Hiaeshulter-Rice, D. (2025a) Climate Change Advocacy and Engagement on Social Media. *Science Communication*, 47(3): 385–412.

Hart, P.S., Feldman, L., Choi, S., Chinn, S. and Hiaeshutter-Rice, D. (2025b). Climate Change Advocacy and Engagement on Social Media. *Science Communication*, 47(3): 385–412.

IPPCC (2023). Climate Change 2023 Synthesis Report. Retrieved from www.ipcc.ch/report/ar6/syr/downloads/report/IPCC_AR6_SYR_LongerReport.pdf. Accessed 26 May 2027.

Juliadi, R., Murwani, E. and Ariestya, A. (2021). *A Netnography Study of Social media Influencers on Creating Netizens' Climate Change Awareness on Instagram.* The 4th International Conference on Communication and Business. Retrieved from https://papers.ssrn.com/sol3/papers.cfm?abstract_id=3871822. Accessed 20 May 2025.

Kozinets, R.V., Dolbec, P.Y. and Earley, A. (2014). Netnographic Analysis: Understanding Culture Through Social Media Data. In U. Flick (Ed.) *Sage Handbook of Qualitative Data Analysis* (pp. 262–275). London: Sage.

Kuypers, J.A. (2009). Framing Analysis: How to conduct a Rhetorical Framing Study of the News. In Kuypers (Ed.) *Rhetorical Criticism: Perspectives in Action* (pp. 181–203). Lanham, MD: Lexington Books.

Li, N. and Su, S.Y. (2018). Message Framing and Climate Change Communication: A Meta-Analytical Review. *Journal of Applied Communications*, 102(3). Article 4.

Maldonado-Torres, N. (2007). On Coloniality of Being. Contributions to the Concept of Development. *Cultural Studies*, 21(2–3): 240–270.

Mavrodieva, A., Rachman, O., Harahap, V. and Shwa, R. (2019). Role of Social Media in Raising Awareness and Engagement in Addressing Climate Change. *Climate*, 7(122): 2–15.

Nisbet, M. (2009). Communicating Climate Change: Why Frames Matter for Public Engagement. *Environment Science and Policy for Sustainable Development*, 51(2): 12–23.

Olteanu, A., Castillo, C., Diakopoulos, N. and Aberer, K. (2015). *Comparing Events Coverage in Online News and Social Media: The Case of Climate Change.* Proceedings of the 9th International AAAI Conference on Web and Social Media. Association for the Advancement of Artificial Intelligence, 26-29 May, 2015. Oxford. United Kingdom (pp. 288–297). https://doi.org/10.1609/icwsm.v9i1.14626.

Russill, C. and Nyssa, Z. (2009). The Tipping Point Trend in Climate Change Communication. *Global Environmental Change*, 19: 336–344.

Saab, A. (2023). Discourses of Fear on Climate Change in International Human Rights Law. *The European Journal of International Law*, 34(1): 113.

Schafer, M.S. (2025). Social Media in Climate Change Communication: State of the Field, New Developments and The Emergence of Generative AI. *Dialogues on Climate Change*, 2(1): 49–55.

Silva, M. (2024, June 16). *How a Kenyan Farmer became a Champion of Climate Change Denial.* Retrieved from www.bbc.com/news/articles/c133r4gyx1no. Accessed 4 June 2025.

Sultana, B.C., Prodhan, T.R., Alam, E., Sohel, M.S., Bari, A.B.M.M., Pal, S.C., Islam, M.K. and Islam, A.R.M.T. (2024). A Systematic Review of the Nexus Between Climate Change and Social Media: Present Status, Trends and Challenges. *Frontier Communication*, 9: 13014000.

Tanveer, Z. (2024). Leveraging Digital Media for Climate Change Awareness: The Role of Social Platforms in Environmental Advocacy. *Multi-disciplinary Journal of Law for Future Security*, 1(1): 1–9.

Treen, K., Williams, H., O'Neil, S. and Coan, T.G. (2022). Discussion of Climate Change on Reddit: Polarised Discourse or Deliberate Debate? *Environmental Communication*, 16(5): 680–698.

Index

Note: Page numbers with an "n" denote endnotes.

Made in the USA
Monee, IL
12 November 2026

599a72f8-3375-4ef6-ae1b-da8eb29fdf6dR01